# CAMBRIDGE LIBRARY COLLECTION

*Books of enduring scholarly value*

## Technology

The focus of this series is engineering, broadly construed. It covers technological innovation from a range of periods and cultures, but centres on the
technological achievements of the industrial era in the West, particularly in
the nineteenth century, as understood by their contemporaries. Infra-structure
is one major focus, covering the building of railways and canals, bridges and
tunnels, land drainage, the laying of submarine cables, and the construction
of docks and lighthouses. Other key topics include developments in industrial
and manufacturing fields such as mining technology, the production of iron
and steel, the use of steam power, and chemical processes such as photography
and textile dyes.

## Lives of the Engineers

A political and social reformer, Samuel Smiles (1812–1904) was also a noted
biographer in the Victorian period. Following the engineer's death in 1848,
Smiles published his highly successful *Life of George Stephenson* in 1857
(also reissued in this series). His interest in engineering evolved and he began
working on biographies of Britain's most notable engineers from the Roman
to the Victorian era. Originally published in three volumes between 1861 and
1862, this work contains detailed and lively accounts of the educations, careers
and pioneering work of seven of Britain's most accomplished engineers. These
volumes stand as a remarkable undertaking, advancing not only the genre, but
also the author's belief in what hard work could achieve. Volume 1 charts the
engineering of early roads, embankments, bridges, harbours and ferries, as
well as the lives of the engineers Sir Hugh Myddelton (*c.*1560–1631) and
James Brindley (1716–72).

# Lives of the Engineers

*With an Account of Their Principal Works*

VOLUME 1

SAMUEL SMILES

CAMBRIDGE UNIVERSITY PRESS

Cambridge, New York, Melbourne, Madrid, Cape Town,
Singapore, São Paolo, Delhi, Mexico City

Published in the United States of America by Cambridge University Press, New York

www.cambridge.org
Information on this title: www.cambridge.org/9781108052924

© in this compilation Cambridge University Press 2012

This edition first published 1861
This digitally printed version 2012

ISBN 978-1-108-05292-4 Paperback

*Sir Hugh Myddelton, Knight & Baronet.*

Engraved by W. Hell, after the portrait by Cornelius Jansen.

Published by John Murray, Albemarle Street. 1861.

# LIVES

OF

# THE ENGINEERS,

WITH

## AN ACCOUNT OF THEIR PRINCIPAL WORKS;

COMPRISING ALSO

## A HISTORY OF INLAND COMMUNICATION IN BRITAIN.

BY SAMUEL SMILES.

"Bid Harbours open, Public Ways extend;
Bid Temples, worthier of God, ascend;
Bid the broad Arch the dang'rous flood contain,
The Mole projected, break the roaring main;
Back to his bounds their subject sea command,
And roll obedient rivers through the land.
These honours, Peace to happy Britain brings;
These are imperial works, and worthy kings."

POPE.

WITH PORTRAITS AND NUMEROUS ILLUSTRATIONS.

VOL. I.

LONDON:

JOHN MURRAY, ALBEMARLE STREET.

1861.

# PREFACE.

THE object of the following volumes is to give an account of some of the principal men by whom the material development of England has been promoted,— the men by whose skill and industry large tracts of fertile land have been won from the sea, the bog, and the fen, and made available for human habitation and sustenance; who have rendered the country accessible in all directions by means of roads, bridges, canals, and railways; and have built lighthouses, breakwaters, docks, and harbours, for the protection and accommodation of our vast home and foreign commerce.

Notwithstanding the national interest which might be supposed to belong to this branch of literature, it has hitherto received but little attention. When the author first mentioned to the late Mr. Robert Stephenson his intention of writing the Life of his father, that gentleman expressed strong doubts as to the possibility of rendering the subject sufficiently popular to attract the attention of the reading public. "The building of bridges, the excavation of tunnels, the making of roads and railways," he observed, "are mere mechanical matters, possessing no literary interest;" and in proof of this he referred to the 'Life of Telford' as "a work got up at great expense, but which had fallen still-born from the press."

Besides the apparent unattractiveness of the subject, its effective treatment involved the necessity of burrowing through a vast amount of engineering reports, which, next to law papers, are about the driest possible reading, except to those professionally interested in them.

Circumstances such as these have probably concurred in deterring literary men from entering upon this field of biography, which has hitherto remained comparatively unexplored. Hence, most of the Lives and Memoirs contained in the following series are here attempted for the first time. All that has appeared relating to Brindley, Smeaton, and Rennie, is comprised in the brief and unsatisfactory notices contained in Encyclopedias and Biographical Dictionaries. What has been published respecting Myddelton's life is for the most part inaccurate, whilst of Vermuyden no memoir of any kind exists. It is true, a 'Life of Telford' has appeared in quarto, but, though it contains most of that engineer's reports, the history of his private life as well as of his professional career is almost entirely omitted.

Besides the Lives of these more distinguished men, the following volumes will be found to contain memoirs of several meritorious though now all but forgotten persons, who are entitled to notice as amongst the pioneers of English engineering. Such were Captain Perry, who repaired the breach in the Thames embankment at Dagenham; blind John Metcalf, the Yorkshire road-maker; William Edwards, the Welsh bridge-builder; and Andrew Meikle, Rennie's master, the inventor of the thrashing-machine. Although the Duke of Bridgewater was not an engineer, we have included a memoir of him in the Life of Brindley, with whose early history he was so closely identified; and also because of the important influence which he exercised on the extension of the canal system and the development of modern English industry.

The subject, indeed, contains more attractive elements than might at first sight appear. The events in the lives of the early engineers were a succession of individual struggles, sometimes rising almost to the heroic. In one case, the object of interest is a London goldsmith, like Myddelton; in another, he is a retired sea-captain, like Perry; a wheelwright, like Brindley; an attorney's clerk, like Smeaton; a millwright, like Rennie; a working mason, like Telford; or an engine brakesman, like Stephenson. These men were strong-minded, resolute, and ingenious, impelled to their special pursuits by the force of their constructive instincts. In most cases they had to make for themselves a way; for there was none to point out the road, which until then had been untravelled. To our mind, there is almost a dramatic interest in their noble efforts, their defeats, and their triumphs; and their eventual rise, in spite of manifold obstructions and difficulties, from obscurity to fame.

It will be observed from the following pages that the works of our engineers have exercised an important influence on the progress of the English nation. But it may possibly excite the reader's surprise to learn how very modern England is in all that relates to skilled industry, which appears to have been among the very youngest outgrowths of our national life.

Most of the Continental nations had a long start of us in art, in science, in mechanics, in navigation, and in engineering. Not many centuries since, Italy, Spain, France, and Holland looked down contemptuously on the poor but proud islanders, contending with nature for a subsistence amidst their fogs and their mists. Though surrounded by the sea, we had scarcely any navy until within the last three hundred years. Even our fisheries were so unproductive, that our markets were supplied by the Dutch, who sold us the herrings caught upon our own coasts. England was then regarded principally as a magazine for the supply of raw materials, which were carried away in foreign ships and partly returned to us in manufactures worked up by foreign artisans. We grew wool for Flanders, as America grows cotton for England now. Even the little manufactured at home was sent to the Low Countries to be dyed.

Most of our modern branches of industry were begun by foreigners, many of whom were driven by religious persecution to seek an asylum in England. Our first cloth-workers, silk-weavers, and lace-makers were French and Flemish refugees. The brothers Elers, Dutchmen, began the pottery manufacture; Spillman, a German, erected the first paper-mill at Dartford; and Boomen, a Dutchman, brought the first coach into England.

When we wanted any skilled work done, we almost invariably sent for foreigners to do it. Our first ships were built by Danes or Genoese. When the *Mary Rose* sank at Spithead in 1545, Venetians were hired to raise her. On that occasion Peeter de Andreas was employed, assisted by his ship-carpenter and three of his sailors, with "sixty English maryners to attend upon them." When an engine was required to pump water from the Thames for the supply of London, Peter Morice, the Dutchman, was employed to erect it.

Our first lessons in mechanical and civil engineering were principally obtained from Dutchmen, who supplied us with our first wind-mills, water-mills, and pumping-engines. Holland even sent us the necessary labourers to execute our first great works of drainage. The Great Level of the Fens was drained by Vermuyden; and another Dutchman, Freestone, was employed to reclaim the marsh near Wells, in Norfolk. Canvey Island, near the mouth of the Thames, was embanked by Joas Croppenburgh and his company of Dutch workmen.. When a new haven was required at Yarmouth, Joas Johnson, the Dutch engineer, was employed to plan and construct the works; and when a serious breach occurred in the banks of the Witham, at Boston, Matthew Hake was sent for from Gravelines in Flanders; and he brought with him not only the mechanics but the manufactured iron required for the work. The art of bridge-building had sunk so low in England about the middle of the last century, that we were under the necessity of employing the Swiss engineer Labelye to build Westminster Bridge.

In short, we depended for our engineering, even more than we did for our pictures and our music, upon foreigners. At a time when Holland had completed its magnificent system of water communication, and when France, Germany, and even Russia had opened up important lines of inland navigation, England had not cut a single canal, whilst our roads were about the worst in Europe. It was not until the year 1760 that Brindley began his first canal for the Duke of Bridgewater.

After the lapse of a century, we find the state of things has become entirely reversed. Instead of borrowing engineers from abroad, we now send them to all parts of the world. British-built steam-ships ply on every sea; we export machinery to all quarters, and supply Holland itself with pumping engines. During that period our engineers have completed a magnificent system of canals, turnpike-roads, bridges, and railways, by which the internal communications of the country have been completely opened up; they have built lighthouses round our coasts, by which ships freighted with the produce of all lands, when nearing our shores in the dark, are safely lighted along to their destined havens; they have hewn out and built docks and harbours for the accommodation of a gigantic commerce; whilst their inventive genius has rendered fire and water the most untiring workers in all branches of industry, and the most effective agents in locomotion by land and sea. Nearly all this has been accomplished during the last century, much of it within the life of the present generation. How and by whom these great achievements have been mainly effected—exercising as they have done so large an influence upon society, and constituting as they do so important an element in our national history—it is the object of the following pages to relate.

It was the author's original intention to have begun this work with the Life of Brindley, the earliest of our canal engineers. But on mentioning the subject to the late Mr. Robert Stephenson—after the publication of his father's Life had shown that this class of biography was not so unattractive to general readers as he had apprehended—the author was urged by that gentleman to trace the history of English engineering from the beginning, and to include the labours of Vermuyden, and especially of Sir Hugh Myddelton, a person of great merit and boldness, considering the times in which he lived, and whom Mr. Stephenson considered entitled to special notice as being the First English Engineer. Memoirs of these men have accordingly been included in the series; and in preparing them the author has availed himself of the information afforded by the collection of State Papers, and (in the case of Myddelton) the Corporation Records of the City of London. He has also to acknowledge the valuable assistance of W. C. Mylne, Esq., engineer to the New River Company, and the Rev. H. T. Ellacombe, M.A., of Clyst St. George, Devon, a lineal descendant of Sir Hugh Myddelton.

The Life of Brindley has been derived almost entirely from original sources; amongst which may be mentioned the family papers in the possession of Robert Williamson,

Esq., of Ramsdell Hall, Cheshire; the documents relating to the engineer in the possession of Lord Ellesmere, proprietor of the Bridgewater Canal; and the valuable MS. collection of Joseph Mayer, Esq., of Liverpool. The author has also to acknowledge information obtained from Robert Rawlinson, Esq., engineer to the Bridgewater Canal, relative to certain interesting details as to the execution of the works of that undertaking.

The materials for the Life of John Rennie have been mainly obtained from Sir John Rennie, C.E., who has kindly placed at the author's disposal the elaborate MSS. prepared by Sir John, descriptive of his father's great works; of which no consecutive account has been published until the present memoir.

The Life of Telford has been principally derived from a large collection of that engineer's confidential letters to his friends in Eskdale, in the possession of Mr. Little, of Carlesgill, near Langholm, containing Telford's own account of the early part of his career; whilst, in the later part, the author has had the assistance of Joseph Mitchell, Esq., and other gentlemen. In preparing this part of the work, the author has reversed the process adopted in the 'Life of Telford' already published: he has omitted the engineer's reports, but included the biography; by which method he believes the narrative will be found considerably improved.

The author's principal labour has consisted in compressing rather than in expanding the large mass of materials placed at his disposal. It would indeed have been much easier to devote two volumes to each of the following lives than it has been to comprise the whole of them within a like compass; but he believes that labour is well bestowed in condensing biography up to a certain point, provided no essential feature is omitted—the interest and readableness of such narratives being very often in an inverse ratio to their length.

With the object of saving unnecessary verbal descriptions, illustrations, in the shape of maps, plans, and sections, have been introduced wherever practicable; and in those cases where a representation is given of a bridge, lighthouse, aqueduct, or harbour, it will be found set in its appropriate landscape. Although the dimensions of the wood engravings are necessarily small, every attention has been paid to accuracy of detail, most of them being drawn to scale.

The drawings by Mr. Percival Skelton—an excellent and graceful artist—have been made in nearly every case on the spot, for the express purpose of this work. Those by Mr. R. P. Leitch and Mr. Wimperis are mostly after original sketches supplied by distant correspondents; and it is hoped that the illustrations generally will be found to add to the interest of the volumes. The whole of the cuts have been executed by Mr. James Cooper, whose accuracy and carefulness in superintending the illustrative department of the work, the author takes this opportunity of acknowledging.

*London, October,* 1861.

# CONTENTS OF VOL. I.

—⋄—

## PART I.—EARLY WORKS OF EMBANKING AND DRAINING.

### CHAPTER I.

### CHAPTER II.

### CHAPTER III.

### CHAPTER IV.

### CHAPTER V.

## PART V.—LIFE OF JAMES BRINDLEY.

# LIST OF ILLUSTRATIONS.

# EARLY WORKS

## OF

# EMBANKING AND DRAINING.

# EARLY WORKS

OF

# EMBANKING AND DRAINING.

## CHAPTER I.

ROMNEY MARSH AND THE EMBANKMENT OF THE THAMES.

THE numerous ancient earthworks existing in various parts of Britain show that the Navvy is by no means a modern character. The mounds of Old Sarum and Silbury Hill, in Wiltshire, by whatever means and for whatever purpose raised, testify to a large amount of patient industry on the part of those who heaped them together. In Wales, Yorkshire, Devonshire, and the more hilly parts of England, the remains of the formidable ditches and embankments constructed for purposes of defence, afford abundant proof that the former people of this country must have been familiar with the use of the spade and mattock. But it would appear, from the remains of ancient British dwellings still extant in different parts of the country, that the early inhabitants lived in mere wigwams, and that engineering skill was scarcely to be expected of them. Their houses seem to have been formed by digging so many round holes in the earth, and covering them over with the branches of trees. Dr. Young, of Whitby, examined the remains of upwards of forty ancient British villages on the Yorkshire Wolds, from which he inferred that the aborigines, especially the more northern tribes, were no further advanced in civilization than the Caffres or

Bechuanas of the present day.[1] Numerous traces of
ancient habitations of a similar kind have been met
with in the southern counties, of which those at Bowhill,
in Sussex, are probably the most remarkable.[2]

The valleys and low-lying grounds being then mostly
covered with dense forests, the naturally cleared high
lands, where timber would not grow, were selected as
the sites of the old villages. Tillage was not yet understood nor practised; the people subsisted by hunting,
or upon their herds of cattle, which found ample grazing
among the hills of Dartmoor, and on the downs of
Sussex and Wiltshire, where most of these remains have
been found.[3] They are especially numerous along the
skirts of Dartmoor, where the hills slope down to the
watercourses. The heights above them are mostly
crowned by tors, or rude fortifications of earth, which
exhibit no greater engineering skill in construction

---

[1] Their clothing, when they wore
any, consisted of skins; they stained
their bodies with paint or ochre, and
often marked them with figures, in
the way of the South Sea tattooing.
They lived in circular huts nearly in
the shape of bee-hives, like those of
the native Africans, as we may yet
see in the remains of these dwellings
at Eyton Grange, Harewood Dale, &c.
To construct a hut, they dug a round
hole in the ground, and, with the
earth and stones cast out in the digging, made a kind of wall, which was
surmounted with boughs of trees
meeting together at the top to form
a sort of roof, over which there might
be a covering of sods to protect them
from the weather, a hole being left on
one side to serve the triple purpose of
a door, a window, and a chimney.
The fire was placed in the centre of
the floor, and the inhabitants sat or
lay on the ground around it. Remains
of the charcoal of their fires are found
in digging in the middle of the hollows
that mark the sites of these ancient
dwellings. In such wretched huts
large families of men, women, and
children would be promiscuously
huddled together, as is the case with
the South African savages; and this
mode of life might give rise to the
statements of Cæsar and Dion Cassius,
that among the Britons it was customary for every ten or twelve men,
and those the nearest relations, to have
their wives in common.—Dr. Young's
'History of Whitby and its Vicinity.'

[2] See 'Notitia Britannia.' By W.
D. Saul. 1845.

[3] We have undoubted proofs from
history and from existing remains
that the earliest habitations were pits,
or slight excavations in the ground,
covered and protected from the incle-
mency of the weather by boughs of
trees or sods of turf. The high grounds
were pointed out by nature as the
fittest for these early settlements,
being less encumbered by wood, and
affording a better pasture for the numerous flocks and herds, from which
the erratic tribes of the first colonists
drew their means of subsistence.—Sir
R. C. Hoare on the 'Antiquities of
Wiltshire.'

than is to be found in many a New Zealand pah.   But, ignorant though the people then seem to have been of the art of construction, it would appear that they must have possessed some skill in mechanical appliances, to have been enabled to transport those huge blocks of stone to their places on Stonehenge, and to erect the cyclopean bridges over the Teign and Dart in Devonshire, the remains of which are among the greatest curiosities extant of ancient engineering.[1]

The art of embanking and draining was introduced into England by quite another race—the adventurous tribes of Belgium and Friesland, who early landed in great numbers along the south-eastern coasts, and made good their footing by the power of numbers, as well as their superior civilization.   These men were tillers of the soil, and wherever they went they settled down to the arts of agriculture, clearing the ground of its primitive forest, and more especially occupying the rich arable lands along the valleys and by the seaside.   The early settlement of Britain by the races which at present occupy it, is usually spoken of as an invasion and a conquest; but there is good reason to believe that it was principally effected by a system of immigration and colonization, such as is going forward under our own eyes at this day in America, Australia, and New Zealand; and that the people who swarmed into the country in early times from Friesland, Belgium, and Jutland, secured their settlement by the spade far more than by the sword.   The Celts were a pastoral race, whilst the immigrants were tillers of the ground. Wherever the new men came, they settled themselves down on their several bits of land, which became their holdings; and they bent their backs over the stubborn soil, watering it with their sweat, and delved, and drained, and cultivated it, until it became fruitful.

---

[1] See subsequent chapter, on Old Bridges.

Thus these agricultural colonists spread themselves over the richer arable lands of the country, and became the dominant race, as is shown by the dominancy of their language in the districts which they occupied, the older population gradually receding before them to the hunting and pastoral grounds of the north and west. The process was slow, but it was continuous. The settlers made the land their own by their labour; and what they recovered by toil from the waste, the forest, and the moor, they held by the strength of their right arms. But the whole proceeding was one of simple persevering industry rather than of war. The men of Teutonic race thus gradually occupied the whole of the reclaimable land, until they were stopped by the hills of Cumberland, of Wales, and of Cornwall. The same process seems to have gone on in the arable districts of Scotland, into which a swarm of colonists from Northumberland poured in the reign of David I.,[1] and quietly settled upon the soil, which they proceeded to cultivate. It is a remarkable confirmation of this view of the early settlement of the country by its present races that the modern English language extends over the whole of the arable land of England and Scotland, and the Celtic tongue only begins where the plough ends.[2]

---

[1] See Cosmo Innes's 'Sketches of Early Scottish History,' 1861.

[2] This was formerly the case in the hill country of Cumberland and Cornwall, where the ancient language has now entirely disappeared. But in Wales, the Scotch Highlands, and the western parts of Ireland, the English traveller still finds himself amongst a race of people who can neither read his language nor understand him when he speaks to them. If they reply, it is obvious that they only partially understand the language; for they speak in broken English, like foreigners. Are they foreigners? No; these are the descendants of the early inhabitants of the soil, speaking the language which was spoken all over Britain long before the English language had been formed or English literature created. Yet there is every reason to believe that even those Celtic races were at one time but foreigners in Britain, and drove forth, if they did not exterminate, some previous race—the men who lived in caves, pits, and holes in the ground, such as are still to be found under Blackheath Point, at Crayford, Dartford Heath, Tilbury, and various other places in Kent, Essex, and the southern counties of England. Probably the remains of the very oldest race in the British Islands are now to be found in the least accessible districts of Galway

One of the most extensive districts along the English coast which lay the nearest to the country from which the continental immigrants first landed was the tract of Romney Marsh,[1] containing about 60,000 acres of land, lying along the south coast of Kent. The reclamation of this tract is supposed to be due to the Frisians, who were familiar with embanking, their own country being in a great measure the result of laborious industry in reclaiming and preserving it from the inland as well as the outland waters. English history does not reach so far back as the period at which Romney Marsh was first reclaimed, but doubtless the work is one of great antiquity. The district is about fourteen miles long and eight broad, divided into Romney Marsh, Wallend Marsh, Denge Marsh, and Guildford Marsh. The tract is a dead uniform level, extending from Hythe, in Kent, westward to Winchelsea, in Sussex; and it is to this day held from the sea by a continuous wall or bank, on the solidity of which the preservation of the

---

in Ireland, where the people exhibit features altogether different from the more modern Milesian Celts of Munster, whose fine physical and moral characteristics remind one of the often-quoted description of them by O'Connell, as "the finest peasantry in the world"; nor was the description by any means exaggerated. The same process of colonization to which we have above referred is even now going forward in the western parts of Ireland, where the old Galway race is being gradually submerged by the wave of modern Irish flowing over them from the northern province of Ulster, and driving them to emigration in large numbers. It may further be observed that the same qualities which enabled the Teutonic races in early times to colonize the arable lands of England, continue to render the modern Englishman the best of all colonists. His self-dependence fits him for enduring the solitude of a wilderness until he has reclaimed it by his

industry. He builds a house in the midst of his clearing, takes up his dwelling there, and his house becomes his castle. This remarkable and inherent difference between the Celt and the Teuton is curiously exemplified by the actual state of things in modern France and England. In the former the agricultural population live in villages, often far from the land they cultivate; in the latter they live in hamlets and detached dwellings, directly upon the soil on which they work. The same characteristic is illustrated in another way. When a Frenchman makes a fortune, he settles in Paris; when an Englishman does so, he retires to live in the country.

[1] Rumen-ea, *Sax.*—*i. e.*, the large watery place. Mr. Holloway is, however, of opinion that the word Roman-ea means "the Isle of the Romans," and was applied to the town of Romney, originally situated upon an island reclaimed by that people.—Holloway's 'History of Romney Marsh.'

district depends, the surface of the marsh being under
the level of the sea at the highest tides.   The following
descriptive view of the marsh, taken from the high
ground above the ancient Roman fortress of Portus
Limanis, near the more modern but still ancient castle
of Lymne, will give an idea of the extent and geographical relations of the district.

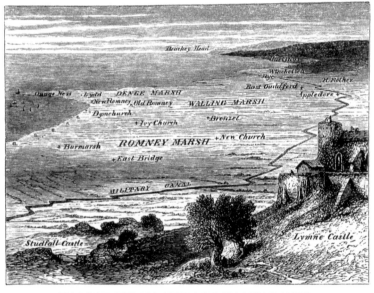

DESCRIPTIVE VIEW OF ROMNEY MARSH, FROM LYMNE CASTLE.

[By Percival Skelton, after his original design.]

The tract is so isolated, that the marshmen say the
world is divided into Europe, Asia, Africa, America,
and Romney Marsh.   It contains few or no trees, its
principal divisions being formed by dykes and watercourses.   It is thinly peopled, but abounds in cattle and
sheep of a peculiarly hardy breed, which are a source
of considerable wealth to the marshmen ; and it affords
sufficient grazing, in genial years, for more than half a
million of sheep, besides numerous herds of cattle.

The first portion of the district reclaimed was an
island, upon which the town of Old Romney now stands ;

and embankments were extended southward as far as New Romney, where an accumulation of beach took place, forming a natural barrier against further encroachments of the sea at that point. The old town of Lydd[1] also originally stood upon another island, as did Ivychurch, Old Winchelsea, and Guildford; the sea sweeping round them and rising far inland at every tide. Burmarsh, and the districts thereabout, were reclaimed at a more recent period; and by degrees the islands disappeared, the sea was shut out, and the whole became firm land. Large additions were made to it from time to time by the deposits of shingle along the coast, which left several towns, formerly important seaports, stranded upon the beach far inland. Thus the ancient Roman port at Lymne, past which the Limen or Rother is supposed originally to have flowed, is left high and dry more than three miles from the sea, and sheep now graze where formerly the galleys of the Romans rode. West Hythe, one of the Cinque Ports, originally the port for Boulogne, is silted up by the wide extent of shingle used by the modern School of Musketry as their practising-ground. Old Romney, about the centre of the marsh, past which the Rother afterwards flowed, was one of the ancient ports of the district, but it is now about two miles from the sea. The marshmen seem to have followed up the receding waters, and then founded the town of New Romney, which also became a Cinque Port; but a storm which occurred

---

[1] Somner, in his 'Treatise of the Roman Ports and Forts in Kent,' cites the Charter of Offa, king of the Angles, by which he grants the Manor of Lydd to Archbishop Janibert in the year 774; and the boundaries are thus described :—" The sea on the north and east, and on the south the territory of King Edwy. It is called Dengemarsh as far as the stone which is placed at the extreme point of the land; and to the west and north, the confines of the kingdom, as far as to Bleechinge." " From whence," adds Somner, " clear enough it is, that the sea, with a large and spacious inlet, arm, and estuary, in those days flowed in between Lydd and Romney, and was there met with the river Limen." Up this river a Danish fleet of 250 vessels sailed many miles inland in the year 893, and the Danes built a castle at Appledore, where they for some time held their rendezvous.

in the reign of Edward I. so blocked up the Rother
with shingle, at the same time breaching the wall, that
the river took a new course, and flowed thenceforward
by Rye into the sea; and the port of New Romney
became lost.   The point of Dungeness, running almost
due south, gains accumulations of shingle so rapidly
from the sea, that it is said to have extended more
than a mile seaward within the memory of persons living.
Rye was founded on the ruins of the Romneys, and also
became a Cinque Port; but notwithstanding the advantage of the river Rother flowing past it, that port has
also become nearly silted up, and now stands about two
miles from the sea.   New Winchelsea, the Portsmouth
and Spithead of its day, is left stranded like the rest
of the old Cinque Ports, and is now but a village surrounded by the remains of its ancient grandeur.   All
this ruin, however, wrought by the invasions of the
shingle upon the seacoast towns, has only served to increase the area of the rich grazing ground of the marsh,
which continues year by year to extend itself seaward.

The colonists who first reclaimed the district must
have found it necessary at once to organize some method
of maintaining the lands won from the sea.   Accordingly we find a very ancient local usage existing
in Romney Marsh, which, though at first unwritten,
eventually acquired the force of law, and was afterwards extensively applied in other districts.   Indeed,
" the law and custom of Romney Marsh" to this day
lies at the bottom of all English legislation on the subject of embanking and draining.   Twenty-four of the
chief men or elders were chosen by the inhabitants to
take all such measures as might be necessary to maintain the sea-banks, and their custom was to levy a
rate upon the occupiers of marsh lands in proportion to
their holdings, for the purpose of executing the necessary repairs.   As long ago as the reign of Henry III.,
or more than six hundred years since, when complaint

was made to him by the twenty-four "jurats"—as the conservators of the marshes were called—that certain holders of lands in the marshes refused to pay their rateable proportion of expenses for maintaining the banks, the King, referring to "the ancient and approved custom" of the district, ordered the sheriff to execute warrants of distress upon the defaulters, and thus support the jurats in the exercise of their customs and liberties for the defence of themselves and others against the sea, and that no peril might by their neglect in so doing come to the King or his kingdom. In due course of time the custom became embodied in a written law, confirmed by the letters patent of successive monarchs; and for many hundred years the successive embankments of land in Romney Marsh have been continued under this ancient jurisdiction, which, in all essential respects, remains in force at the present day.

From the very earliest times the tendency to "scamp" work seems to have existed; and not only so, but the tendency to job on the part of those who had public moneys to disburse. Thus, in the reign of Edward II., we find that the sea broke through the bank near Denge Marsh, and inflicted great injury on the marshmen. On inquiry it was found that the maintenance had been neglected, that the banks had been imperfectly repaired, and that the whole mischief had been caused "through the pravity of ill-disposed men, who chiefly mind their particular gain, though it be by cheating the public; that were it not for a strict watch over them, all good order would be subverted, and little else but cosenage, if not rapine, would be practised."[1]

The same custom of Romney Marsh with respect to the embankment of lands, prevailed all over Kent; and in the Isle of Thanet, at Sandwich, and along the low marsh lands in the valley of the Stour, the like practice

---

[1] Dugdale, 'History of Imbanking and Draining,' p. 29 — a work containing a great deal of curious information on this subject.

was extensively adopted.[1]  But by far the most interesting and important early work of this class was the embankment of the Thames, now become the great highway between the capital of Britain and the world.  It
may not be generally known, but it is nevertheless
true, that the Thames is an artificial river almost from
Richmond to the sea.  Before human industry confined
it within its present channel, it was a broad estuary, in
many parts between London and Gravesend several
miles wide.  The higher tides covered Plumstead and
Erith Marshes on the south, and Plaistow, East Ham,
and Barking Levels on the north ; the river meandering
in many devious channels at low water, leaving on either
side vast expanses of rich mud and ooze.  Opposite the

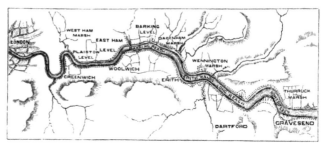

MAP OF THE VALLEY OF THE THAMES (WESTERN PART).

[Ordnance Survey.]

City of London, the tides washed over the ground now
covered by Southwark and Lambeth ; the district called
Marsh still reminding us of its former state, as Bankside
informs us of the mode by which it was reclaimed by
the banking out of the tidal waters.

A British settlement is supposed to have been formed at

---

[1] There is a tradition extant to the
effect that Goodwin Sands were once
dry land protected by embankments,
and that in consequence of a rate levied
for their repair having been diverted
towards the building of Tenterden
Steeple, the sea burst in and swallowed
up the land.  Hence the well-known
proverb, otherwise inexplicable, of
" Tenterden Steeple the cause of Goodwin Sands ;" though, if the tradition be
founded on fact, it possesses the usual
pertinence of most old proverbs.

an early period on the high ground on which St. Paul's Cathedral stands, by reason of its natural defences, being bounded on the south by the Thames, on the west by the Fleet, and on the north and east by morasses, Moorfields Marsh having only been reclaimed within a comparatively recent period.   The natural advantages of the situation were great, and the City seems to have acquired considerable importance even before the Roman period.   The embanking of the river has been attributed to that indefatigable people ; but on this point no evidence exists.   The numerous ancient British camps found in all parts of the kingdom afford sufficient proof that the early inhabitants of the country possessed a knowledge of the art of earthwork ; and it is not im-

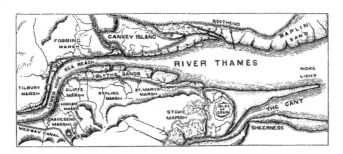

MAP OF THE VALLEY OF THE THAMES (EASTERN PART).

N.B. The dotted line represents the embankments raised along the banks of the river.

probable that the same Belgian tribes who reclaimed Romney Marsh were equally quick to detect the value for agricultural purposes of the rich alluvial lands along the valley of the Thames, and proceeded accordingly to embank them after the practice of the country from which they had come.   The work was carried on from one generation to another, as necessity required, until the Thames was confined within its present limits, the process of embanking serving to deepen the river and greatly improve it for purposes of navigation, while large tracts of fertile land were at the same time added to the food-producing capacity of the country.

It was long, however, before the banks effectually resisted the higher tides in the Thames, and scarcely a season passed without some breach occurring, followed by an inundation of the marsh-lands which had been won. There were frequent burstings of the bank on the south side between London Bridge and Greenwich ; the district of Bermondsey, then green fields, being especially liable to be submerged.  Commissions were appointed on such occasions, with full powers to distrain for rates and to impress labourers in order that the requisite repairs might at once be carried out.  In some cases the waters for a long time proved the victors, and carried everything before them.  Thus, in the reign of Henry VIII., the marshes of Plumstead and Lesnes, now used as a practising-ground by the Woolwich garrison, were completely drowned by the waters which had burst through Erith Breach, and for a considerable time all measures taken to reclaim them proved ineffectual.  The low lands lying immediately to the east of the royal palace at Greenwich, called Combe Marshes, were also inundated, and they were only reclaimed by the help of one Acontius, an Italian ; in reward for which an allotment was made to him of some six hundred acres of the recovered land. Another Italian, Baptista Castilione, with other undertakers, won back a further portion ; and at length the banks were all securely raised again along the south side of the river.

The maintenance of the embankment along the north shore seems to have been a work of still greater difficulty, and destructive inundations were frequent down to a comparatively recent period.  Thus we hear of breaches occurring at Wapping and Limehouse as late as the sixteenth century, and the Isle of Dogs was often overflowed and recovered with difficulty.  Lower down the river, the long bank which protects the Dagenham and Barking Levels was particularly liable to be burst through, by which the whole valley of the Lea, as

well as the rich lands along the south boundary of
Essex, were frequently laid under water. The drowned
state of these lands led to the employment, in 1621, of
the distinguished Dutch engineer Cornelius Vermuyden,
who succeeded in repairing the breaches and recovering
a large extent of inundated country; although we shall
afterwards find the waters from time to time breaking
through the banks, and temporarily re-establishing their
ascendency.

In many other parts of England, similar works of
reclamation were carried out according to the law and
custom of Romney Marsh. There was an extensive
tract of marsh in Somerset to which the same practice
was successfully applied—the well-known district of
Sedgemoor; so called because of the rushes, reeds, and
sedge with which it was overgrown in summer, whilst
in winter it was covered with water and could only be
traversed in boats. In ancient times, an abbot of Glas-
tonbury made a causeway across the level, which is still
named after him Graylock's Fosse. But it was not
until the reign of James I. that that rich tract of
about thirteen thousand acres of land, now covered with
orchards and cornfields, was reclaimed by drainage and
embankment; and it was the dykes or "rhines" cut in
the moor which afterwards threw Monmouth's untrained
troops into confusion during their night-attack on the
royal army, and in a great measure caused their dis-
comfiture.

For many centuries the English people were thus
engaged in slowly subduing the stubborn soil, reclaiming
the waste places, and making the land pleasant and
healthy to live in. While the population remained
comparatively scanty, the urgency to labour was not
excessive, and the progress made, although steady, was
but slow. Englishmen were, for the most part, satisfied
in those olden times to live like respectable country
folks; not burdened with many younger sons; enjoying

leisure and enjoying liberty; not very enterprising nor very laborious; taking things easy. But the growth of a healthy population, confined within a limited territory, gradually stimulated them to increased effort, and brought out their full working energy. The climate was favourable to labour, and patient industry was the inheritance of the race which held possession of the soil. By degrees its native riches were brought to light: it was abundantly stored with the best raw materials,—stone, lime, clay, coal, iron, and the useful metals; but, best of all, it was peopled with strong, hardy, willing men.

The land was too full of natural wealth, and lay too near the powerful military nations of Europe, to be held by a weak or idle people. It was large enough for independence, and, though limited by its coast-line on all sides, contained within it the elements of almost boundless expansion. It lay anchored by the side of Europe, in the watery highway between north and south, with its chief navigable river, the Thames, offering a convenient access to ships from every port. By persistent industry the land was gradually transformed. Year by year, and century by century, the people went on improving it. From a pasture-range it became a corn-farm and a garden; and alongside of agriculture there grew up handicrafts, trades, and manufactures, by which its rich raw materials were worked up in all manner of tools, machines, and fabrics. The powers of nature were laid under contribution, and wind, water, and steam became the allies and servants of man. Bridle-paths were superseded by wheel-roads; rivers were dug out and deepened, or used to feed artificial channels of water-communication; and from a land of horse-vehicles it became one of steam-trains and iron railroads. All branches of agriculture, trade, and manufacture, advanced with accelerated pace, until England has become the world's school of industry and the world's workshop.

# CHAPTER II.

## The Great Level of the Fens.

Situated as Britain is in a vapoury climate, and surrounded by the sea, water was, from the first, the chief element which English skill and industry had to fight against; and in effectively resisting it, or in subjugating and controlling it, the engineer's talent was first displayed. We have seen that to reclaim and hold the land against the violence of the ocean, embankments and sea-walls were built around the low-lying grounds on the coast. A subsequent stage of operations was to conduct the surplus rain and river waters descending from the interior across long stretches of level land, sometimes through fens and marshes, into the sea; for which purpose wide drains had to be dug, and powerful sluices erected at the point of junction of the fresh water with the salt. Then, still contending with the powers of water, the engineer raised lighthouses on solitary rocks, far out at sea, capable of resisting the most violent force of the waves; whilst around the coast he built piers and formed harbours for the accommodation of the ships thus safely lighted to port. To connect county with county, the opposite banks of rivers were bound together by means of bridges, the engineer still fighting against water in securing the foundations for his piers. Then he skilfully contrived to subdue the same element, and convert it into his most docile servant; leading it through new channels to drive mills and machinery, or along aqueducts for the supply of towns, or in canals upon which boats and ships laden with merchandise might be safely floated.

In the first place, water had to be contended against as a fierce enemy. This was the case at Romney Marsh and along the banks of the Thames, as we have already described. But perhaps the most interesting district of the kind in England, where for centuries the struggle has been persistently maintained by ingenuity and industry against the powers of water both within and without the land, is the extensive low-lying tract of country, situated at the junction of the counties of Lincoln, Huntingdon, Cambridge, and Norfolk, commonly known as the Great Level of the Fens. The area of this district presents almost the dimensions of a province, being from sixty to seventy miles from north to south, and from twenty to thirty miles broad, the high lands of the interior bounding it somewhat in the form of a horse-shoe. It contains about 680,000 acres of the richest land in England, and is as much the product of art as the kingdom of Holland, opposite to which it lies. It has been reclaimed and drained by the labour of successive generations of engineers, and it is only preserved for purposes of human habitation and culture by continuous watchfulness from day to day. As presenting a series of some of the finest works which energy and perseverance have ever achieved, we regard these great Fen districts—flat and unattractive though they be to the lovers of the picturesque—as among the most interesting parts of England.

Not many centuries ago, this vast tract of about two thousand square miles of land was entirely abandoned to the waters, forming an immense estuary of the Wash, into which the rivers Witham, Welland, Glen, Nene, and Ouse discharged the rainfall of the central counties of England. It was an inland sea in winter, and a noxious swamp in summer, the waters expanding in many places into settled seas or meres, swarming with fish and screaming with wild-fowl. The more elevated parts were overgrown with tall reeds, which appeared at a distance like fields of waving corn; and they were haunted by immense flocks of starlings, which, when disturbed, would rise in such numbers as almost to darken the air. Into this great dismal swamp the floods descending from the interior were carried, their waters mingling and winding by many devious channels before they reached the sea. They were laden with silt, which became deposited in the basin of the Fens. Thus the river-beds were from time to time choked up, and the intercepted waters forced new channels through the ooze, meandering across the level, and often winding back upon themselves, until at length the surplus waters, through many openings, drained away into the Wash. Hence the numerous abandoned beds of old rivers still traceable amidst the Great Level of the Fens—the old Nene, the old Ouse, and the old Welland. The Ouse, which in past times flowed into the Wash at Wisbeach (or Ouse Beach), now enters at King's Lynn, near which there is another old Ouse. But the probability is that all the rivers flowed into a lake, which existed on the tract known as the Great Bedford Level, from thence finding their way, by numerous and frequently shifting channels into the sea.

Along the shores of the Wash, where the fresh and salt waters met, the tendency to the deposit of silt was the greatest; and in the course of ages, the land at the outlet of the inland waters was raised above the level of the interior. Accordingly, the first land reclaimed in the district was the rich fringe of deposited silt lying along the shores of the Wash, now known as Marshland and South Holland. This was the work of the Romans, a hard-working, energetic, and skilful people; of whom the Britons are said to have complained[1] that they wore out and consumed their hands and bodies in clearing

---

[1] Tacitus, 'Life of Agricola.'

the woods and banking the fens. The bulwarks or causeways which they raised to keep out the sea are still traceable at Po-Dyke in Marshland, and at various points near the old coast-line.

On the inland side of the Fens the Romans are supposed to have constructed another great work of drainage, still known as Carr Dyke, extending from the Nene to the Witham. It means Fen Dyke, the fens being still called Carrs in certain parts of Lincoln. This old drain is about sixty feet wide, with a broad, flat bank on each side; and originally it must have been at least forty miles in extent, winding along under the eastern side of the high land, which extends in an irregular line up the centre of the district from Stamford to Lincoln. It was calculated to receive all the high-land and flowing waters, preventing them flooding the lower grounds; and was thus of the nature of an intercepting or " catch-water " drain. [1]

The same people also laid several causeways across the Fens for military purposes. Thus Herodian [2] alludes to the construction of such causeways for the purpose of enabling the Roman soldiers to pass over them and fight on dry land, the Britons having taken refuge from them by swimming. Such was probably the origin of the causeway made of gravel—still traceable, though in most places covered over with moor-soil—extending

---

[1] Mr. Rennie had the highest opinion of this work; and in one of his reports, written about fifty years since, he thus described it:—" This great Roman work extended originally from the river Nene below Peterborough to the city of Lincoln, and perhaps even to the river Trent at Torksey. I have traced its course for the greatest part of the way, and a more judicious and well-laid-out work I have never seen. But through neglect it has been suffered in many places to be encroached upon, and in others to be silted up, to the great injury of the country originally benefited by it; and, although the part between the Nene and the Welland has been kept in better repair than any other part I have seen, it is far short of what it ought to be; and to the bad condition of this drain much of the injury done by the floods to the first district of the North Level is to be attributed."

[2] ' Life of Severus,' lib. iii.

from Denver in Norfolk over the Great Wash to Charke, and from thence to Marsh and Peterborough, a distance of nearly thirty miles.[1]

The eastern parts of Marshland and Holland were thus the first lands reclaimed in the district, and they were available for purposes of agriculture long before any attempts had been made to drain the lands of the interior. Indeed, it is not improbable that these early embankments thrown up along the coast had the effect of increasing the inundations of the lower-lying lands of the level; for, whilst they dammed the salt water out, they also held back the fresh, no provision having been made for improving and deepening the outfalls of the rivers flowing through the Level into the Wash. The Fen lands in winter were thus not only flooded by the rainfall of the Fens themselves, and by the upland waters which flowed from the interior, but also by the daily flux of the tides which drove in from the German Ocean, holding back the fresh waters, and even mixing with them far inland.[2]

The Fens, therefore, continued flooded with water down to the period of the Middle Ages, when there seems to have been water enough in the Witham to float the ships of the Danish sea rovers as far inland as Lincoln, where ships' ribs and timbers have recently

---

[1] The causeway was about sixty feet broad, and laid with gravel about three feet thick. A cutting made across it at Eldernell shows the permanent manner in which the Romans did their work. It is laid upon the moor, the lowest layer being of oak-branches, then a considerable thickness of Northamptonshire rough flagstone, and then alternate layers of gravel with a small layer of clay, which, together, have formed a cement that nothing but the vigorous application of the pick can remove.

[2] The tides in the Wash are about the highest and strongest on the east coast. They rush with great vehemence through the harbours of Lynn, Wisbeach, and Boston,—the resistance caused by their meeting with the ebb-waters being called the Aegar, which rises to its greatest impetuosity during the equinoxes. This word Aegar, or Higre, it may be added, is merely the name of the old North-men's god, applied, like Neptune, to the sea itself.

been found deep sunk in the bed of the river. The first reclaimers of the Fen lands seem to have been the religious recluses, who settled upon the islands overgrown with reeds and rushes which rose up at intervals amidst the Fen level, and where, amidst the waste, they formed their solitary settlements. One of the first of the Fen islands thus occupied was the Isle of Ely, or Eely—so called, it is said, because of the abundance and goodness of the eels caught in the neighbourhood, and in which rents were paid in early times. It stood solitary amidst a waste of waters, and was literally an island. Etheldreda, afterwards known as St. Audrey, the daughter of the King of the East Angles, retired thither, secluding herself from the world and devoting herself to a recluse life. A nunnery was built, then a town, and the place became famous in the religious world. The pagan Danes, however, had no regard for Christian shrines, and a fleet of their pirate ships, sailing across the Fens, attacked the island and burnt the nunnery. It was again rebuilt and a church sprang up, the fame of which so spread abroad that Canute, the Danish king, determined to visit it. It is related that as his ships sailed towards the island his soul rejoiced greatly, and on hearing the chanting of the monks in the quire wafted across the waters, the king joined in the singing and ceased not until he had come to land. Canute more than once sailed across the Fens with his ships, and the tradition survives that on one occasion, when passing from Ramsey to Peterborough, the waves were so boisterous on Whittlesea Mere (now a district of fruitful cornfields), that he ordered a channel to be cut through the body of the Fen westward of Whittlesea to Peterborough, which to this day is called by the name of the "King's Delph."

The Fen islands were also the refuge of that loose and lawless portion of the population, which even in civilized societies invariably hovers about the margin of the moor, the waste, and the fen. The islands were long the haunts of marauders and banditti, and when plunder failed them, the abundance of fish and fowl offered them a ready means of subsistence. But about the period of the Conquest a new class of refugees swarmed into them. The defeated and still resisting Saxons fled thither for shelter against the mailed men-at-arms of the Norman. The situation of Ely at the junction of the ancient branch of the Ouse (called the West River) with the Cam on its course from Essex and Cambridgeshire, surrounded by morasses and fens, and accessible only by one entrance at Aldreth High Bridge, rendered it of great strength. It became a camp of refuge for the Saxons, who, led by Hereward, maintained for many years their last desperate but unavailing struggle for independence.

The other Fen islands which acquired a similar celebrity in those ancient times were Crowland, Ramsey, Thorney, and Spinney. They rose up at intervals far apart amidst the dead watery level of the Fens, grown over with rushes, flags, and sedge. The atmosphere which hung over them was moist and putrid, and "full of rotten harrs." But the very desolation and horror which enveloped the district seem to have proved attractions in the eyes of the recluse Guthlac—the saint of the Fen islands. Having journeyed towards the margin of the Fens, he inquired of the borderers what they knew thereof; and they told him many things of the dreadfulness and solitude of these places, but especially that in the remote and secret parts of the Fen there lay a certain island which no one dared to inhabit because of the strange and uncouth monsters with which it abounded. Whereupon Guthlac earnestly entreating that he might be shown that place, a fisherman proceeded to row him thither in his boat, and landing him at the spot now known as Croyland, there left him. Guthlac built for himself "a hut in a hollow, on the side of a heap of turf;" and what with worship and what with work, Guthlac gradually converted the little island into a green oasis amidst the waste.[1] As his fame spread abroad, other worshippers and labourers gathered around him, and Croyland soon became the most flourishing of all the islands in the Fens. At first the soil was so rotten and boggy, that a pole might be thrust down into it for thirty feet; but by digging and embankment, by tillage and culture, the land was converted into a garden of plenty. On the site of Guthlac's wooden oratory a new and stately stone structure was built on oak and alder piles driven deep into the bog, and the abbey of Croyland became the resort of pilgrims from far and near. A village and then a town sprang up—causeways and embankments were extended farther into the Fens—drains and sluices were dug to let off water from the standing pools—more land was reclaimed and tilled—until the monastery grew richer and richer, and increasing numbers of people resorted to Croyland for purposes of devotion, employment, and subsistence. Other islands near at hand were gradually subdued in like manner, to which the Croy-

---

[1] The horrors first encountered by Guthlac in his desolate island are graphically described in the following metaphorical account, contained in the Life of the Saint in the Cottonian Library. Not long after his landing, the legend says, — " St. Guthlac, being awake in the night time, betwixt his hours of prayer, as he was accustomed, of a sudden he discerned his cell to be full of black troops of unclean spirits, which crept in under the door, as also at chinks and holes; and, coming in both out of the sky and from the earth, filled the air, as it were, with dark clouds. In their looks they were cruel, and of form terrible, having great heads, long necks, lean faces, pale countenances, ill-favoured beards, rough ears, wrinkled foreheads, fierce eyes, stinking mouths, teeth like horses, spitting fire out of their throats, crooked jaws, broad lips, loud voices, burnt hair, great cheeks, high breasts, rugged thighs, bunched knees, bended legs, swoln ankles, preposterous feet, open mouths, and hoarse cries; who with such mighty shrieks were heard to roar, that they filled almost the whole distance from heaven with their bellowing noises; and by and by rushing into the house, first bound the holy man, then drew him out of his cell, and cast him over head and ears into the dirty fen; and, having so done, carried him through the most rough and troublesome parts thereof, drawing him amongst brambles and briars for the tearing of his limbs." It would appear from this graphic report as if the horrid stagnancy of the fens had afflicted the saint with an intolerable nightmare.

land men went in their boats or skerries to milk the
cows—the boats being so small that they could carry
only two men and their milk-pails.   As yet no corn
grew within five miles of Croyland, and there was an
old proverb of the district which said that "all the carts
that come to Croyland are shod with silver," for the
good reason that the ground all about it was so boggy
that neither horse nor cart *could* approach it; and hence
the proverb.

Thorney and Ramsey were other Fen islands, each
the seat of an abbey.   Both stood solitary amidst the
dead level waste around them.   Deep and boggy quagmires separated Ramsey from the high lands on the
west, whilst several large meres, abounding in eels,
pikes, hakedes, and other fish, stretched away towards
the east.   Like Croyland and Thorney, it was approach-
able only by boats, until a causeway was made to it
across the marsh — the monks being the engineers.
Another of these causeways was made from Soham to
Ely, which was considered in its day a work so wonderful, that it was afterwards attributed to a miracle
performed by the monk who constructed it; and Egelric,
a Peterborough monk, made a firm causeway of wood
and gravel through the Fens between Deeping and
Spalding, for the convenience of foot-passengers.   A
considerable inducement to the industry of the churchmen was, no doubt, the increased value given to the
Fen lands thus reclaimed, which were added from time
to time to the endowments of their respective establishments.   Hence we find serious disputes occurring
between the Bishops of Ely and the Abbots of Ramsey
as to the boundaries of their Fen lands, and the controversy became so hot amongst the brethren on one occasion, that it is related that "on the feast-day of Saint
Peter ad Vincula, two of the canons of the priory of the
Holy Trinity, disputing thereof, grew to such high words
as contracted an implacable hatred betwixt them, that,

studying a revenge, the one took an opportunity to murther the other." [1]

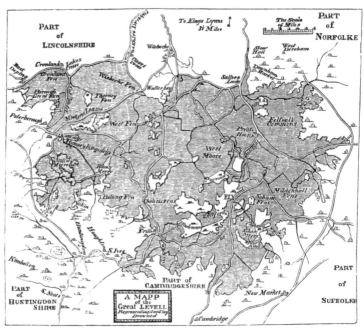

MAP OF THE FENS AS THEY LAY DROWNED.[2]
[After Dugdale.]

These attempts at reclaimment, however, made comparatively small impression on the vast extent of drowned lands forming the great Fen Level. They touched only the higher points, and being conducted on no defined plan or system, the efforts made to drain one spot often had the effect of only sending a flood of water upon another, or perhaps diverting in some new

---

[1] Dugdale, 'History of Draining and Imbankment.'

[2] The map represents the Fens as they lay drowned at a much later period than that above described. The parts dark shaded were covered with water as late as the reign of James I. The map is corrected from Dugdale's 'History of Drainage and Imbankment,' and rendered more intelligible to modern readers. In Dugdale's book, the south is at the top of the map, and the west to the right hand—the reverse of our present arrangement. But the old map-makers were by no means particular; and there are other maps in Dugdale in which the north is to the right of the map, and the west is at the top.

direction the water which before had driven a mill, or formed a channel for purposes of navigation. The rivers also were constantly liable to get silted up, and to form for themselves new courses; sometimes in a night undoing all that it had taken years to accomplish. Hence we find perpetual local litigation prevailing in the district from a very early date; and numerous inquisitions were held for the purpose of determining in what way the waters were best to· be dammed back, or in any way got rid of. In these inquisitions the Bishop of Ely, the Abbots of Croyland, Thorney, and Ramsey, and the Prior of Spalding, took a leading part—being concerned for the large Fen property attached to their respective establishments. A few of the large landowners attempted experiments in drainage on a small scale; but in those days the enterprise of the barons was mostly in a warlike direction. In 1427 Gilbert Halloft, a Baron of the Exchequer, residing at Well, in what is called the North Level, drained and cultivated a small tract of country with tolerable success; and another baron, Richard de Rulos, Lord of Burne and Deeping, by diverting the waters of the Welland and building them back by strong embankments, succeeded in reclaiming the very rich lands of Market Deeping, " out of the very pits and bogs thereby making a garden of pleasure."

Deep Ing, or Low Meadow, is one of the lowest parts of the Fens, being below the level of the sea at high-water, but rich and fertile as any land in England. Indeed many of the richest Fen lands lie considerably beneath the sea-level—those inland being even lower than the marsh lands which fringe the sea-coast. The floor of the old church at Wigenhall, St. Germans, was at least seven feet below high-water mark of the river Ouse. The same river, after one of its burstings, washed away the churchyard at Old Lynn, and compelled the removal of the church farther inland. The violence of the sea was often felt by the poor inhabitants of these remote districts; for sometimes, in a single night, was undone the tedious industry of centuries. But the district then lay far apart from the highways of intercourse in England. Marshland was cut off from the inland counties by the impassable Fens which lay in the hollow of the Great Level; and the troubles and sufferings of the marshmen and fenmen excited but little interest. On one occasion, however, a royal army had nearly been cut off by the fury of the sea driving up the Wash, impelled by one of the north-east winds formerly so destructive along that coast. It was the army of King John when crossing the marshes between King's Lynn and Sleaford. They had nearly reached the north shore when they heard the terrible roar of the Aegar. Pressing on, impelled by terror, the King, with his immediate followers, succeeded in reaching the firm land, but not a minute too soon, for the carriages and sumpter horses, which bore the military chest, were swallowed up in the whirlpool caused by the furious meeting of the flowing tide with the waters of the Welland. Hence Shakespeare, in his tragedy of ' King John,' makes Falconbridge say to Hubert—

> " I tell thee, Hubert, half my power this night,
> Passing these flats, is taken by the tide;
> These Lincoln washes have devoured them;
> Myself, well-mounted, hardly have escaped."

Each suffering locality, acting for itself, did what it could to preserve the land which had been won from the sea, and to check the recurrence of inundations. Dyke-reeves were appointed along the sea-borders, with a force of shore-labourers at their disposal, to see to the security of the embankments; and fen-wards were constituted inland, over which commissioners were set, for the purpose of keeping open the drains, maintaining the dykes, and preventing destruction of life and property by floods, whether descending into the Fens from the high lands or bursting in upon them from the sea. Where lands became suddenly drowned, the Sheriff was authorised to impress diggers and labourers for raising embankments; and commissioners of sewers were afterwards appointed, with full powers of local action, after the law and usage of Romney Marsh.   In one district we find a public order made that every man should plant with willows the bank opposite his portion of land towards the fen, "so as to break off the force of the waves in flood times;" and swine were not to be allowed to go upon the banks unless they were ringed, under a penalty of a penny (equal to a shilling in our money) for every hog found unringed.

One of the first works attempted on a large scale, with a view to the thorough drainage of part of the North Level, was that carried out by John Morton, Bishop of Ely, in the reign of Henry VII.   He caused a forty-foot cut or canal to be dug from near Peterborough to Guyhirne, continuing it eastward, through Wisbeach, to the sea, the distance being forty miles.   Its object was, to enable the overflowings of the river Nene, into which the drainage of many thousands of acres of land flowed, to be more quickly evacuated, and at the same time to enable navigation to be carried on between Peterborough and the sea.   The Bishop took great pleasure in superintending the construction of the work, which is still called by his name, 'Morton's Leam.'   He had a lofty brick tower built at Guyhirne, where the waters met, and "up into that tower he would often go to oversee and set out the works." [1]   This Bishop was the first to introduce into the district the practice of making straight cuts and artificial rivers for the purpose of more rapidly voiding the waters of the Fens—a practice which has been extensively adopted by the engineers who succeeded him.

---

[1] Atkyns's Report, anno 1618.

At the dissolution of the monasteries, in the reign of Henry VIII., the drainage of the Fens, which, up to that time, had been conducted principally by the churchmen, suffered a serious check. The embankments were neglected; the rivers were allowed to silt up; and the Fen lands, covered with water, were abandoned to fishes and water-fowl. The sea on the one hand, and the inland waters on the other, required the exercise of unceasing vigilance. Not only had the sea to be held back, but the inland waters must be got rid of; and of the two, the latter were usually the most dangerous. Where the river outfalls were neglected, and allowed to be choked up by silt, the first heavy rainfall in the interior brought down a flood, for whose escape there was no adequate outlet; hence devastating inundations from time to time occurred, and the waters rapidly regained their dominion over the land. Thus, in Elizabeth's reign, there occurred a great drowning of the East Fen, near Boston, by which some five thousand acres of land, under partial cultivation, were completely submerged; and the poor fenmen who had formed their miserable settlements on the islets rising above the waste, saw their little holdings swept away. Most of them barely escaped with their lives, but many were drowned. Appeals were sent up to the Government respecting the deplorable state of the district; and renewed efforts were made to keep out the waters. A commission was appointed (20th Elizabeth) with the object of devising and adopting measures for the drainage of the North Level; but the work was of so formidable a character as to deter the commissioners from taking any steps beyond the simplest defensive measures to protect the land already reclaimed. A General Drainage Act was passed some years later, by the advice of Lord Burghley (43rd Elizabeth), which was of so comprehensive a character that it embraced the drainage of all the marshes and drowned lands in England. But comparatively little was done to carry its provisions into effect, although measures were about that time adopted, under concessions made to the Cecils and the Fitz-williams, which enabled them to drain, and that very ineffectually, several portions of land near Clough's Cross, to the eastward of Peterborough.

A few years later James I. ascended the throne, and shortly after, in 1607, a series of destructive floods burst in the embankments along the eastern coast of England, and, sweeping away many farms and vil-lages, did immense damage. One of the worst occurred through the bursting of Terington Dyke, near King's Lynn, by reason of the violence of a storm from the north-east, which drove the spring-tide through the sea-banks. The jury for the hundred of Freebridge held an inquest upon the bodies of those who were drowned, and the following brief extract from their presentment will give an idea of the horrors of the situation :—" In their distress the people of the town fled to the church for refuge ; some to haystacks ; some to the baulks in the houses, till they were near famished ; poor women leaving their children swimming in their beds, till good people, adventuring their lives, went up to the breast in the water to fetch them out at the windows ; whereof Mr. Browne, the minister, did fetch divers to the church upon his back ; and had it not pleased God to move the hearts of the mayor and aldermen of King's Lynn with compassion, who sent beer and victual thither by boat, many had perished ; which boats came the direct way over the soil from Lynne to Terington." [1]

---

[1] The following description is given in a curious little black-letter book of the period, entitled ' More strange Newes of Flouds in England :'—" In the danger every man layed first handes on what he loued best; some made away with his wife, some his children, some, careles both of wife and child, hurried away his goods. He that had seene this troublesome night's work would have thought upon the miserable night of Troy. Here waded one up to the middle loaded with wealth, when noting how the water increased, and calling to mind his haplesse children, with a sigh, as

When the King was informed of this great calamity which had befallen the inhabitants of the Fens, principally through the decay of the old works of drainage and embankment, he is said to have made the right royal declaration, that "for the honour of his kingdom, he would not any longer suffer these countries to be abandoned to the will of the waters, nor to let them lie waste and unprofitable; and that if no one else would undertake their drainage, he himself would become their undertaker." A commission was in the first place appointed to ascertain the extent of the evil, and from the schedule which was shortly afterwards prepared and sent in of the drowned lands lying more particularly along the river Ouse, it appeared that there were not less than 307,242 acres lying outside the Fen dykes which required drainage and protection. A bill was brought into Parliament for the purpose of enabling rates to be levied and measures to be taken for the drainage of this land, but the Bill was summarily rejected. Two years later, the "little bill" for draining 6000 acres in Waldersea county was passed—the first district act for Fen drainage sanctioned by Parliament. The King then called Chief-Justice Popham to his aid, and sent him down to the Fens to undertake a portion of the work; and he induced a company of Londoners to undertake another portion, the adventurers receiving two-thirds of the reclaimed lands as a recompense. "Popham's Eau," and "The Londoners' Lode," still mark the scene of these operations. The works, however, did not prove very successful, not having been carried out with sufficient practical knowledge on the

---

loath to part from what he so dearly loued, he throwes it downe, runs to bedde, wakens his wife, and from her sides snatches the sleeping infants. Here comes a husband with his wife on his backe, and under either arme an infant. The sonne carries the father, the brother the sister, the daughter the mother, whilest the unmerciful conqueror breakes downe the walles of the houses, taking pittie neither of aged nor sere, findes some at play, some a sleepe in chayres, many in their beds, that never dreamed of misfortune till it waked them."

part of the adventurers, nor after any well-devised plan. Indeed, so long as the river outfalls were neglected, the drainage of one district only had the effect of drowning some other; and hence arose perpetual quarrels amongst the fen-owners, with constant appeals to the law. The England of that day was very weak in engineering ability; and it was natural that the King, in this emergency, should bethink him of resorting for help to the skilled drainers of Holland, then the great country of water engineers. Out of this necessity arose the employment of Cornelius Vermuyden, the Dutchman, whose career in England will form the subject of succeeding chapters.

The need of skilled engineering for the rescue of the drowned lands in the Fens was at this time certainly most imminent. It would be difficult to imagine anything more dismal than the aspect which they presented. In winter, a sea without waves; in summer, a dreary mud-swamp. The atmosphere was heavy with pestilential vapours, and swarmed with insects. The meres and pools were, however, rich in fish and wildfowl. The Welland was noted for sticklebacks, a little fish about two inches long, which appeared in dense shoals near Spalding, every seventh or eighth year, and used to be sold during the season at a halfpenny a bushel, for field manure. Pikes were plentiful near Lincoln; hence the proverb, " Witham pike, England hath none like." Fen-nightingales, or frogs, especially abounded. The birds-proper were of all kinds; wildgeese, herons, teal, widgeons, mallards, grebes, coots, godwits, whimbrels, knots, dottrels, yelpers, ruffs, and reeves, many of which have long since been banished from England. Mallards were so plentiful that 3000 of them, with other birds in addition, have been known to be taken at one draught. Round the borders of the fens there lived a thin and haggard population of " Fen-slodgers," called " yellow-bellies" in other districts, who derived a precarious subsistence from fowling and fishing. They were described by writers of the time as "a rude and almost barbarous sort of lazy and beggarly people." Disease always hung over the district, ready to pounce upon the half-starved fenmen. Camden spoke of the country between Lincoln and Cambridge as "a vast morass, inhabited by fenmen, a kind of people, according to the nature of the place where they dwell, who, walking high upon stilts, apply their minds to grazing, fishing, or fowling." The proverb of "Cambridgeshire camels" doubtless originated in this old practice of stilt-walking in the Fens; the fenmen, like the inhabitants of the Landes, mounting upon high stilts to spy out their flocks across the dead level. But the flocks of the fenmen consisted principally of geese, which were called the "fenmen's treasure;" the fenman's dowry being "three-score geese and a pelt," or sheep-skin used as an outer garment. The geese throve where nothing else could exist, being equally proof against rheumatism and ague, though lodging with the natives in their sleeping-places. Even of this poor property, however, the slodgers were liable at any time to be stripped by sudden inundations.

In the oldest reclaimed district of Holland, containing many old village churches, the inhabitants, in wet seasons, were under the necessity of rowing to church in their boats. In the other less reclaimed parts of the Fens the inhabitants were much worse off. "In the winter time," said Dugdale, "when the ice is only strong enough to hinder the passage of boats, and yet not able to bear a man, the inhabitants upon the hards and banks within the Fens can have no help of food, nor comfort for body or soul; no woman aid in her travail, no means to baptize a child or partake of the Communion, nor supply of any necessity saving what these poor desolate places do afford. And what expectation of health can there be to the bodies of men, where there is no element good? the air being for the most part cloudy, gross, and full of rotten harrs; the water putrid and muddy, yea, full of loathsome vermin; the earth spungy and boggy, and the fire noisome by the stink of smoaky hassocks."[1]

The wet character of the soil at Ely may be inferred from the circumstance that the chief crop grown in the neighbourhood was willows; and it was a common saying there, that "the profit of willows will buy the owner a horse before that by any other crop he can pay for his saddle."[2]    There was so much water constantly lying above Ely, that in olden times the Bishop of Ely was accustomed to go in his boat to Cambridge.    When the outfalls of the Ouse became choked up by neglect, the surrounding districts were subject to severe inundations; and after a heavy fall of rain, or after a thaw in winter, when the river swelled suddenly, the alarm spread abroad, "the bailiff of Bedford is coming!" the Ouse passing by that town.    But there was even a more terrible bailiff than he of Bedford; for when a man was stricken down by the ague, it was said of him, " he is arrested by the bailiff of Marsh-land;" this disease extensively prevailing all over the district when the poisoned air of the marshes began to work.

---

[1] Dugdale, 'History of Imbanking and Draining.' In this curious old book a great deal of interesting matter is to be found relating to the drainage works of early times, though overlaid with considerable 'Dryasdust' citation.    Dugdale seems to have ransacked all literature for any information bearing, however remotely, on his subject.    He was employed under 'The Adventurers' when a young man, as early as 1643, and afterwards published his book at the request of Lord Georges, for some time Surveyor-General of the Great Bedford Level.

[2] 'Anglorum Speculum; or, the Worthies of England in Church and State:' London, 1684.

## CHAPTER III.

### Drainage of Hatfield Chase—Sir Cornelius Vermuyden.

Cornelius Vermuyden, the Dutch engineer, was invited over to England, about the year 1621, to stem a breach in the embankment of the Thames near Dagenham, which had been burst through by the tide. He was a person of good birth and education, the son of Giles Vermuyden, by Sarah his wife, who was the daughter of Sir Cornelius Wordendyke, a gentleman of some importance in his time. His birthplace was at St. Martin's Dyke, in the isle of Tholen, in Zeeland. He had been trained as an engineer, and having been brought up in a country where embanking was studied as an art and afforded employment to a considerable proportion of its inhabitants, he was familiar with the most approved methods of defending land against the encroachments of the sea. He was so successful in his operations at Dagenham, that when it was found necessary to drain the royal park at Windsor, he was employed to direct the labourers in that work, by which he became known to James I., who took a peculiar interest in works of internal improvement.[1] Among the several public undertakings promoted by that monarch, were the reclamation of Canvey Island, at the mouth of the Thames; Sedgemoor, in Somersetshire; Brading Haven, in the Isle of Wight; and the drainage of Hatfield Level and the Great Bedford Level; as well as the construction of the New River, hereafter to be described.

---

[1] From an order on the Exchequer, dated 13th Feb., 1623, it appears that the trenching and drainage of Windsor Great Park were executed under the direction of Vermuyden, at an expense of 300*l.*—' State Paper Office—Issues of the Exchequer.'

The extensive district of Axholme, of which the Level of Hatfield Chase formed only a part, resembled the Great Level of the Fens in many respects, being a large fresh-water bay formed by the confluence of the rivers Don, Went, Ouse, and Trent, which brought down into the Humber almost the entire rainfall of Yorkshire, Derbyshire, Nottingham, and North Lincoln, and into which the sea also washed. The uplands of Yorkshire bounded this watery tract on the west, and those of Lincolnshire on the east. Rising up about midway between them was a single hill, or rather elevated ground, formerly an island, and still known as the Isle of Axholme. There was a ferry between Sandtoft and that island in times not very remote, and the farmers of Axholme were accustomed to attend market at Doncaster in their boats, though the bottom of the sea over which they then rowed is now amongst the most productive corn-land in England. The waters extended to Hatfield, which lies along the Yorkshire edge of the level on the west; and it is recorded in the ecclesiastical history of that place that a company of mourners, with the corpse they carried, were once lost when proceeding by boat from Thorne to Hatfield. When Leland visited the county in 1607, he went by boat from Thorne to Tudworth, over what at this day is rich ploughed land. The district was marked by numerous merestones, and many fisheries are still traceable in local history as having existed at places now far inland.

The Isle of Axholme was in former times a stronghold of the Mowbrays, being unapproachable save by water. In the reign of Henry II., when Lord Mowbray held it against the King, it was taken by the Lincolnshire men, who attacked it in boats; and, down to the reign of James I., the only green spot which rose above the wide waste of waters was this solitary isle. In early times the whole of the south-eastern part of the county of York, from Conisborough Castle to the sea, belonged

to the royal domains; but one estate after another was
alienated, until at length, when James I. succeeded to
the crown of England, there only remained the manor
of Hatfield, which, watery though it was, continued to
be dignified with the appellation of a Royal Chase.
There was, however, plenty of deer in the neighbourhood, for De La Pryme says that in his time they were
as numerous as sheep on a hill, and that family venison
was as abundant as mutton in a poor man's kitchen.[1]
But the principal sport which Hatfield furnished was in
the waters and meres adjacent to the old timber manor-
house.   Prince Henry, the King's eldest son, on the
occasion of a journey to York, rested at Hatfield on his
way, and had a day's sport in the Royal Chase, which is
curiously described by De La Pryme :—" The prince and
his retinue all embarked themselves in almost a hundred
boats that were provided there ready, and having fright-
ened some five hundred deer out of the woods, grounds,
and closes adjoining, which had been drawn there the
night before, they all, as they were commonly wont,
took to the water, and this little royal navy pursuing
them, soon drove them into that lower part of the level,
called Thorne Mere, and there, being up to their very
necks in water, their horned heads raised themselves so
as almost to represent a little wood.   Here being encompassed about with the little fleet, some ventured
amongst them, and feeling such and such as were fattest,
they either immediately cut their throats, or else tying
a strong long rope to their heads, drew them to land
and killed them."

Such was the last battue in the Royal Chase of Hatfield; for shortly after, King James brought the subject
of the drainage of the tract under the notice of Cornelius
Vermuyden, who, on inspecting it, declared the project
to be quite practicable.   The level of the Chase con-

---

[1] De La Pryme, ' History of the Level of Hatfield Chase.'

tained about 70,000 acres, the waters of which found
their way to the sea through many changing channels,
like the rivers of the Fens.   There were numerous
places in the level deeper than others; some of them
meres abounding in fish; hence Fishlake and Fishtoft,
which were famous for this commodity.  Various attempts had been made to diminish the flooding of the
lands.  In the fourteenth century several deep trenches
were dug, to let off the water, but they probably
admitted as much as they allowed to escape, and the
drowning continued.  Commissioners were appointed,
but they did nothing.  The country was too poor, and
the people too unskilled, to undertake so expensive and
laborious an enterprise.

A local jury was then summoned by the King to
consider the question of the drainage, but they broke
up, after expressing their opinion of the utter impracticability of carrying out any effective plan for
the withdrawal of the waters.  Vermuyden, however,
declared that he would undertake and bind himself to
do that which the jury had pronounced to be impossible.
The Dutch had certainly been successful beyond all
other nations in projects of the same kind.  No people
had fought against water so boldly, so perseveringly,
and so successfully.  They had made their own land
out of the mud of the rest of Europe, and, being rich
and prosperous, were ready to enter upon similar enterprises in other countries.  On the death of James I.,
his successor confirmed the preliminary arrangement
which had been made with Vermuyden, with a view
to the drainage of Hatfield Manor; and on the 24th of
May, 1626, after a good deal of negotiation as to terms,
articles were drawn up and signed between the Crown
and Vermuyden, by which the latter undertook to
reclaim the drowned lands, and make them fit for
tillage and pasturage.  It was a condition of the contract
that Vermuyden and his partners in the adventure were

to have granted to them one entire third of the lands so recovered from the waters.

Vermuyden was a bold and enterprising man, full of energy and resources. He also seems to have possessed the confidence of capitalists in his own country, for we find him shortly after proceeding to Amsterdam to raise the money, of which England was then so deficient; and a company was formed, composed almost entirely of Dutchmen, for the purpose of carrying out the necessary works of reclamation. Amongst those early speculators in English drainage we find the names of the Valkenburgh family, the Van Peenens, the Vernatti, Andrew Boccard, and John Corsellis. Of the whole number of shareholders amongst whom the lands were ultimately divided, the only names of English sound are those of Sir James Cambell, Knight, and Sir John Ogle, Knight, who were amongst the smallest of the participants.

Many of the Dutch capitalists came over to look after their own interest, and Vermuyden collected from different parts the skilled labour of a large number of Dutch and Flemish workmen. It so happened that there were then scattered up and down over England numerous foreign labourers—Dutchmen who had been brought from Holland to embank the lands at Dagenham and Canvey Island on the Thames, and others who had been driven from their own countries by religious persecution — French Protestants from Picardy, and Walloons from Flanders. The countries in which those people had been born and bred resembled in many respects the marsh and fen districts of England, and they were practically familiar with the reclamation of such lands, the digging of drains, the raising of embankments, and the cultivation of marshy ground. Those immigrants had already settled down in large numbers in the eastern counties, and along the borders of the Fens, at Wisbeach, Whittlesea, Thorney, Spalding, and the neighbourhood.[1]   The poor foreigners readily answered Vermuyden's call, and many of them took service under him at Hatfield Chase, where they set to work with such zeal, and laboured with such diligence, that before the end of the second year the work was so far advanced that a commission was issued for the survey and division amongst the participants of the reclaimed lands.[2]   The plan of drainage adopted seems to have been, to carry the waters of the Idle by direct channels into the Trent, instead of allowing them to meander at will through the level of the Chase.   Deep drains were cut, through which the water was drawn from the large pools standing near Hatfield and Thorne.   The Don also was blocked out of the level by embankments, and forced through its northern branch, by Turn-bridge, into the river Aire.   But this last attempt proved a mistake, for the northern channel was found insufficient for the discharge of the waters, and floodings of the old lands about Fishlake, Sykehouse, and Snaith took place; to prevent which, a wide and deep channel, called the Dutch River, was afterwards cut, and the waters of the Don were sent directly into the Ouse, near Goole. This great and unexpected addition to the cost of the undertaking appears to have had a calamitous effect,

---

[1] It has been observed that the buildings in many of the old Fen towns to this day have a Flemish appearance, as the names of many of the inhabitants have evidently a foreign origin.   Those of Descow, Le Plas, Egar, Bruynne, &c., are said to be still common.   Among the settlers in the level of Hatfield was Mathew de la Pryme, who emigrated from Ypres in Flanders during the persecutions of the Duke of Alva.   Professor Pryme, of Cambridge, has been mentioned as a lineal descendant of the family. Tablets to several of the name are still to be found in Hatfield Church.

[2] The following document in the State Paper Office relates to the grant of the reclaimed lands:—"July 11, 1628.  Grant to Cornelius Vermuyden, of the manors of Hatfield, Fishlake, Thorpe, Stainforth, and Dowesthorpe, county York, subject to a rent of 150l. per annum, and to a covenant for the grant to be void if his Majesty repay to the grantee 10,000l. with interest on Septr. 25th."—Docquet. We also find in the same papers (vol. cvii. 14) a grant to Vermuyden for the lives of himself, his son, and his two daughters, Sarah and Catherine, of a moiety of divers wastes and surrounded grounds belonging to Misen, in the county of Nottingham, consisting of 2600 acres, reclaimed by the same works.

and brought distress and ruin on many who had engaged in it. The people who dwelt on the northern branch of the Don complained loudly of the adventurers, who

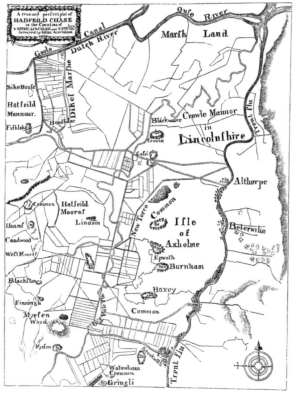

MAP OF THE LEVEL OF HATFIELD CHASE.
[Corrected, after Dugdale.]

were denounced as foreigners and marauders; and they were not satisfied with mere outcry, but took the law into their own hands; broke down the embankments, assaulted the Flemish workmen, and several persons lost their lives in the course of the riots which ensued.[1]

---

[1] R. Ansbie writes the Duke of Buckingham from Tickill Castle, under date the 21st August, 1628, as follows:—" What has happened betwixt Mr. Vermuyden's friends and workmen and the people of the Isle of Axholme these inclosed will give a taste. Great riots have been committed by the people, and a man killed by the Dutch party, the killing

Vermuyden did what he could to satisfy the inhabitants. He employed large numbers of native workmen, at considerably higher wages than had before been paid there; and he strenuously exerted himself to relieve those who had suffered from the changes he had effected, so far as could be done without incurring a ruinous expense.[1] Dugdale relates that there could be no question about the great benefits which the execution of the drainage.works conferred upon the labouring population; for whereas, before the reclamation, the country round about had been " full of wandering beggars," these had now entirely disappeared, and there was abundant employment for all who would work, at good wages. An immense tract of rich land had been completely recovered from the waters, but it could only be made valuable and productive after long and diligent cultivation. Vermuyden was supported by the Crown, and on the 6th of January, 1629, he received the honour of knighthood at the hands of Charles I., in recognition of the skill and energy he had displayed in adding so large a tract to the cultivable lands of England. In the same year he took a grant from the Crown of the

of whom is conceived to be murder in all who gave direction for them to go armed that day. These outrages will produce good effects. They will procure conformity in the people, and enforce Vermuyden to sue for favour at the Duke's hands,—if not for himself, for divers of his friends, especially for Mr. Saines, a Dutchman, who has an adventure of 13,000*l.* in this work. Upon examination of the rest of Vermuyden's people, thinks it will appear that he gave them orders to go armed. Expected to have heard from Mr. Fortherley about Vermuyden's defeazance, but he having fallen short therein, it has been given out by Vermuyden's party that he has bought the lordship; but the writer, with strong assurances, has removed that stumbling-block."—' State Papers,' vol. cxiii. 38.

[1] F. Vernatti, one of the Dutch capitalists who had contributed largely towards the cost of the works, writes to Monsieur St. Gillis, in October, 1628:—" The absence of Mr. Vermuyden, and the great interest the writer takes in the business of embankment at Haxey, has led him to engage in it with eye and hand. The mutinous people have not only desisted from their threats, but now give their work to complete the dyke, which they have fifty times destroyed and thrown into the river. A royal proclamation made by a serjeant-at-arms in their village, accompanied by the sheriff and other officials, with fifty horsemen, and an exhortation mingled with threats of fire and vengeance, have produced this result."—' State Papers,' vol. cxix. 73.

whole of the reclaimed lands in the manor of Hatfield, amounting to about 24,500 acres, agreeing to pay the Crown the sum of 16,080*l.*, an annual rent of 193*l.* 3*s.* 5½*d.*, one red rose ancient rent, and an improved rent of 425*l.* from Christmas, 1630.[1] Power was also granted him to erect one or more chapels wherein the Dutch and Flemish settlers might worship in their own language;[2] and they then settled down peacefully to cultivate the soil which their labours had won. They built houses, farmsteads, and windmills; and even now the Dutch look of the villages, and the Dutch character of many of the inhabitants in the district, are still observable.

It was long, however, before the hostility and jealousy of the native population could be appeased. The idea of foreigners settling as colonists upon lands over which, though waste and swamp, their forefathers had enjoyed rights of common, was especially distasteful to them, and bred bitterness in many hearts. All over the Fen district the dispossessed fenmen were almost in a state of revolt, and they had numerous sympathisers among the rest of the population. Thus, on one occasion, we find the Privy Council sending down a warrant to all postmasters to furnish Sir Cornelius Vermuyden with horses and a guide to enable him to ride post from London to Boston, and from thence to Hatfield.[3] But at Royston " Edward Whitehead, the constable, in the absence of the postmaster, refused to provide horses, and on being told he should answer for his neglect, replied, 'Tush! do your worst: you shall have none of my horses in spite of your teeth.' "[4] Complaints were made to the Council as to the injury done to the surrounding districts by the drainage works; and an inquisition was

---

[1] 'State Papers,' vol. cxlvii. 21.
[2] 'State Papers,' vol. cxxiii. 26.
[3] Warrant of Council, dated Whitehall, May 12th, 1630.—'State Papers,' vol. clxvi. 56.
[4] Affidavit of George Johnson, servant of Sir Cornelius Vermuyden.—'State Papers,' vol. clxx. 17.

held on the subject before the Earls of Clare and New-castle, and Sir Gervase Clifton. Vermuyden was heard in defence, and a decision was given in his favour; but he seems to have acted with precipitancy in taking out subpœnas against many of the old inhabitants for damage said to have been done to him and his agents. Several persons were apprehended and confined in York gaol, and the feeling of bitterness between the native population and the Dutch settlers grew more intense from day to day. Lord Wentworth, President of the North, at length interfered; and after surveying the lands, he ordered that all suits should cease, and the restoration of the old rights of common, which had in some cases been interfered with. Vermuyden was also directed to assign to the tenants certain tracts of moor and marsh ground, to be enjoyed by them in common. He attempted to evade the decision, holding it to be unjust; but the Lord President was too powerful for him, and he therefore felt that, as opposition was of no use, it was better that he should altogether withdraw from the undertaking, which he did; first conveying his lands to trustees, and afterwards disposing of his interest in them.[1]

The necessary steps were then taken to relieve the old lands which had been flooded, by the cutting of the Dutch River at a heavy expense. Great difficulty was experienced in raising the requisite funds; the Dutch capitalists now holding their hand, or transferring their

[1] The Dutch settlers lived for the most part in single houses, dispersed through the newly-recovered country. A house built by Vermuyden remains. It was chiefly of timber, and what is called *stud-bound*. It was built round a quadrangular court. The eastern front was the dwelling-house. The other three sides were stables and barns. Another good house was built by Mathew Valkenburgh, on the Middle Ing, near the Don, which afterwards became the property of the Boynton family. Sir Philibert Vernatti and the two De Witts erected theirs near the Idle. A chapel for the settlers was also erected at Sandtoft, in which the various ordinances of religion were performed; and the public service was read alternately in the Dutch and French languages.— The Rev. Joseph Hunter's 'History and Topography of the Deanery of Doncaster,' 1828, vol. i. 165-6.

interest to other proprietors, at a serious depreciation in the value of their shares. The Dutch River was, however, at length cut, and all reasonable ground of complaint, so far as respected the lands along the North Don, was removed. For some years the new settlers cultivated their lands in peace; when suddenly they were reduced to the greatest distress, through the troubles arising out of the wars of the Commonwealth. In 1642 a committee sat at Lincoln to watch over the interests of the Parliament in that county. The Yorkshire royalists were very active on the other side of the Don, and the rumour went abroad that Sir Ralph Humby was about to march into the Isle of Axholme with his forces. To prevent this the committee at Lincoln gave orders to break the dykes, and pull up the flood-gates at Snow-sewer and Millerton-sluice, which was done. Thus in one night the results of so many years' labour were undone, and the greater part of the level again lay under water. The damage inflicted on the Hatfield settlers in that one night was estimated at not less than 20,000l. The people who carried out these orders were, no doubt, glad to have the opportunity of taking their full revenge upon the foreigners, who, they alleged, had robbed them of their commons. They levelled the houses of the settlers, destroyed their growing corn, and broke down the fences; and, when some of them tried to stop the destruction of the sluices at Snow-sewer, the rioters stood by with loaded guns, and swore they would stay until the whole levels were drowned again, and the foreigners forced to swim away like ducks.

After this mischief had been done, the commoners set up their claims as participants in the lands which had not been drowned, and from which the foreigners had been driven. In this they were countenanced by Colonel Lilburne, who, with a force of Parliamentarians, occupied Sandtoft, driving the Protestant minister out of his house, and stabling their horses in the chapel. A bargain was actually made between the Colonel and the commoners, by which 2000 acres of Epworth Common were to be assigned to him, on condition of their right being established as to the remainder, and that they were to be held harmless on account of the cruelties which they had perpetrated on the poor settlers of the level.[1]     When the injured parties attempted to obtain redress by law, Lilburne, by his influence with the Parliament, the army, and the magistrates, parried their efforts for eleven years.[2]     He was, however, eventually compelled to disgorge; and though the original settlers at length got a decree of the Council of State in their favour, and those of them who survived were again permitted to occupy their holdings, the nature of the case rendered it impossible that they should receive any adequate redress for their losses and sufferings.[3]

In the mean time Sir Cornelius Vermuyden had not been idle.     He was as eagerly speculative as ever. Before he had parted with his interest in the reclaimed lands at Hatfield, he was endeavouring to set on foot his scheme for the reclamation of the drowned lands in the Cambridge Fens; for we find the Earl of Bedford,

---

[1] In the course of the riots not fewer than eighty-two dwelling-houses of the foreign settlers were destroyed, and their chapel at Sandtoft was defaced, with circumstances which distinctly mark the vulgar and brutal character of the assailants. For ten days the isle-men were in a state of open rebellion.—Hunter's 'History of the Deanery of Doncaster,' vol. i. p. 166.

[2] Colonel Lilburne attempted an ineffectual defence of himself in the tract entitled 'The Case of the Tenants of the Manor of Epworth truly stated by Col. Jno. Lilburne,' Nov. 18th, 1651.

[3] For a long time after this, indeed, the commoners continued at war with the settlers, and both were perpetually resorting to the law—of the courts as well as of the strong hand. One Reading, a counsellor, was engaged to defend the rights of the drainers or participants, but his office proved a very dangerous one. The fen-men regarded him as an enemy, and repeatedly endeavoured to destroy him. Once they had nearly burned him and his family in their beds. Reading died in 1716, at a hundred years old, fifty of which he had passed in constant danger of personal violence, having fought "thirty-one set battles" with the fen-men in defence of the drainers' rights.—See the Rev. W. B. Stonehouse's 'History and Antiquities of the Isle of Axholme,' 4to., London, 1839.

in July, 1630, writing to Sir Harry Vane, recommending him to join Sir Cornelius and himself in the enterprise.[1]  Before the end of the year we find Vermuyden entering into a contract with the Crown for the purchase of Malvern Chase, in the county of Worcester, for the sum of 5000*l.*,[2] which he forthwith proceeded to reclaim and enclose.  Shortly after he took a grant of 4000 acres of waste land on Sedgemoor, with the same object, for which he paid 12,000*l.*  Then in 1631 we find him, in conjunction with Sir Robert Heath, taking a lease for thirty years of the Dovegang lead-mine, near Wirksworth, reckoned the best in the county of Derby.  But from this point he seems to have become involved in a series of lawsuits, from which he never altogether shook himself free.  Legal troubles accumulated about him with reference to the Hatfield estates, and he appears for some time to have suffered imprisonment.[3]  He was also harassed by the disappointed Dutch capitalists at the Hague and Amsterdam, who had suffered heavy losses by their investment at Hatfield, and took legal proceedings against him.  He had no sooner, however, emerged from confinement than we find him fully occupied with his new and grand project for the drainage of the Great Level of the Fens.

---

[1] 'State Papers,' vol. clxxi. 30.
[2] 'State Papers,' vol. clxxiv. 1.
[3] Feb. 25th, 1634. Petition of Sir Cornelius Vermuyden to the King, for reparation for his imprisonment and unjust prosecution, and for compensation for 216*l.* 8*s.*, which he overpaid. —'State Papers,' Chas. I., No. 242, May 1634; Report on the Petition, Ibid., No. 769.

# CHAPTER IV.

DRAINAGE OF THE GREAT LEVEL—SIR CORNELIUS VERMUYDEN.

THE outfalls of the numerous rivers flowing through the Fen Level having been neglected, and the old drains suffered to be silted up, the waters were rapidly regaining their old dominion. Districts which had been partially reclaimed were again becoming drowned, and the waters even threatened with ruin the older settled farms and villages situated upon the islands of the Fens. The Commissioners of Sewers at Huntingdon attempted to raise funds for the purposes of drainage by levying a tax of six shillings an acre upon all marsh and fen lands, but not a shilling of the tax was collected. This measure having failed, the Commissioners of Sewers of Norfolk, at a session held at King's Lynn, in 1629, determined to call to their aid Sir Cornelius Vermuyden. At an interview, to which he was invited, he offered to find the requisite funds to undertake the drainage of the Level, and to carry out the works after the plans submitted by him, on condition that 95,000 acres of the reclaimed lands were granted to him as a recompense. A contract was entered into on those terms; but so great an outcry was immediately raised against such a contract being made with a foreigner that it was abrogated before many months had passed.

Then it was that Francis, Earl of Bedford, the owner of many of the old church-lands in the Fens, was induced to take the place of Vermuyden, and become chief undertaker in the drainage of the extensive tract of fen country now so well known as the Great Bedford Level. Several others of the large adjoining landowners

entered into the project with the Earl, contributing
sums towards the work, in return for which a propor-
tionate acreage of the reclaimed lands was to be allotted
them. The new undertakers, however, could not dis-
pense with the services of Vermuyden. He had, after
long study of the district, prepared elaborate plans for
its drainage, and, besides, had at his command an
organized staff of labourers, mostly Flemings, who were
well accustomed to this kind of work. Westerdyke,
also a Dutchman, prepared and submitted plans, but
Vermuyden's were preferred, and he accordingly proceeded with the enterprise.

The difficulties encountered in carrying on the
works were very great, arising principally from the
want of funds. The Earl of Bedford became seriously
crippled in his resources; he raised money upon his
other valuable landed property until he could raise no
more, and many of the smaller undertakers were completely ruined. Vermuyden, with much determination,
took measures to provide the requisite means to pay the
workmen and prosecute the drainage to completion;
until the undertakers became so largely his debtors that
they were under the necessity of conveying to him many
thousand acres of the reclaimed lands, even before the
works were completed, as security for the large sums
which he had advanced.

The most important of the new works executed at
this stage were as follows :—Bedford River (now known
as Old Bedford River), extending from Erith on the Ouse
to Salter's Lode on the same river : this cut was 70 feet
wide and 21 miles long, and its object was to relieve and
take off the high floods of the Ouse.[1]  Bevill's Leam was
another extensive cut, extending from Whittlesea Mere

---

[1] We insert the annexed map at this
place, although it includes the drain-
age-works subsequently constructed,
in order that the reader may be enabled
more readily to follow the history of
the various cuts and drains executed
in the Fen country from about the
middle of the sixteenth century down
to about the year 1830.

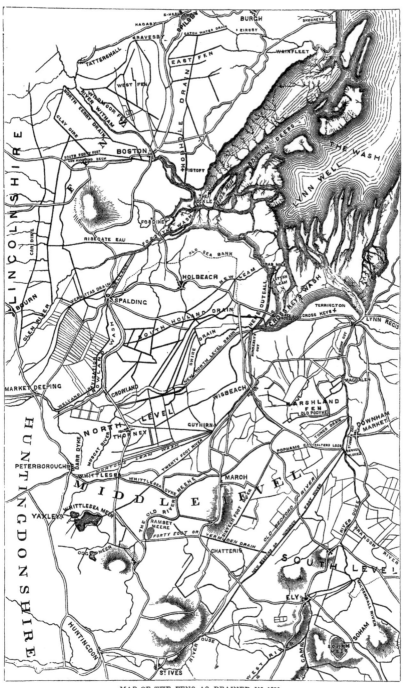

MAP OF THE FENS AS DRAINED IN 1830.
[After Telford's Plan and the Ordnance Survey.]

E 2

to Guyhirne, 40 feet wide and 10 miles long; Sam's Cut, from Feltwell to the Ouse, 20 feet wide and 6 miles long; Sandy's Cut, near Ely, 40 feet wide and 2 miles long; Peakirk Drain, 17 feet wide and 10 miles long; with other drains, such as Mildenhall, New South Eau, and Shire Drain. Sluices were also erected at Tydd upon Shire Drain, at Salter's Lode, and at the Horseshoe below Wisbeach, together with a clow,[1] at Clow's Cross, to keep out the tides; while a strong fresh-water sluice was also provided at the upper end of the Bedford River.

These works were not permitted to proceed without great opposition on the part of the Fen men, who frequently assembled to fill up the Cuts which the labourers had dug, and to pull down the banks which they had constructed. They also abused and maltreated the foreigners when the opportunity offered, and sometimes mobbed them while employed upon the drains, so that in several places they had to work under a guard of armed men. Difficult though it was to deal with the unreclaimed bogs, the unreclaimed men were still more impracticable. Although their condition was very miserable, they nevertheless enjoyed a sort of wild liberty amidst the watery wastes, which they were not disposed readily to give up. Though they might alternately shiver and burn with ague, and become prematurely bowed and twisted with rheumatism, still the Fens were their " native land," such as it was, and their only source of subsistence, precarious though it might be. The Fens were their commons, on which their geese grazed. They furnished them with food, though the finding thereof was full of adventure and hazard. What cared the Fen men for the drowning of the land? Did not the water bring them fish, and the fish attract wild-fowl, which they could snare and shoot? Thus the proposal to drain the Fens and to convert them into

---

[1] A clow is a sluice regulated by being lifted or dropped perpendicu- larly, like a portcullis. The other sluices open and shut like gates.

wholesome and fruitful lands, however important in a
national point of view, as enlarging the resources and
increasing the wealth of the country, had no attraction
whatever in the eyes of the Fen men.   They muttered
their discontent, and everywhere met the " adventurers,"
as the reclaimers were called, with angry though ineffectual opposition.   But their numbers were too few,
and they were too widely scattered, to make any combined effort at resistance.   They could only retreat to
other fens where they thought they might still be safe,
carrying their discontent with them, and complaining that
their commons were taken from them by the rich, and,
what was worse, by foreigners—Dutch and Flemings.
The jealous John Bull of the towns became alarmed at
this idea, and had rather that the water than these
foreigners had possession of the land.   " What ! " asked
one of the objectors, " is the old activitie and abilities
of the English nation grown now soe dull and insufficient that wee must pray in ayde of our neighbours to
improve our own demaynes ?   For matter of securitie,
shall wee esteem it of small moment to put into the
hands of strangers three or four such ports as Linne,
Wisbeach, Spalding, and Boston, and permit the countrie
within and between them to be peopled with over-
thwart neighbours ; or, if they quaile themselves, must
wee give place to our most auncient and daungerous enemies, who will be readie enough to take advantage of
soe manie fair inlets into the bosom of our land, lying
soe near together that an army landing in each of them
may easily meet and strongly entrench themselves with
walls of water, and drowne the countrie about them at
their pleasure ? " [1]

---

[1] ' The Drayner Confirmed,' tract,
1629.   Another   violent   pamphlet
against the drainage was published
about the same time, entitled ' The
Anti-Projector ; or, the History of the
Fenn Project;' small 4to., London,
(*circa*) 1628. The writer was opposed
to Vermuyden, for his interference
with the prescriptive rights of the
poor and others settled in the fenny
districts.

Thus a great agitation against the drainage sprang up in the Fen districts, and a wide-spread discontent prevailed, which, as we shall afterwards find, exercised an important influence on the events which culminated in the Great Rebellion of a few years later. Among the other agencies brought to bear against the Fen drainers was the publication of satirical songs and ballads—the only popular press of the time; and the powtes or poets doubtless accurately enough represented the then state of public opinion, as their ballads were sung with great applause about the streets of the Fen towns. One of these, entitled 'The Powte's Complaint,' was among the most popular.[1]  In another—a

---

[1] This poem is quoted in Badeslade's 'History of the Ancient and Present State of King's Lynn and of Cambridge,' London, 1725, and runs as follows:—

### THE POWTE'S COMPLAINT.

Come, Brethren of the water, and let us all assemble,
To treat upon this Matter, which makes us quake and tremble;
For we shall rue, if it be true that Fens be undertaken,
And where we feed in Fen and Reed, they'll feed both Beef and Bacon.

They'll sow both Beans and Oats, where never Man yet thought it;
Where Men did row in Boats, ere Undertakers bought it;
But, *Ceres*, thou behold us now, let wild Oats be their Venture,
Oh, let the Frogs and miry Bogs destroy where they do enter.

Behold the great Design, which they do now determine,
Will make our Bodies pine, a prey to Crows and Vermine;
For they do mean all Fens to drain, and Waters overmaster,
All will be dry, and we must die—'cause Essex calves want pasture.

Away with Boats and Rudder, farewel both Boots and Skatches,
No need of one nor t'other, Men now make better Matches;
Stilt-Makers all, and Tanners, shall complain of this Disaster,
For they will make each muddy lake for Essex Calves a Pasture.

The feather'd Fowls have Wings, to fly to other Nations;
But we have no such things, to help our Transportations;
We must give place, O grevious Case! to horned Beasts and Cattle,
Except that we can all agree to drive them out by Battel.

Wherefore let us intreat our antient Water-Nurses
To shew their Power so great as t help to drain their Purses;
And send us good old Captain Flood to lead us out to Battel,
Then Two-penny Jack, with Scales on 's Back, will drive out all the Cattle.

This Noble Captain yet was never know to fail us,
But did the conquest get of all that did assail us;
His furious Rage none could assuage; but, to the World's great Wonder,
He tears down Banks, and breaks their Cranks and Whirligigs assunder.

God *Eolus*, we do thee pray that thou wilt not be wanting;
Thou never said'st us nay—now listen to our canting;
Do thou deride their Hope and Pride that purpose our Confusion,
And send a Blast that they in haste may work no good Conclusion.

Great

drinking song—we find the Dutch pointed at as the chief offenders.  The following stanzas may serve as a specimen :—

> " Why should we stay here, and perish with thirst ?
> To th' new world in the moon away let us goe,
> For if the Dutch colony get thither first,
>     'Tis a thousand to one but they'll drain that too !
>
> Then apace, apace drink, drink deep, drink deep,
>     Whilst 'tis to be had let's the liquor ply;
> The drainers are up, and a coile they keep,
>     And threaten to draine the kingdom dry." [1]

The Fen drainers might, however, have outlived these attacks, had the works executed by them been successful ; but unhappily they failed in many respects. Notwithstanding the numerous deep cuts made across the Fens in all directions at such great cost, the waters still retained their dominion over the land.   The Bedford River and the other drains merely acted as so many additional receptacles for the surplus water, without relieving the drowned districts to any appreciable extent.   This arose from the engineer confining his attention almost exclusively to the inland draining and embankments, while he neglected to provide any sufficient outfalls for the waters themselves into the sea.   Vermuyden committed the error of adopting the Dutch method of drainage, in a district where the circumstances were different in many respects from those which prevailed in Holland.   In Zeeland, for instance, the few rivers passing through it were easily banked up and carried

---

Great *Neptune*, God of Seas, this Work must needs provoke ye;
They mean thee to disease, and with Fen-Water choak thee;
But with thy Mace do thou deface, and quite confound this matter,
And send thy Sands to make dry lands when they shall want fresh Water.

And eke we pray thee, *Moon*, that thou wilt be propitious,
To see that nought be done to prosper the Malicious;
Tho' Summer's Heat hath wrought a Feat, whereby themselves they flatter,
Yet be so good as send a Flood, lest Essex Calves want Water.

[1] This song is preserved in the collection entitled ' Witt and Drollery,' 12mo., London, 1655.  A poetical pamphlet was, however, published in favour of the drainage, entitled ' A True and Natural Description of the Great Level of the Fenns,' in 216 heroic lines.

out to sea, whilst the low lying lands were kept clear of surplus water by pumps driven by windmills. There, the main object of the engineer was to build back the river and the ocean; whereas in the Great Level the problem to be solved was, how to provide a ready out-fall to the sea for the vast body of fresh water falling upon as well as flowing through the Fen districts. This essential point was unhappily overlooked by the early drainers; and it has thus happened that one of the principal labours of the modern engineers has been to rectify the errors of Vermuyden and his followers; more especially by providing efficient outlets for the dis-charge of the fen waters, deepening and straightening the rivers, and compressing the streams in their course through the level, so as to produce a more powerful current and scour, down to their point of outfall into the sea.

This important condition of successful drainage having been overlooked, it may readily be understood how un-satisfactory was the result of the works first carried out in the Bedford Level. In some districts the lands were no doubt improved by the additional receptacles provided for the surplus waters, but the great extent of fen land still lay for the most part wet, waste, and unprofitable. Hence, in 1634, a Commission of Sewers held at Hun-tingdon pronounced the drainage to be defective, and the 400,000 acres of the Great Level to be still subject to inundation, especially in the winter season. The King, Charles I., then resolved himself to undertake the reclamation, with the object of converting the Level, if possible, into "winter grounds." He took so much per-sonal interest in the work that he even designed a town to be called Charleville, which was to be built in the midst of the Level, for the purpose of commemorating the undertaking. Sir Cornelius Vermuyden was again employed to carry out the King's design. He had many enemies, but he could not be dispensed with; being the

only man of recognised ability in works of drainage at
that time in England.

The works constructed in pursuance of this new
design were these :—an embankment on the south side
of Morton's Leam, from Peterborough to Wisbeach; a
navigable sasse, or sluice, at Standground ; a new river
cut between the stone sluice at the Horse-shoe and the
sea below Wisbeach, 60 feet broad and 2 miles long,
embanked at both sides; and a new sluice in the marshes
below Tydd, upon the outfall of Shire Drain.   These
and other works were in full progress, when the political
troubles of the time came to a height, and brought all
operations to a stand still for many years.   The discontent caused throughout the Fens by the drainage operations had by no means abated ; but, on the contrary,
considerably increased.   In other parts of the kingdom,
the attempts made about the same time by Charles I.
to levy taxes without the authority of Parliament, gave
rise to much agitation.   In 1637 occurred Hampden's
trial, arising out of his resistance to the payment of ship-
money : by the end of the same year the King and Parliamentary party were mustering their respective forces,
and a collision between them seemed imminent.   At this
juncture the discontent which prevailed throughout the
Fen counties was an element of influence not to be
neglected.   It was adroitly represented that the King's
sole object in draining the Fens was merely to fill his
impoverished exchequer, and enable him to govern
without a Parliament.   The discontent was thus fanned
into a fierce flame; on which Oliver Cromwell, the
member for Huntingdon, until then comparatively unknown, availing himself of the opportunity which offered,
of increasing the influence of the Parliamentary party
in the Fen counties, immediately put himself at the head
of a vigorous agitation against the further prosecution
of the scheme.   He was very soon the most popular man
in the district; he was hailed 'Lord of the Fens' by the

Fen men ; and he went from meeting to meeting, stirring
up the public discontent, and giving it a suitable direction. "From that instant," says Mr. Forster, "the
scheme became thoroughly hopeless. With such desperate
determination he followed up his purpose—so actively
traversed the district, and inflamed the people everywhere—so passionately described the greedy claims of
royalty, the gross exactions of the commission, nay, the
questionable character of the improvement itself, even
could it have gone on unaccompanied by incidents of
tyranny,—to the small proprietors insisting that their
poor claims would be merely scorned in the new distribution of the property reclaimed,—to the labouring
peasants that all the profit and amusement they had
derived from *commoning* in those extensive wastes were
about to be snatched for ever from them,—that, before
his almost individual energy, King, commissioners, noble-
men-projectors, all were forced to retire, and the great
project, even in the state it then was, fell to the ground."

The success of the Cambridge Fen men in resisting
the reclamation of the wastes, encouraged those in the
more northern districts to take even more summary
measures to get rid of the drainers, and restore the lands
to their former state. The Earl of Lindsey had succeeded
at great cost in enclosing and draining about 35,000
acres of the Lindsey Level, and induced numerous farmers
and labourers to settle down upon the land, to plough
and sow it. They erected dwellings and farm-buildings,
and were busily at work, when the Fen men suddenly
broke in upon them, destroyed their. buildings, killed
their cattle, and let in the waters again upon the land.
So, too, in the West and Wildmore Fen district, between
Tattershall and Boston in Lincolnshire, where considerable progress had been made by a body of " adventurers"
in reclaiming the wastes. After many years' labour and

---

[1] Lives of Eminent British Statesmen (Lardner's ' Cabinet Cyclopædia,'
vol. vi. p. 60).

much cost, they had succeeded in draining, enclosing, and cultivating an extensive tract of rich land, and they were peaceably occupied with their farming pursuits, when a mob of Fen men collected from the surrounding districts, and under pretence of playing at football, levelled the enclosures, burnt the corn and the houses, destroyed the cattle, and even killed many of the people who occupied the land. They then proceeded to destroy the drainage works, by cutting across the embankments and damming up the drains, by which the country was inundated, and restored to its original condition.

The greater part of the Level thus again lay waste, and the waters were everywhere extending their dominion over the dry land through the choking up of the drains and river outfalls by the deposit of silt. Matters were becoming even worse than before, but could not be allowed thus to continue. In 1641 the Earl of Bedford and his participants made an application to the Long Parliament then sitting, for permission to re-enter upon the works; but the civil commotions which still continued prevented any steps being taken, and the Earl himself shortly after died, in a state of comparative penury, to which he had reduced himself by his devotion to this great work. Again, however, we find Sir Cornelius Vermuyden upon the scene. Undaunted by adversity, and undismayed by the popular outrages committed upon his poor countrymen in Lincolnshire and Yorkshire, he still urged that the common weal of England demanded that the rich lands lying under the waters of the Fens should be reclaimed, and made profitable for human uses. He saw a district almost as large as the whole of the Dutch United Provinces remaining waste and worse than useless, and he gave himself no rest until he had set on foot some efficient measure for its drainage and reclamation. What part he took in the political discussions of the time, we know not; but we find the eldest of his sons, Cornelius, a colonel in the Parliamentary army[1] stationed in the Fens under Fairfax, shortly before the battle of Naseby. Vermuyden himself was probably too much engrossed by his drainage project to give much heed to political matters; and besides, he could not forget that Charles, and Charles's father, had been his fast friends.

In 1642, whilst the civil war was still raging, appeared Vermuyden's 'Discourse' on the Drainage of the Fens,[2] wherein he pointed out the works which still remained to be executed in order effectually to reclaim the 400,000 acres of land capable of growing corn, which formed the area of the Great Level. His suggestions formed the subject of much pamphleteering discussion,[3] for several years, during which also numerous petitions were presented to Parliament, urging the necessity for perfecting the works. At length, in 1649, authority was granted to William, Earl of Bedford, and other participants, to prosecute the undertaking which

---

[1] "The party under Vermuyden waits the King's army, and is about Deeping; has a command to join with Sir John Gell, if he commands him."—Cromwell's Letter to Fairfax, 4th June, 1645. This Vermuyden resigned his commission a few days before the battle of Naseby, having, as he alleged, special reasons requiring his presence beyond the seas, whence he does not seem to have returned until after the Restoration. In 1665 we find him a member of the Corporation of the Bedford Level.

[2] The title of Vermuyden's tract is—'A Discourse touching the Drayning the Great Fennes lying within the severall Covnties of Lincolne, Northampton, Huntingdon, Norfolke, Suffolke, Cambridge, and the Isle of Ely, as it was presented to his Majestie by Sir Cornelivs Vermviden, Knight; whereunto is annexed the Designe or Map. Published by Authority. London: Printed by Thomas Fawcet, dwelling in Grub Street, neere the lower pumpe. 1642.' (Small 4to. pp. 32.)

[3] Among the numerous replies were the following:—'Exceptions against Sr. Cornelius Vermuden's Discourse for the Draining of the Great Fennes, &c., which, in January, 1638, he presented to the King for his designe. Wherein his Majesty was misinformed and abused in regard it wanteth all the essentiall parts of a Designe. And the greate advantageous Workes made by the late Earle of Bedford slighted, and the whole adventure disparaged. Published by Andrewes Burrell, Gentleman.' (4to. Lond., 1642, pp. 19.) [This curious tract is by one who was himself a drainer, and bitterly opposed to Vermuyden and all other foreign adventurers in England.] 'Andrewes Burrell's Briefe Relation.' (Small 4to. Lond. 1642.) [This is another and smaller tract, in which the author exposes the draining views of Vermuyden, and himself suggests a more practicable course.]

his father had begun, and steps were shortly after taken
to commence the works. Again was Westerdyke, the
Dutch engineer, called in to criticise Vermuyden's
plans; and again was Vermuyden triumphant over all
his opponents. He was selected, once more, to direct
the drainage, which, looking at the defects of the works
previously executed by him, and the difficulties in which
the first Earl had thereby become involved, must be
regarded as a marked proof of the man's force of purpose,
as well as his recognised integrity of character.

Vermuyden again collected his Dutchmen about him,
and vigorously began operations. But they had not
proceeded far before they were again almost at a stand
still for want of funds; and throughout their entire
progress they were hampered and hindered by the same
great difficulty. Some of the participants sold and
alienated their shares in order to get rid of further
liabilities; others held on to the last, but were reduced
to the lowest ebb. Means were, however, adopted to
obtain a supply of cheaper labour; and application was
made by the adventurers for a supply of men from
amongst the Scotch prisoners who had been taken at
the battle of Dunbar. A thousand of them were granted
for the purpose, and employed on the works to the north
of Bedford River, where they continued to labour until
the political arrangements between the two countries
enabled them to return home.[1] When the Scotch labourers had left, some difficulty was again experienced
in carrying on the works. The local population were
still hostile, and occasionally interrupted the labourers
employed upon them; a serious riot at Swaffham having
only been put down by the help of the military. Fortunately, Blake's victory over Van Tromp, in 1652,
supplied the Government with a large number of Dutch

---

[1] Wells, in his 'History of the Bedford Level,' adds :—" Many, however,
settled in the Fens, and were the
origin of most of the Scotch families
and names that now exist in the
Great Level." (Vol. i. p. 244.)

prisoners, five hundred of whom were at once forwarded to the Level, where they proved of essential service in prosecuting the works.

The most important of the new rivers, drains, and sluices included in this second undertaking, were the following :—The New Bedford River, cut from Erith on the Ouse to Salter's Lode on the same river, reducing its course between these points from 40 to 20 miles : this new river was 100 feet broad, and ran nearly parallel with the Old Bedford River. A high bank was raised along the south side of the new cut, and an equally high bank along the north side of the old river, a large space of land, of about 5000 acres, being left between them, called the Washes, for the floods to " bed in," as Vermuyden termed it. Then the river Welland was defended by a bank, 70 feet broad and 8 feet high, extending from Peakirk to the Holland bank. The river Nene was also defended by a similar bank, extending from Peterborough to Guyhirne ; and another bank was raised between Standground and Guyhirne, so as to defend the Middle Level from the overflowing of the Northampton-shire waters. The river Ouse was in like manner restrained by high banks extending from Over to Erith, where a navigable sluice was provided. Smith's Leam was cut, by which the navigation from Wisbeach to Peterborough was opened out. Among the other cuts and drains completed at the same time, were Vermuyden's Eau, or the Forty Feet Drain, extending from Welch's Dam to the river Nene near Ramsey Mere ; Hammond's Eau, near Somersham, in the county of Huntingdon ; Stonea Drain and Moore's Drain, near March, in the Isle of Ely ; Thurlow's Drain, extending from the Forty Feet to Popham's Eau ; and Conquest Lode, leading to Whittlesea Mere. And in order to turn the tidal waters into the Hundred Feet River, as well as to prevent the upland floods from passing up the Ten Mile River towards Littleport, Denver Sluice, that great bone of after contention, was erected. Another important work in the South Level was the cutting of a large river called St. John's, or Downham Eau,[1] 120 feet wide, and 10 feet deep, from Denver Sluice to Stow Bridge on the Ouse, with sluices at both ends, for the purpose of carrying away with greater facility the flood waters descending from the several rivers of that level. Various new sluices were also fixed at the mouths of the rivers, to prevent the influx of the tides, and most of the old drains and cuts were at the same time scoured out and opened for the more ready flow of the surface waters.

At length, in March, 1652, the works were declared to be complete, and the Lords Commissioners of Adjudication appointed under the Act of Parliament proceeded to inspect them. They embarked upon the New River, and sailing over it to Stow Bridge, surveyed the new eaus and sluices executed near that place, after which they returned to Ely. There Sir Cornelius Vermuyden read to those assembled a discourse, in which he explained the design he had carried out for the drainage of the district; in the course of which he stated as one of the results of the undertaking, that in the North and Middle Levels there were already 40,000 acres of land "sown with cole seed, wheat, and other winter grain, besides innumerable quantities of sheep, cattle, and other stock, where never had been any before. These works," he added, "have proved themselves sufficient, as well by the great tide about a month since, which overflowed Marshland banks, and drowned much ground in Lincolnshire and other places, and a flood by reason of a great snow, and rain upon it following soon after, and yet never hurt any part of the whole Level; and the view of them, and the consideration of what hath previously been said, proves a clear

---

[1] The St. John's Eau, being a straight cut, is known in the district as "The Poker;" and Marshland Cut, being in the shape of a pair of tongs, is commonly called "Tongs Drain."

draining according to the Act." He concluded thus,— "I presume to say no more of the work, lest I should be accounted vain-glorious; although I might truly affirm that the present or former age have done nothing like it for the general good of the nation. I humbly desire that God may have the glory, for his blessing and bringing to perfection my poor endeavours, at the vast charge of the Earl of Bedford and his participants."[1]

A public thanksgiving took place to celebrate the completion of the undertaking; and on the 27th of March, 1653, the Lords Commissioners of Adjudication of the Reclaimed Lands, accompanied by their officers and suite,—the Company of Adventurers, headed by the Earl of Bedford,—the magistrates, and leading men of the district, with a vast concourse of other persons,—attended public worship in the cathedral of Ely, when the Rev. Hugh Peters, chaplain to the Lord-General Cromwell, preached a sermon on the occasion.

Vermuyden's perseverance had thus far triumphed. He had stood by his scheme when all others held aloof from it. Amidst the engrossing excitement of the civil war, the one dominating idea which possessed him was the drainage of the Great Level. While the nation was divided into two hostile camps, and the deadly struggle was proceeding between the Royalists and Parliamentarians, Vermuyden's sole concern was how to raise the funds wherewith to pay his peaceful army of Dutch labourers in the Fens. To carry on the works he sold every acre of the lands he had reclaimed; he first sold the allotment of lands won by him from the Thames at Dagenham in 1621; then he sold his interest in his lands at Sedgemoor and Malvern Chase; and in 1654 we find him conveying the remainder of his lands in Hatfield Level. He was also under the necessity of selling all the lands apportioned to him in the Bedford

---

[1] Wells's 'History of the Bedford Level,' vol. i., 275.

Level itself, in order to pay the debts incurred in their drainage.   But although he lost all, it appears that the company in the end preferred heavy pecuniary claims against him which he had no means of meeting ; and in 1656 we find him appearing before Parliament as a suppliant for redress.   Thenceforward he entirely disappears from public sight; and it is supposed that, very shortly after, he went abroad and died, a poor, broken down old man, the extensive lands which he had reclaimed and owned having been conveyed to strangers.[1]

The drainage of the Fens, however, was not yet complete.   The district was no longer a boggy wilderness, but much of it in fine seasons was covered with waving crops of corn.   As the swamps were drained, farm buildings, villages, and towns gradually sprang up, and the toil of the labourer was repaid by abundant harvests.   The anticipation held forth in the original charter granted by Charles I.[2] to the reclaimers of the Bedford Level was more than fulfilled.   " In those places which lately presented nothing to the eyes of the beholders but great waters and a few reeds thinly scattered here and there, under the Divine mercy might be seen pleasant pastures of cattle and kine, and many houses belonging to the inhabitants."   But the tenure by which the land continued to be held was unremitting vigilance and industry ; the difficulties interposed by nature tending to discipline the skill, to stimulate the enterprise, and evoke the best energies of the people who had rescued the fields from the watery waste.   There was still the ten-

---

[1] Vermuyden's second daughter, Catherine, married Thomas Babington, Esq., of Somersham, Huntingdonshire, son and heir of Thomas Babington, Esq., of Rothley Temple, Leicestershire. It may be remembered that Zachary Macaulay married into the Babington family, and that the late Thomas Babington Macaulay was born at

Rothley Temple.—There is a tradition at Hatfield that Vermuyden died in the poor-house at Belton, but Dr. Hunter, in his ' Deanery of Doncaster,' says this is incorrect.

[2] The Charter, commonly known as " The Lynn Law," was granted in 10 Car. I.

dency of the river outfalls to silt up and fall into decay.
At any time neglect was certain to be followed by inundation,[1] and the "Bailiff of Bedford" for nearly another
century continued to pay the dwellers in the Fens many
a rude visit, sometimes in a night sweeping away the
polder works of many homesteads.

In 1713 a violent tide rushed up the Ouse, and, encountering high floods descending from the uplands,
their combined force was such that Denver Sluice
was blown up and destroyed. The ebb of the Bedford
waters flowed nearly two hours up the Cambridge River,
carrying with them additional silt and sand every spring
tide. The state of the South Level gradually became
much deteriorated, until the year 1748, when the sluice
was reconstructed under the direction of Labelye, the
Swiss architect who built Westminster Bridge. Various
engineers were employed at different times during the
last century in correcting the defects of the early works,
or in carrying out further improvements; the most pro-
minent being Perry, Golborne, and Kinderley. The
great scheme of Kinderley, proposed in 1751, was a
suggestion of genius. He designed to convey the con-
joined waters of the Ouse and the Nene into the centre
of the Wash, there to unite with the Welland and the
Witham. By this measure the navigation of the Wash
would have been greatly improved, and its shifting sands
avoided, whilst as much new land would have been
drained and reclaimed as almost to have justified the
addition of a new county. Every one of his cuts
was proposed on the same principle that has governed
later improvements—that of avoiding broad channels
with shifting sand banks, and confining the rivers to
*narrow* channels, in order to secure *depth* by force and
weight of current. But Kinderley's grand idea was not

---

[1] The account of a destructive inundation is given in a publication entitled 'A true and impartial Relation
of the great damages done by the late
great Tempest, and overflowing of the
Tide upon the coast of Lincolnshire
and Norfolk,' &c. 1671.

carried out; it was too large for the narrow views and
the still narrower purses of the landed proprietors in the
Level a hundred years since. It was reserved for Mr.
Rennie to evoke the enterprise of the Fen lords, and
induce them to carry out the thorough drainage of the
low lying lands on sound scientific principles. He clearly
pointed out that this could only be satisfactorily effected
by cutting down the outfalls to low water of spring
tides, and thus facilitating the escape of the waters to
the utmost extent,—a course of action which, as we shall
hereafter point out, he carried into effect with remarkable success.

Meanwhile, improvements of all kinds went steadily
on, until all the rivers flowing through the Level were
artificially altered and diverted into new channels, excepting the Nene, which is the only natural river in
the Fen district remaining comparatively unaltered.
New dykes, causeways, embankments, and sluices were
formed; many droves, leams, eaus, and drains were
cut, furnished with gowts or gates at their lower ends,
which were from time to time dug, deepened, and
widened. Mills were set to work to pump out the water
from the low grounds; first windmills, sometimes with
double-lifts, as practised in Holland; and more recently
powerful steam-engines, as first recommended by Mr.
Rennie. Sluices were also erected to prevent the inland
waters from returning; strong embankments extending
in all directions, to keep the rivers and tides within
their defined channels. To protect the land from the
sea waters as well as the fresh,—to build and lock
back the former, and to keep the latter within due
limits,—was the work of the engineer; and by his skill,
aided by the industry of his contractors and workmen,
water, instead of being the master and tyrant as of old,
became man's servant and pliant agent, and was used as
an irrigator, a conduit, a mill-stream, or a water-road
for extensive districts of country. In short, in no part

of the world, except in Holland, have more industry and skill been displayed in reclaiming and preserving the soil, than in Lincolnshire and the districts of the Great Bedford Level. Six hundred and eighty thousand acres of the most fertile land in England, or an area equal to that of North and South Holland, have been converted from a dreary waste into a fruitful plain, and fleets of vessels traverse the district itself, freighted with its rich produce. Taking its average annual value at 4l. an acre, the addition to the national wealth and resources may be readily calculated.

The prophecies of the decay that would fall upon the country, if "the valuable race of Fenmen" were deprived of their pools for pike, and fish, and wild-fowl, have long since been exploded. The population has grown in numbers, in health, and in comfort, with the progress of drainage and reclamation. The Fens are no longer the lurking places of disease,[1] but as salubrious as any other parts of England. Dreary swamps are supplanted by pleasant pastures, and the haunts of pike and wild-fowl have become the habitations of industrious farmers and husbandmen. Even Whittlesea Mere and Ramsey Mere,—the only two lakes, as we were told in the geography books of our younger days, to be found in the south of England,—have been blotted out of the map, for they have been drained by the engineer, and are now covered with smiling farms and pleasant homesteads.

---

[1] It is stated in a recent report of the Registrar-General that, whilst the mortality of Pau in the Pyrenees, a place resorted to by British invalids on account of its salubriousness, is 23 in 1000, that of Ely is only 17 in 1000.

# CHAPTER V.

## STOPPAGE OF DAGENHAM BREACH—CAPTAIN PERRY.

BEFORE dismissing from consideration those early under-takings of embankment and drainage, we may briefly allude to further works which were rendered necessary by the neglected embankment of the Thames down to a comparatively recent period.

The banks first raised seemed to have been in many places of insufficient strength; and when a strong north-easterly wind blew down the North Sea, and the waters became pent up in that narrow part of it lying between the Belgian and the English coasts,—and especially when this occurred at a time of the highest spring tides,— the strength of the river embankments became severely tested throughout their entire length, and breaches often took place, occasioning destructive inundations.

Thus, in the year 1676, a serious breach took place at Limehouse, by which a number of houses was destroyed, and it was with great difficulty the waters could be banked out again. The wonder is that sweeping, as the new current did, over the Isle of Dogs, in the direction of Wapping, and in the line of the present West India Docks, the channel of the river was not then permanently altered. But Deptford was already established as a royal dockyard, and probably the diversion of the river would have inflicted as much local injury, judging by comparison, as it unquestionably would do at the present day. The breach was accordingly stemmed, and the course of the river held in its ancient channel by Dept-ford and Greenwich. Another destructive inundation shortly after occurred through a breach made in the

embankment of the West Thurrock Marshes, in what is
called the Long Reach, nearly opposite Greenhithe;
where the lands remained under water for seven years,
and it was with great difficulty the breach could be
closed.   But at length the tides were shut out, leaving
a large lake upon the land in the direction in which
the waters had rushed; and the breach and lake are to
be found marked on the maps to this day.

But the most destructive and obstinate breach of all
was that made by the river in the north bank a little
to the south of the village of Dagenham, by which
the whole of the Dagenham and Havering Levels lay
drowned at every tide.   It will be remembered that a
similar breach had occurred about 1621, which Vermuyden had succeeded in stopping; and at the same
time he embanked or "inned" the whole of Dagenham
Creek, through which the little rivulet flowing past
the village of that name found its way to the Thames.
Across the mouth of this rivulet Vermuyden had erected
a sluice, of the nature of a "clow," being a strong gate
suspended by hinges, which opened to admit of the
egress of the inland waters at low tide, and closed
against the entrance of the Thames when the tide rose.
It happened, however, that a heavy inland flood, and
an unusually high spring tide, occurred simultaneously
during the prevalence of a strong north-easterly wind,
in the year 1707; when the united force of the waters
meeting from both directions blew up the sluice, the
repairs of which had been neglected, and in a very short
time nearly the whole area of the above Levels was
covered by the waters of the Thames.

At first the gap was so slight as to have been easily
closed, being only from 14 to 16 feet broad.   But having
been neglected, the tide ran in and out of the opening for
years, and every tide wore the channel deeper, and made
the stoppage of the breach more difficult.   At length
the channel was upwards of 30 feet deep at low water,

and about 100 feet wide; and a lake more than a mile and a half in extent was formed inside the line of the river embankment. Above a thousand acres of rich lands were spoiled for all useful purposes, and by the scouring of the waters out and in at every tide, about a hundred and twenty acres were completely washed away. The soil was carried into the channel of the Thames, where it formed a bank of about a mile in length, reaching half way across the river. This state of things could not be allowed to continue, for the navigation of the Thames was seriously interrupted by the obstruction, and there was no knowing where the mischief would stop.

Various futile attempts were made by the adjoining landowners to stem the breach. They filled old ships with chalk and stones, and had them scuttled and sunk in the hole, throwing in baskets of chalk and earth outside them, together with bundles of straw and hay to stop up the interstices; but when the full tide rose, it washed them away like so many chips, and the opening was again bored clean through. Then the expedient was tried of sinking into the hole gigantic trunks made expressly for the purpose, fitted tightly together, and filled with chalk. Power was obtained to lay an embargo on the cargoes of chalk and ballast contained in passing ships, for the purpose of filling these machines, as well as damming up the gap; and as many as from ten to fifteen freights of chalk a day were thrown in, but without effect. One day when the tide was on the turn, the force of the water lifted one of the monster trunks sheer up from the bottom, when it toppled round, the lid opened, out fell the chalk, and, righting again, the immense box floated out into the stream and down the river. One of the landowners interested in the stoppage ran along the bank, and shouted out at the top of his voice " Stop her ! oh stop her !" But the unwieldy object being under no guidance was carried down stream towards the shipping lying at Gravesend, where its unusual appearance, standing so high out of the water, excited great alarm amongst the sailors. The empty trunk, however, floated safely past, down the river, until it reached the Nore, where it stranded upon a sandbank.

The Government next lent the undertakers an old royal ship called the *Lion*, for the purpose of being sunk in the breach, which was done, with two other ships; but the *Lion* was broken in pieces by a single tide, and at the very next ebb not a vestige of her was to be seen. No matter what was sunk, the force of the water at high tide bored through underneath the obstacle, and only served to deepen the breach. After the destruction of the *Lion*, the channel was found deepened to 50 feet at low water, at the very place where she had been sunk.

All this had been merely tinkering at the breach, and every measure that had been adopted merely proved the incompetency of the undertakers. The obstruction to the navigation through the deposit of earth and sand in the river being still on the increase, and after the bank had been open for a period of seven years, an Act was passed in 1714, enabling it to be repaired at the public expense. But it is an indication of the very low state of engineering ability in the kingdom at the time, that several more years passed before the measures taken with this object were crowned with success, and the opening was only closed after a fresh succession of failures. The works were first let to one Boswell, a contractor. He proceeded very much after the method which had already failed so egregiously, sinking two rows of caissons or chests across the breach, between which he proposed to erect the piles and drift work; but his chests were blown up again and again. Then he tried pontoons of ships, which he loaded and sunk in the opening; but the force of the tide, as before, rushed under and around them, and broke them all to pieces, the only result being to make the gap in the bank considerably bigger than before. Boswell at length abandoned all further attempts to close it, after suffering a heavy loss; and the engineering skill of England seemed likely to be completely baffled by this hole in a river's bank.

The competent man was, however, at length found in Captain Perry, who had just returned from Russia, where, having been able to find no suitable employment for his abilities in his own country, he had for some time been employed by the Czar Peter in carrying on extensive engineering works.

John Perry was born at Rodborough, in Gloucestershire, in 1669, and spent the early part of his life at sea. In 1693 we find him a lieutenant on board the royal ship the *Montague*. The vessel having put into harbour at Portsmouth to be refitted, Perry is said to have displayed considerable mechanical skill in contriving an engine for throwing out a large quantity of water from deep sluices (probably for purposes of dry docking) in a very short space of time. The *Montague* having been repaired, she put to sea, and was shortly after lost. As the English navy had suffered greatly during the same year, partly by mismanagement, and partly by treachery, the Government was in a very bad temper, and Perry was tried for alleged misconduct. The result was, that he was sentenced to pay a fine of 1000*l.*, and to undergo ten years' imprisonment in the Marshalsea. This sentence must, however, have been subsequently mitigated, for we find him in 1695 publishing a "Regulation for Seamen," with a view to the more effectual manning of the English navy; and in 1698 the Marquis of Caermarthen and others recommended him to the notice of the Czar Peter, then resident in England, by whom he was invited to go out to Russia, to superintend the establishment of a royal fleet, and the execution of several gigantic works which he contemplated for the purpose of opening up the resources of his empire. Perry was engaged by the Czar at a salary of 300*l.* a year, and shortly after accompanied him to Holland, from whence he proceeded to Moscow to enter upon the business of his office.

One of the Czar's grand designs was to open up a system of inland navigation, to connect his new city of Petersburgh with the Caspian Sea, and also to place Moscow upon another line, by forming a canal between the Don and the Volga. In 1698 the works had been begun by one Colonel Breckell, a German officer in the Czar's service. But though a good military engineer, it turned out that he knew nothing of canal making; for the first sluice which he constructed was immediately blown up. The water, when let in, forced itself under the foundations of the work, and the six months' labour of several thousand workmen was destroyed in a night. The Colonel, having a due regard for his personal safety, immediately fled the country in the disguise of a servant, and was never after heard of. Captain Perry entered upon this luckless gentleman's office, and forthwith proceeded to survey the work he had begun, some seventy-five miles beyond Moscow. Perry had a vast number of labourers placed at his disposal, but they were altogether unskilled, and therefore comparatively useless. His orders were to have no fewer than 30,000 men at work, though he seldom had more than from 10,000 to 15,000; but one twentieth the number of skilled labourers would have better served his purpose. He had many other difficulties to contend with. The local nobility or boyars were strongly opposed to the undertaking, declaring it to be impossible; and their observation was, that God had made the rivers to flow one way, and it was presumption in man to think of attempting to turn them in another.

Shortly after the Czar had returned to his dominions, he got involved in war with Sweden, and was defeated by Charles XII. at the battle of Narva, in 1701.

Although the Don and Volga canal was by this time half dug, and many of the requisite sluices were finished, the Czar sent orders to Perry to let the works stand, and attend upon him immediately at St. Petersburgh. Leaving one of his assistants to take charge of what had been done, Perry waited upon his royal employer, who had a great new design on foot of an altogether different character. This was the formation of a royal dockyard on one of the southern rivers of Russia, where he contemplated building a fleet of war ships, wherewith to act against the Turks in the Black Sea. Perry immediately entered upon the office to which he was appointed, of Comptroller of Russian Maritime Works, and proceeded to carry out the new project. The site of the Royal Dockyard was fixed at Veronize on the Don, and there Perry was occupied for several years, with a vast number of workmen under him, in building a dockyard, with storehouses, ship sheds, and workshops. He also laid down and superintended the building of numerous vessels, one of them of eighty guns; and the slips on which he built them are said to have been very ingenious and well contrived.

The creation of this dockyard was far advanced, when Perry received a fresh command to appear before the Czar at St. Petersburgh. Peter had now founded his new capital there, and desired to connect it with the Volga by means of a canal, to enable provisions, timber, and building materials to flow freely to the city from the interior of the empire. Perry forthwith entered upon an extensive survey of the intervening country, tracing to their respective heads the rivers flowing into Lake Ladoga. He surveyed three routes, and recommended for execution, as the most easy, that by the river Svir from Lake Ladoga to Lake Onega, from thence by the river Kovja to Lake Biela, then down into the Volga by a cut from Bielozersk to Schneska. The fall on the Petersburgh side of the navigation was 445 feet from the summit level, and 110 feet to the Volga. The works were immediately begun, and carried on, though with occasional interruptions, caused by the war in which Peter was for some time longer engaged with his formidable enemy of Sweden, but whom he eventually routed at Pultawa, in 1709.

Before the works were completed, however, Perry fled from Russia like his predecessor, but not for the same reason. During the whole of his stay in the kingdom he had been unable to get paid for his work, valuable although his services had been. His applications for his stipulated salary were put off with excuses from year to year. Proceedings in the courts of law were out of the question in such a country; he could only dun the Czar and his ministers; and at length his arrears had become so great, and his necessities so urgent, that he could no longer endure his position, and threatened to quit the Czar's service. It came to his ears that the Czar had threatened on his part, that if he did, he would have Perry's head; and the engineer immediately took refuge at the house of the British minister, who shortly after contrived to get him safely conveyed out of the country, but without being paid. He returned to England in 1712, as poor as he had left it, though he had so largely contributed to create the navy of Russia, and to lay the foundations of its afterwards splendid system of inland navigation. Shortly after his return to this country he published his description of Russia,[1] which,

---

[1] The title of the book, now little known, is—' The State of Russia under the present Czar : In relation to the great and remarkable things he has done, as to his naval preparations, the regulating his army, the reforming his people, and improvement of his country; particularly those works on which the author was employed, with the reasons of his quitting the Czar's service, after having been fourteen years in that country. London : Printed for Benjamin Tooke, at the Middle Temple Gate in Fleet Street, 1716.' The work was translated into French, under the title—' Etat présent de la Grande Russie ou Moscovie,' &c. Paris and Brussels, 12mo. 1717; The Hague, 12mo. 1717 ; Amsterdam, 12mo. 1720. It was also published in German, entitled—' Der Jetzige Staat von Russland oder Moskau,' &c. 2 vols. Leipzig, 1724.

as the first authentic account of the extraordinary progress of that new empire, was read with great avidity in
England, and was shortly after translated into nearly
all the languages of Europe.

It will be remembered that all attempts made to stop
the breach at Dagenham had thus far proved ineffectual;
and it threatened to bid defiance to the engineering
talent of England. Perry seemed to be one of those men
who delight in difficult undertakings, and he no sooner
heard of the work than he displayed an eager desire to
enter upon it. He went to look at the breach shortly
after his return, and gave in a tender with a plan for its
repair; but on Boswell's being accepted, which was
the lowest, he held back until that contractor had tried
his best, and failed. The road was now clear for Perry,
and again he offered to stop the breach and execute the
necessary works for the sum of 25,000l.[1] His offer was
this time accepted, and the works were commenced in the
beginning of the year 1715. The opening was now of
great width and depth, and a lake had been formed on
the land from 400 to 500 feet broad in some places, and
extending nearly 2 miles in length. Perry's plan of
operations may be briefly explained with the aid of his
own map. (*See* next page.)

In the first place he sought to relieve the tremendous
pressure of the waters against the breach at high tide,
by making other openings in the bank through which
they might more easily flow into and out of the inland
lake, without having exclusively to pass through the gap
which it was his object to stop. He accordingly had
two openings, protected by strong sluices, made in the
bank a little below the breach, and when these had been
opened and were in action, he commenced operations at
the breach itself. He began by driving in a row of
strong timber piles across the channel; and these piles

---

[1] Boswell's price had been 16,300l., and he undertook to do the work in
fifteen months.

he dovetailed one into the other so as to render them
almost impervious to water. He also threw in large
quantities of clay outside the piling, and formed a sort of
puddle, which served at the same time to resist the force
of the water, and to protect the foundation of the piles.

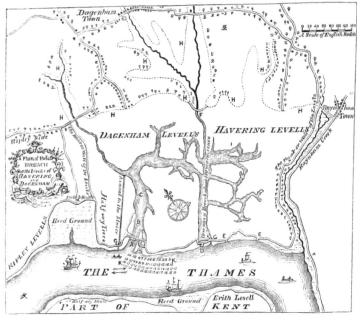

PERRY'S PLAN OF DAGENHAM BREACH.

A.—The Dam whereby the Breach was stopped.
B.—The site of Boswell's works.
C.—The site of the Landowners' works.
D.—The site of Perry's Sluices.
E.—The site of Boswell's Sluices.
F.—A Dam and Sluice made for recovery of the
    Meadows shortly after the Breach had
    occurred.

G.—Small Sluice for drainage of the land waters.
HH.—The dotted line represents the extent of
    the inundation caused by the Breach.
I.—Places where stags' horns were dug up.
K.—Parallel lines, showing the depth at low
    water at every 60 yards dstancie from
    the shore.

Scarcely had Perry commenced this part of the work,
and proceeded so far as to exhibit his general design, than
Boswell, the former contractor, presented a petition to
Parliament against the engineer being allowed to go on,
alleging that his scheme was utterly impracticable. The
work being of great importance, and executed at the public expense, a Parliamentary Committee was appointed,
when Perry was called before them and examined at

great length.   His answers were so explicit, and, on
the whole, so satisfactory, that at the close of the examination one of the members thus spoke the sense of the
Committee :—" You have answered us like an artist, and
like a workman ; and it is not only the scheme, but the
man, that we recommend."

Perry was then allowed to proceed, and the work
went steadily forward.   About three hundred men were
employed in stopping the breach, and it occupied them
about five years to accomplish it.   Pile driving is slow
work, but it gradually advanced, and every foot of ground
secured was made good.   As the piling approached from
both sides towards the centre, the water rushed through
the narrowed aperture with increased violence, pouring
in and out like a cascade.   It was even feared that the
channel in the centre would be worn so deep as not to
be reached by the longest piling.   Numerous accidents
incidental to such an undertaking occurred, and on one
occasion it was feared that after all the waters would
be the victor, spouting through the crevices in the
timbers, and pouring over the top of the work like
a waterfall.   But at last the centre pile was driven ;
a stout clay bank was heaped all round the breach, as
it is still to be seen ; the opening was effectually stopped,
and the waters drained away by the sluices, leaving the
extensive inland lake, which is to this day used by the
Londoners as a place for fishing and aquatic recreation.

A good idea of the formidable character of the embankments extending along the Thames may be obtained by a
visit to this place.   Standing on the top of the bank,
which is from 40 to 50 feet above the river level at low
water,[1] we see on the one side the Thames, with its

---

[1] The banks themselves are from 17
to 25 feet high in the neighbourhood
of Dagenham, and from 25 to 30 feet
at the base.   The marks of the old
breach are still easily traceable, and at
one point the rotting timbers of a
large square box are still seen sticking
out of the bank, which may possibly
be one of the machines filled with
stones and chalk sunk by the unfortunate contractor Boswell.

shipping passing and repassing, high above the inland level when the tide is up, and the still lake of Dagenham and the far extending flats on the other. Looking from the lower level on these strong banks extending along the stream as far as the eye can reach, we can only see the masts of sailing ships and the funnels of large steamers leaving behind them long trails of murky smoke,—at once giving an idea of the gigantic traffic that flows along this great watery highway, and the

DAGENHAM LAKE.
[By Edmund M. Wimperis, after a Sketch by the Author]

enormous labour which it has cost to bank up the lands and confine the river within its present artificial limits. We do not exaggerate when we state that these formidable embankments, winding along the river side, up creeks and tributary streams, round islands and about marshes, from London to the mouth of the Thames, are not less than 300 miles in extent.

It is to be regretted that Perry gained nothing but fame by his great work. The expense of stopping the breach far exceeded his original estimate; he required more materials than he had calculated upon ; and frequent strikes amongst his workmen for advances of wages greatly increased the total cost. These circumstances seem to have been taken into consideration by the Government in settling with the engineer, and a grant of 15,000*l.* was voted to him in consideration of his extra outlay. The landowners interested also made him a present of a sum of 1000*l.* But even then he was left a loser; and although the public were so largely benefited by the complete success of the work, which restored the navigation of the river, and enabled the drowned lands within the embankment again to be effectually reclaimed, he did not receive a farthing's remuneration for his five years' great anxiety and labour.[1]

After this period Perry seems to have been employed on harbour works,—more particularly at Rye and Dover; but none of these were of great importance, the enterprise of the country being as yet dormant, and its available capital for public undertakings comparatively limited. It appears from the Corporation Records of Rye, that in 1724 he was appointed engineer to the proposed new harbour works there. The port had become very much silted up, and for the purpose of restoring the navigation it was designed to cut a new channel, with two pier-heads, to form an entrance to the harbour. The plan further included a large stone sluice and draw-bridge, with gates, across the new channel, about a quarter of a mile within the pier-heads; a wharf constructed of timber along the two sides of the channel, up to the sluice; together with other well-designed improvements. But the works had scarcely been begun before the Commissioners displayed a strong disposition to job, one of them withdrawing for the purpose of

---

[1] Perry afterwards published a history of the above undertaking in a little book entitled—'An Account of the Stopping of Daggenham Breach; and proposals for rendering the Ports of Dover and Dublin commodious for entertaining large ships; to which is prefixed a Plan of the Levels which were overflowed by the Breach.' 8vo. London, 1721.

supplying the stone and timber required for the new
works at excessive prices, and others forming what was
called "the family compact," or a secret arrangement
for dividing the spoil amongst them.  The plan of Perry
was only carried out to a limited extent; and though
the pier-heads and stone sluice were built, the most important part of the work, the cutting of the new channel,
was only partly executed, when the undertaking was
suspended for want of funds.

From that time forward, Perry's engineering ability
was very much confined to making reports as to what
things should be done, rather than in being employed to
do them.   In 1727 he published his "Proposals for
Draining the Fens in Lincolnshire;" and he seems to
have been employed there as well as in Hatfield Level,
where "Perry's Drain" still marks one of his works.
He was acting as engineer for the adventurers who
undertook the drainage of Deeping Fen, in 1732, when
he was taken ill and died at Spalding, in the sixty-third
year of his age; and he lies buried in the churchyard of
that town.

THE THAMES FROM DAGENHAM BANK, LOOKING UP THE RIVER
[By R. P. Leitch, after a Sketch by the Author.]

VIEW OF BRADING HAVEN, TEMPORARILY RECLAIMED BY SIR HUGH MYDDELTON,
AS SEEN FROM THE VILLAGE OF BRADING.

[By Percival Skelton, after his original Drawing.]

# L I F E

OF

# SIR HUGH MYDDELTON.

## CHAPTER I.

### WATER SUPPLY OF LONDON IN EARLY TIMES.

WHILE the engineer has so often to contend with all his
skill against the powers of water, and to resist it as a
fierce enemy, he has also to deal with it as a useful
agent, and treat it as a friend. Water, like fire, though
a bad master, is a most valuable servant; and it is
the engineer's business, amongst other things, to render
the element docile, tractable, and useful. Even in the
Fens, water was not to be entirely got rid of. Were
this possible, the Great Level, instead of a boggy reed
swamp, would be merely converted into an arid, dusty
desert. Provision had, therefore, to be made for the
accommodation and retention of sufficient water to serve
for irrigation and the watering of cattle, at the same
time that the lines of drains or cuts were so laid out
as to be available for purposes of navigation.

But water is also one of the indispensable necessaries
of life for man himself, an abundant supply of it being
essential for human health and comfort. Hence all
the ancient towns were planted by the banks of rivers,
principally because the inhabitants required a plentiful
supply of water for their daily use. Old London had
not only the advantage of its pure, broad stream flowing

along its southern boundary, so useful as a water road, but it also possessed an abundance of wells, from which a supply of pure water was obtained, adequate for the requirements of its early population. The river of Wells, or Wallbrook, flowed through the middle of the city; and there were numerous wells in other quarters, the chief of which were Clerke's Well, Clement's Well, and Holy Well, the names of which still survive in the streets built over them.

As London grew in size and population, these wells were found altogether inadequate for the wants of the inhabitants; besides, the water drawn from them became tainted by the impurities which filter into the soil wherever large numbers are congregated. Conduits were then constructed, through which water was led from Paddington, from James's Head, Mewsgate, Tyburn, Highbury, and Hampstead. There were sixteen of such public conduits about London, and the Conduit Streets which still exist throughout the metropolis mark the sites of several of these ancient works.[1] The copious supply of water by these conduits was all the more necessary at that time, as London was for the most part built of timber, and liable to frequent fires, to extinguish which promptly, every citizen was bound to have a barrel full of water in readiness outside his door. The corporation watched very carefully over their protection, and inflicted severe punishments on such as interfered with

---

[1] The conduits used, in former times, to be yearly visited with considerable ceremony. For instance, we find that —"On the 18th of September, 1562, the Lord Mayor (Harper), the Aldermen, with many worshipful persons, and divers of the Masters and Wardens of the twelve companies, rode to the Conduit's-head [now the site of Conduit Street, New Bond Street], for to see them after the old custom. And afore dinner they hunted the hare and killed her, and thence to dinner at the head of the Conduit. There was a good number entertained with good cheere by the Chamberlain, and, after dinner, they hunted the fox. There was a great cry for a mile, and at length the hounds killed him at the end of St. Giles's. Great hallooing at his death, and blowing of hornes; and thence the Lord Mayor, with all his company, rode through London to his place in Lombard Street."—Stowe's 'Survey of London.' It would appear that the ladies of the Lord Mayor and Aldermen attended on these jovial occasions, riding *in waggons*.

the flow of water through them. We find a curious instance of this in the City Records, from which it appears that, on the 12th November, 1478, one William Campion, resident in Fleet Street, had cunningly tapped the conduit where it passed his door, and conveyed the water into a well in his own house, "thereby occasioning a lack of water to the inhabitants." Campion was immediately had up before the Lord Mayor and Aldermen, and after being confined for a time in the Comptour in Bread Street, the following further punishment was inflicted on him. He was set upon a horse with a vessel like unto a conduit placed upon his head, which being filled with water running out of small pipes from the same vessel, he was taken round all the conduits of the city, and the Lord Mayor's proclamation of his offence and the reason for his punishment was then read. When the conduit had run itself empty over the culprit, it was filled again. The places at which the proclamation was read were the following,—at Leadenhall, at the pillory in Cornhill, at the great conduit in Chepe, at the little conduit in the same street, at Ludgate and Fleet Bridge, at the Standard in Fleet Street, at Temple Bar, and at St. Dunstan's Church in Fleet Street; from whence he was finally marched back to the Comptour, there to abide the will of the Lord Mayor and Aldermen.[1]

The springs from which the conduits were supplied in course of time decayed; perhaps they gradually diminished by reason of the sinking of wells in their neighbourhood for the supply of the increasing suburban population. Hence a deficiency of water began to be experienced in the city, which in certain seasons almost amounted to a famine. There were frequent contentions at the conduits for "first turn," and when water was scarce, these sometimes grew into riots. The water carriers came prepared for a fight, and at length

---

[1] 'Corporation Records.' Index No. I., fo. 184 b.

the Lord Mayor had to interfere, and issued his proclamation forbidding persons from resorting to the conduits armed with clubs and staves. This, however, did not remedy the deficiency. It is true the Thames,—"that most delicate and serviceable river," as Nichols terms it,[1] was always available; but an increasing proportion of the inhabitants lived at a distance from the river. Besides, the attempt was made by those who occupied the lanes leading towards the Thames to stop the thoroughfare, and allow none to pass without paying a toll. A large number of persons then obtained a living as water carriers,[2] selling the water by the "tankard" of about three gallons; and they seem to have formed a rather unruly portion of the population.

The difficulty of supplying a sufficient quantity of water to the inhabitants by means of wells, conduits, and water carriers, continued to increase, until the year 1582, when Peter Morice, a Dutchman, undertook, as the inhabitants could not go to the Thames for their water, to carry the Thames to them. With this object he erected an ingenious pumping engine in the first arch of London Bridge, worked by water wheels driven by the rise and fall of the tide, which then rushed with great velocity through the arches. This machine forced the water through leaden pipes, which were laid into the houses of the citizens; and the power with which Morice's forcing pumps worked was such, that he was enabled

---

[1] 'Progresses of James I.,' vol. ii., 699. The Corporation records contain numerous references to the preservation of the purity of the water in the river. The Thames also furnished a large portion of the food of the city, then abounding in salmon and other fish. The London fishermen constituted a large class, and we find numerous proclamations made relative to the netting of the "salmon and porpoises"—wide nets and wall nets being especially prohibited. Fleets of swans on the Thames were a pictu-resque feature of the river down even to the time of James II.

[2] The water carrier was commonly called a "Cob," and Ben Jonson seems to have given a sort of celebrity to the character by his delineation of "Cob" in his 'Every Man in his Humour.' Gifford, in a note on the play, pointed out that there is an avenue still called "Cob's Court," in Broadway, Blackfriars; not improbably (he adds) from its having formerly been inhabited principally by the class of water carriers.

to throw the water over St. Magnus's steeple, greatly
to the wonderment of the Mayor and Aldermen, who
assembled to witness the experiment. The machinery
succeeded so well that a few years later we find the
corporation empowering the same engineer to. use the
second arch of London Bridge for a similar purpose.[1]

But even this augmented machinery for pumping
was found inadequate for the supply of London. The
town was extending rapidly in all directions, and the
growing density of the population along the river banks
was every year adding to the impurity of the water,
and rendering it less and less fit for domestic purposes. Hence the demand for a more copious and ready
supply of pure water continued steadily to increase.
Where was the new supply to be obtained, and how
was it to be rendered the most readily available for
the uses of the citizens? Water is by no means a
scarce element in England; and no difficulty was experienced in finding a sufficiency of springs and rivers of
pure water at no great distance from the metropolis.
Thus, various springs were known to exist in different
parts of Hertfordshire and Middlesex; and many inde-
finite plans were proposed for conveying their waters to
London. To enable some plan or other to be carried
out, the corporation obtained an Act towards the end of
Queen Elizabeth's reign,[2] empowering them to cut a
river to the city from any part of Middlesex or Hertfordshire; and ten years were specified as the time allowed
for carrying out the necessary works. But, though many
plans were suggested and discussed, no steps were taken
to cut the proposed river. The enterprise seemed too
large for any private individual to undertake; and
though the corporation were willing to sanction it, they

---

[1] The river pumping-leases continued in the family of the Morices
until 1701, when the then owner sold
his rights to Richard Soams for
38,000l., and by him they were afterwards transferred to the New River
Company at a still higher price.

[2] Act 13 Eliz. c. 18.

were not disposed to find any part of the requisite means for carrying it out. Notwithstanding, therefore, the necessity for a large supply of water, which became more urgent in proportion to the increase of population, the powers of the Act were allowed to expire without anything having been done to carry them into effect.

In order, however, to keep alive the parliamentary powers, another Act was obtained in the third year of James I.'s reign (1605),[1] to enable a stream of pure water to be brought from the springs of Chadwell and Amwell, in Hertfordshire; and the provisions of this Act were enlarged and amended in the following session.[2] From an entry in the journals of the corporation, dated the 14th October, 1606, it appears that one William Inglebert petitioned the court for liberty to bring the water from the above springs to the northern parts of the city " in a trench or trenches of brick." The petition was "referred," but nothing further came of it; and the inhabitants of London continued for some time longer to suffer from the famine of water—the citizens patiently waiting for the corporation to move, and the corporation as patiently waiting for the citizens.

The same difficulty of water-supply had been experienced in other towns, but more especially at Plymouth, where the defect had been supplied by the public spirit and enterprise of one of the most distinguished of English admirals—no other than the great Sir Francis Drake. It appears from the ancient records that water was exceedingly scarce in Plymouth, and the inhabitants had to send their clothes more than a mile from the town to be washed, and that the water used for domestic purposes was mostly fetched from Plympton, about five miles distant. Sir Francis Drake, who was born within ten miles of Plymouth, and settled in the neighbourhood of the town after having realized a considerable fortune

---

by his adventures on the Spanish Main, observing the great inconvenience suffered by the population from this want of water, as well as the difficulty of furnishing the ships frequenting the port with that indispensable necessary, conceived the project of supplying the deficiency by leading a store of water to the town from one of the numerous springs on Dartmoor. Accordingly, in 1587, when he represented Bossiney (Tintagel) in Cornwall, he obtained an Act enabling him to convey a stream from the river Mew or Meavy; and in the preamble to the Act it was expressed that its object was not only to ensure a continual supply of water to the inhabitants, but to obviate the inconvenience hitherto sustained by seamen in watering their vessels. It would appear, from documents still extant, that the town of Plymouth contributed 200*l.* towards the expenses of the works, Sir Francis being at the remainder of the cost; and on the completion of the undertaking the corporation agreed to grant him a lease of the aqueduct for a term of twenty years, at a nominal rental. Drake lost no time in carrying out the work, which was finished in four years after the passing of the Act; and its completion in 1591, on the occasion of the welcoming of the stream into the town, was attended with great public rejoicings.[1]

The " Leet," as it is called, is a work of no great magnitude, though of much utility. It was originally nothing more than an open trench cut along the sides of the moor, in which the water flowed by a gentle inclination into the town and through the streets of Plymouth. The distance between the head of the aqueduct at Sheep's

---

[1] The tradition survives to this day that Sir Francis Drake did not cut the Leet by the power of money and engineering skill, but by the power of magic. It is said of him that, calling for his horse, he mounted it and rode about Dartmoor until he came to a spring sufficiently copious for his design, on which, pronouncing some magical words, he wheeled round, and, starting off at a gallop, the stream formed its own channel, and followed his horse's heels into the town.

Tor and Plymouth, as the crow flies, is only seven miles; but the length of the Leet—so circuitous are its windings—is nearly twenty-four miles. After its completion Drake presented the Leet to the inhabitants of Plymouth " as a free gift for ever," and it has since remained vested in the corporation,—who might, however, bestow more care than they do on its preservation against impurity. Two years after the completion of the Leet, the burgesses, probably as a mark of their gratitude, elected Drake their representative in parliament. The water proved of immense public convenience, and Plymouth, instead of being one of the worst supplied, was rendered one of the best watered towns in the kingdom. Until a comparatively recent date the water flowed from various public conduits, and it ran freely on either side of the streets—as is still observed at Salisbury and other southern towns—that all classes of the people might enjoy the benefit of a full and permanent supply throughout the year.[1] One of the original conduits still remains at the head of Old Town-street, bearing the inscription, " Sir Francis Drake first brought this water into Plymouth, 1591."

The example of Plymouth may possibly have had an

---

[1] Westcote, who was contemporary with Drake, thus alludes to his undertaking :—" The streets [of Plymouth] are fairly paved and kept clean and sweet, much refreshed by the fresh stream running through it plenteously, to their great ease, pleasure, and profit; which was brought into the town by the skill and industrious labours of the ever to be remembered with due respect and honourable regard, Sir Francis Drake, Knight; who, when it was a dry town, fetching their water and drying their clothes some miles thence, by a composition made with the magistracy he brought in this fair stream of fresh water. The course thereof from the head is seven miles, but by indenting and circling through hills, dales, and waste bogs, but with greatest labour and cost through a mighty rock generally supposed impossible to be pierced, at least thirty. But in this his undaunted spirit and bounty (like another Hannibal making way through the impassable Alps) had soon the victory, and finished it to the great and continual commodity of the town, and his own commendation."—Westcote's ' Devonshire in 1630.' 4to. Exeter, 1845, p. 378. For the site of the " mighty rock, generally supposed," &c., see the Ordnance map of Devon, a little to the north of Plymouth, where the word " Tunnel " is found marked on the course of the old Dartmoor granite tramway, parallel with which, in that locality, the Leet runs.

influence upon the Corporation of London in obtaining
the requisite powers from Parliament to enable them
to bring the springs of Chadwell and Amwell to the
thirsty population of the metropolis; but unhappily
they had as yet no Drake to supply the requisite capital
and energy.   In March, 1608, one Captain Edmond
Colthurst petitioned the Court of Aldermen for permission to enter upon the work;[1] but it turned out that
the probable cost was far beyond the petitioner's means,
without the pecuniary help of the corporation, and that
being withheld, the project fell to the ground.   After
this, one Edward Wright is said to have actually begun
the works;[2] but they were suddenly suspended, and the
citizens of London were as far from their supply of
pure water as ever.   At this juncture, when all help
seemed to fail, and when men were asking each other
" who is to do this great work, and how is it to be done ?"
citizen Hugh Myddelton, impatient of further delay,
came forward and boldly said, " If no one else will undertake this work, I will do so, and execute it at my own
cost."   Yet Hugh Myddelton was no engineer, not even
an architect or a builder, but only a goldsmith; possessing, however, an amount of energy of character and
enterprising public spiritedness, in which the Londoners
of those days seem to have been generally wanting.

---

[1] ' Records of the City of London,'
6th James I.

[2] Art. ' Canal,' in Addenda to Hutton's ' Mathematical and Philosophical
Dictionary.'   Mr. Wright was the
author of a celebrated treatise on Navigation, entitled ' Certain Errors in
Navigation Detected and Corrected,'
originally published in 1599, and reprinted, with additions, in 1657.

## CHAPTER II.

### HUGH MYDDELTON, GOLDSMITH AND MERCHANT ADVENTURER.

THE Myddeltons are an ancient family in North Wales, and have at various times held large possessions in the

MYDDELTON'S NATIVE DISTRICT.
[Ordnance Survey.]

vale of Clwyd and the neighbourhood. They trace their origin to a noted chieftain of the twelfth century, one Blaidd, lord of Penllyn, from whom also descended the Mostyn family, the Vaughans of Nannau, and the Salusburies of Llanrwst. One of Blaidd's descendants married Cecilia, the daughter and heiress of Sir Alexander Myddelton of Myddelton, Shropshire, whose name he assumed, and various branches of Myddeltons sprang from the union. Those of Gwaenynog, near Denbigh, are the elder branch, and the estate is still held by their lineal descendant. Ystrad was another patrimony of the Myddeltons in the time of the Tudors, and there are monuments of the family still to be seen in Llanrhaiadr church. Nearer Denbigh is a third estate which belonged to the Myddeltons, called Galch-hill : it is situated between Gwaenynog and the town of Denbigh, within sight of the old castle, which commands a view of one of the richest and most beautiful valleys in the kingdom. Three brothers held the several estates of

Gwaenynog, Ystrad, and Galch-hill about the end of the sixteenth century—Robert, John, and Richard;[1] the last being the father of Sir Hugh Myddelton, the subject of our present memoir.

MYDDELTON'S HOUSE AT GALCH-HILL, DENBIGH.[2]
[By E. M. Wimperis, after an original Sketch.]

Richard Myddelton, of Galch-hill, was governor of Denbigh Castle in the reigns of Edward VI., Mary, and Elizabeth. He seems to have been a man eminent for uprightness and integrity, and is supposed to have been

[1] Williams's 'Ancient and Modern Denbigh.' From this book we take the following incident relating to the Myddelton family in early times :— "David Myddelton, Receiver of Denbigh in the 19th Edward IV., and Valectus Coronæ in the 2nd Richard III., paid his addresses to Elyn, daughter of Sir John Donne, of Utkinton, in Cheshire, and gained the lady's affections. But the parents preferred their relative, Richard Donne, of Croton. The marriage was accordingly celebrated; but David Myddelton watched the bridegroom leading his bride out of church, killed him on the spot, carried away his widow, and married her forthwith. So that she was maid, widow, and a wife twice in one day. From Roger, the eldest son of this match, the Myddeltons of the above branch are descended."

[2] The old-fashioned whitewashed house, the back of which is represented in the above engraving, is said to have been the house in which Hugh Myddelton was born. It has, however, undergone numerous alterations since his time, though some portions of it, on the lower story, are very ancient.

the first member who sat in Parliament for Denbigh.[1]
His wife was one Jane Dryhurst, the daughter of an
alderman of Denbigh, by whom he had a very numerous family. He was buried, with his wife, in the parish

FAC SIMILE OF THE MYDDELTON BRASS IN WHITCHURCH PORCH.

church of Denbigh, called Whitchurch or St. Marcellus,
where a small monumental brass placed within the
porch, represents Richard Myddelton and Jane his wife,
with their sixteen children, all kneeling. Behind him
are nine sons, and behind her seven daughters. He died
in 1575; she in 1565. The tablet rehearses his virtues
in the quaint lines inscribed underneath.[2]

---

[1] The privilege was first granted about the year 1536, in the 27th Henry VIII. The first name in the list of representatives of the borough which has been preserved, is that of Richard Myddelton, 1542, in the 33rd Henry VIII. But it is also pro-. bable that he represented the town in the preceding Parliament, which sat only forty days.—Williams's ' Records of Denbigh.'

[2] In vayn we bragg and boast of blood, in vayne of sinne we vaunte,
Syth flesh and blood must lodge at last where nature did us graunte.
So where he lyeth that lyved of late with love and favour muche,
To fynde his friend, to feel his foes, his country skante had suche.
Whose lyff did well reporte his death, whose death hys lyff doth trye,
And poyntes with fynger what he was that here in claye doth lye.
His vertues shall enroll his actes, his tombe shall tell his name,
His sonnes and daughters left behinde shall blaze on earth his fame.
Look under feete and you shall finde, upon the stone yow stande,
The race he ranne, the lyff he led, each with an upright hande.

The epitaph was more truthful than epitaphs usually are ; and as respects the fame of Richard Myddelton's offspring, it might even be regarded as prophetic.  The third son, William, was one of Queen Elizabeth's famous sea captains.  He was educated at Oxford, but, inflamed with a love of adventure, he early went to sea, and eventually rose to distinction.  In 1591 we find him with the small English fleet sent to intercept the Spanish galleons off the Azores.  Philip II., having received intelligence of the design, had equipped and sent to sea a much more powerful squadron for the purpose of effectually frustrating it.  Captain Myddelton first sighted the enemy, and kept company with them for three days, until he had obtained full intelligence of their strength, when he rejoined the fleet under Admiral Howard.  The vigilance of Myddelton on the occasion is said to have saved the English squadron, though Sir Richard Greville, the Vice-Admiral, got entangled with the enemy, and his ship (the only one taken) was captured by the Spaniards, after resisting their whole force for twelve hours, and repulsing their boarders fifteen times.[1]  While engaged on his various cruises, Myddelton occupied his leisure hours in translating the Book of Psalms into Welsh.  He finished the work in the West Indies, and it was published in 1603, shortly after his death.[2]  He was also the author of ' Barddoniaeth, or the Art of Welsh Poetry,' a work for some time held in considerable estimation.  The fourth son of Richard Myddelton was Thomas, an eminent citizen and grocer of London.  He served the office of sheriff in 1603, when

---

[1] Greville died, and his ship went down two days after the fight.  His death was as noble as his life had been.  Shortly before his death he said :— "Here I, Richard Greville, die, with a joyful and quiet mind ; for that I have ended my life as a true soldier ought to do, fighting for his country, Queen, religion, and honour ; my soul willingly departing from this body, leaving behind the lasting fame of having behaved as every valiant soldier is in his duty bound to do."

[2] William Myddelton's version of the Psalms was reprinted at Llanfair, Caereinion, in 1827, with a preface by that eminent Welsh scholar and poet, Walter Davies.

he was knighted; and he was elected Lord Mayor in
1613. He was the founder of the Chirk Castle family,
now represented by Mr. Myddelton Biddulph.[1] Charles,
the fifth son, succeeded his father as governor of Denbigh Castle, and when he died bequeathed numerous
legacies for charitable uses connected with his native
town. The sixth son was Hugh Myddelton, the illustrious goldsmith and engineer. Robert, the seventh,
by trade a skinner, was, like two of his brothers, a
London citizen, and afterwards a member of parliament.
Foulk, the eighth son, served as high sheriff of the
county of Denbigh. This was certainly a large measure
of worldly prosperity and fame to fall to the lot of one
man's offspring.

The precise date of Hugh Myddelton's birth is unknown; but it was probably about the year 1555.[2] We
have no record of his early life, and have no desire to
invent anything to supply the defect. All that we know
is, that he was bred to business in London, under the
eye of his elder brother Thomas, the grocer and merchant

---

[1] Sir Thomas realised considerable
wealth by trade, and occasionally
helped King James with loans of
money during that monarch's pecuniary difficulties. Thus we find his
name appearing in the ' Pell Records'
(24 June, 1609) as the recipient of
3000l., together with 290l. 15s., being
the interest thereon, which he had advanced as a loan to King James for
one year. The interest was at the
then current rate of between nine and
ten per cent. Sir Thomas contributed
500l. towards the Free Schools of the
Grocers' Company, of which he was a
member; and he also left 7l. a year to
the poor of the same Company, as
well as the rent of two tenements in
Baynard's Castle, which they enjoy to
this day. Nor did he forget his
Welsh countrymen, for he provided
the Welsh " nation " with a new edition of the Scriptures at his own expense. He is the same Sir Thomas

Myddelton of whom it is related that,
having married a young wife in his
old age, the famous song, " Room for
cuckolds, here comes my Lord Mayor,"
was composed in his honour on the
occasion.

[2] His mother died in 1565, after
having given birth, as we have seen,
to sixteen children; Hugh being the
sixth of nine sons. He was thus, probably, at least ten years old at his
mother's death. This surmise as to
the probable period of his birth is confirmed by a passage which occurs in a
letter written by Myddelton to his
cousin, Sir John Wynne, in 1625, in
which he declined entering upon any
new undertakings because of the in-
firmities of age. His words were, " I
am growne into years." At that time
he would probably be about seventy,
though we find him alive in 1631,
six years later.

adventurer. In those days country gentlemen of moderate income were accustomed to bind their sons apprentices to merchants, especially where the number of younger sons was large, as it certainly was in the case of Richard Myddelton of Galch-hill. There existed at that time in the metropolis numerous exclusive companies or guilds, the admission into which was regarded as a safe road to fortune. The merchants were then few in number, and they constituted almost an aristocracy in themselves; indeed, they were not unfrequently elevated to the peerage because of their wealth as well as public services, and not a few of our present noble families can trace their pedigree back to some wealthy skinner, mercer, or tailor, of the reigns of James or Elizabeth.

Hugh Myddelton was entered an apprentice of the guild of the Goldsmiths' Company. Having thus set his son in the way of well-doing, Richard Myddelton left him to carve out his own career, and rely upon his own energy and ability. He had done the same with Thomas, whom he had helped until he could stand by himself; and William, whom he had educated at Oxford as thoroughly as his means would afford. These sons having been fairly launched upon the world, he bequeathed the residue of his property to his other sons and daughters.[1]

The goldsmiths of that day were not merely dealers in plate, but in money. They had succeeded to much of the business formerly carried on by the Jews and Venetian merchants established in or near Lombard-street. They usually united to the trade of goldsmith that of banker, money-changer, and money-lender, dealing generally in the precious metals, and exchanging plate and foreign coin for gold and silver pieces of English manufacture, which had become much depreciated by long use as well

---

[1] 'Records of Denbigh,' p. 201-2.

as by frequent debasement.[1] It was to the goldsmiths that persons in want of money then resorted, as they would now resort to money-lenders and bankers; and their notes or warrants of deposit circulated as money, and suggested the establishment of a bank-note issue, similar to our present system of bullion and paper currency. They held the largest proportion of the precious metals in their possession; hence, when Sir Thomas Gresham, one of the earliest bankers, died, it was found that the principal part of his wealth was comprised in gold chains.[2]

The place in which Myddelton's goldsmith's shop was situated was in Bassishaw (now called Basinghall) Street, and he lived in the overhanging tenement above it, as was then the custom of city merchants. Few, if any, lived away from their places of business. The roads into the country, close at hand, were impassable in bad weather, and dangerous at all times. Basing Hall was only about a bow-shot from the City Wall, beyond which lay Finsbury Fields, the archery ground of London, which extended from the open country to the very wall itself, where stood Moor Gate. The London of that day consisted almost exclusively of what is now called The City; and there were few or no buildings east of Aldgate, north of Cripplegate, or west of Smithfield. At the accession of James I. there were only a few rows of thatched cottages in the Strand, along which, on the river's side, the boats lay upon the beach. At the same time there were groves of trees in Finsbury, and green pastures in Holborn; Clerkenwell was a village; St. Pancras boasted only of a little church standing in meadows; and St. Martin's, like St. Giles's, was literally

---

[1] Henry VIII. suffered his coin to be so far debased that no regular exchanges could be made; and the confusion made way for the London goldsmiths to leave off their proper trade of goldsmithrie, and to turn exchangers of plate and foreign money for English coins. — Macpherson's 'Annals of Commerce.' 4to., 1805. Vol. ii., p. 357.

[2] It may be remembered that Rubens was accustomed to be paid for his pictures by so many links of gold chain.

"in the fields." All the country to the west was farm and pasture land; and woodcocks and partridges flew over the site of the future Regent Street, May Fair, and Belgravia.

The population of the city was about 150,000, living in some 17,000 houses,[1] which were nearly all of timber, with picturesque gable-ends, and sign boards swinging over the footways. The upper parts of the houses so overhung the foundations, and the streets were so narrow, that D'Avenant said the opposite neighbours might shake hands without stirring from home. The ways were then quite impassable for carriages, which had not yet indeed been introduced into England; all travelling being on foot or on horseback. When coaches were at length introduced and became fashionable, the aristocracy left the city, through the streets of which their carriages could not pass, and migrated westward to Covent Garden and Westminster.

Those were the days for quiet city gossip and neighbourly chat over matters of local concern; for London had not yet grown so big or so noisy as to extinguish that personal interchange of views on public affairs which continues to characterize most provincial towns. Merchants sat at their doorways in the cool of the summer evenings, under the overhanging gables, and talked over the affairs of trade; whilst those courtiers who still had their residences within the walls, hung about the fashionable shops to hear the city gossip and talk over the latest news. Myddelton's shop appears to have been one of such fashionable places of resort, and the pleasant tradition was long handed down in the parish of St. Matthew, Friday-street, that Hugh Myddelton and Walter Raleigh used to sit together at the door of the goldsmith's shop, and smoke the newly introduced weed, tobacco, greatly to the amazement of the passers by.[2] It is not impro-

---

[1] Strype's Edition of Stowe's 'Survey.'

[2] Malcolm's 'Manners and Customs of London,' p. 115.

bable that Captain William Myddelton, who lived in
London after his return from the Spanish main in 1591,
formed an occasional member of the group; for Pennant
states that he and his friend Captain Thomas Price, of
Plâsgollen, and another, Captain Koet, were the first
who smoked, or as they then called it, " drank " tobacco
publicly in London, and that the Londoners flocked from
all parts to see them.[2]

Hugh Myddelton did not confine himself to the trade
of a goldsmith, but from an early period his enterprising
spirit led him to embark in ventures of trade by sea;
and hence, when we find his name first mentioned in
the year 1597, in the records of his native town of
Denbigh, of which he was an alderman and " capitall
burgess," as well as the representative in Parliament, he
is described as " Cittizen and Gouldsmythe of London,
and one of the Merchant Adventurers of England."[3]
The trade of London was as yet very small, but a
beginning had been made. A charter was granted
by Henry VII., in 1505, to the Company of Merchant
Adventurers of England, conferring on them special
privileges. Previous to that time, almost the whole
trade had been monopolized by the Steelyard Company
of Foreign Merchants, whose exclusive privileges were
formally withdrawn in 1552. But for want of an English
mercantile navy, the greater part of the foreign carrying
trade of the country continued long after to be conducted
by foreign ships.

The withdrawal of the privileges of the foreign merchants in England had, however, an immediate effect in
stimulating the home trade, as is proved by the fact, that
in the year following the suppression of the foreign company the English Merchant Adventurers shipped off

---

[1] He resided at the old Elizabethan
house in Highgate, afterwards occupied as an inn, called the " King's
Head."

[2] 'Tour in Wales,' vol. ii., p. 31.
Ed. 1784.

[3] Williams's ' Ancient and Modern
Denbigh,' p. 105.

for Flanders no less than 40,000 pieces of cloth.[1] Cloth-making now became one of the staple manufactures of England, and instead of allowing the foreigners to export the raw wool, work it up abroad by foreign artisans, and return it for sale in the English markets, the English merchants themselves employed the English artisans, aided by the numerous Protestant refugees who had fled into England from French Flanders and the Low Countries, to work up the raw material, when they became large exporters instead of importers of the manufactured articles. Into this trade of cloth manufacture Hugh Myddelton entered with great energy; and he prosecuted it with so much success, that in a speech delivered by him in the House of Commons on the proposed cloth patent, he stated that he and his partner then maintained several hundred families by that trade.[2]

Besides engaging in this new born branch of manufacture, it is not improbable that Myddelton's enterprising spirit,—encouraged by his intimacy with Raleigh and other sea captains, including his own brother William, who had made profitable captures on the Spanish main,—led him to embark in the maritime adventures which were so common at that period, though they would now be regarded as little better than piracy.[3] Drake sacked the Spanish towns, burnt their ships, and carried off their gold, while England was yet at peace with Spain. Drake's vessels were the property of private persons, who sent them forth upon adventures on the high seas; and the results were so astounding, that it was no wonder the example should be followed, more especially after Spain had declared war against England. The

---

[1] 'Pictorial History of England,' vol. ii., p. 784.

[2] 'House of Commons' Journals,' vol. i., p. 491. (20th May, 1614.)

[3] The proceedings of Drake seem to have been regarded by some in this light, even at that time; for Camden says, " Nothing troubled Drake more than that some of the chief men at court refused to accept the gold which he offered them, as being gotten by piracy."

records of the corporation of London contain some curious entries relative to the fitting out of the ships which were sent to sea for the capture of galleons, and the subsequent division of the spoil. On such occasions the several companies of the city combined with the corporation in forming a common purse, and bound themselves by agreement to share in the loss or gain of their adventures, in proportion to the amounts severally subscribed by them. In 1593 we find a richly laden carrack captured by Sir Walter Raleigh, and brought into the Thames a prize; and on the 15th of November, in that year, a committee was appointed "on behalf of such of the city companies as had ventured in the late fleet, to join with such honourable personages as the Queen hath appointed, to take a perfect view of all such goods, prizes, spices, jewels, pearls, treasures, &c., lately taken in the carrack, and to make sale and division thereof."[1] It appears that 12,000*l.* (or equivalent to about four times the value of our present money) was divided amongst the companies which had adventured; and 8000*l.* was similarly netted by them on another occasion. But the poor of the city were not forgotten in the distribution of the money; two shillings in the pound of the clear gain having been divided among the poor people living within the freedom visited by the plague.[2]

---

[1] 'Corporation of City of London Records,' jor. 23, fol. 156.

[2] The plague was then a frequent visitor in the city. Numerous proclamations were made by the Lord Mayor and Corporation on the subject,—proclamations ordering wells and pumps to be drawn, and streets to be cleaned,—and precepts for removing hogs out of London, and against the selling or eating of pork. Wherever the plague was in a house, the inhabitant thereof was enjoined to set up outside a pole of the length of seven feet, with a bundle of straw at the top, as a sign that the deadly visitant was within. Wife, children, and servants belonging to that house must wear white rods in their hands for thirty-six days before they were considered purged. It was also ordered subsequently, that on the street-door of every house infected, or upon a post thereby, the inhabitant must exhibit imprinted on paper a token of St. Anthony's Cross, otherwise called the sign of the Taw **T**, that all persons might have knowledge that such house was infected.—'Corporation of City of London Records,' jor. 12, fol. 136. No. 1. Years 1590 to 1694.

At a comparatively advanced age, Myddelton took to himself a wife; and the rank and fortune of the lady he married afford some indication of the position he had then attained. She was Miss Elizabeth Olmstead, the daughter and sole heiress of John Olmstead of Ingate-stone, Essex, with whom the thriving goldsmith and merchant adventurer received a considerable accession of property. That he had secured the regard of his neighbours, and did not disdain to serve them in the local offices to which they chose to elect him, is apparent from the circumstance that he officiated for three years as churchwarden for the parish of St. Matthew, to which post he was appointed in the year 1598. But he had public honours offered to him of a more distinguished character. He continued to keep up a friendly con-nection with his native town of Denbigh, and he seems to have been mainly instrumental in obtaining for the borough its charter of incorporation in the reign of Elizabeth. In return for this service the burgesses elected him their first alderman, and in that capacity he signed the first by-laws of the borough in 1597. On the back of the document are some passages in his hand-writing, commencing with "Tafod aur yngenau dedwydd" [A golden tongue is in the mouth of the blessed], followed by other aphorisms, and concluding with some expressions of regret at parting with his brethren, the burgesses of Denbigh, whom he had specially visited on the occasion.

On his next visit to the town he appears to have entered upon a mining enterprise, in the hope of being able to find coal in the neighbourhood. In a letter written by him in 1625 to Sir John Wynne, he thus refers to the adventure :—"It may please you to under-stand that my first undertaking of publick works was amongst my own people, within less than a myle of the place where I hadd my first beinge (24 or 25 years since), in seekinge of coals for the town of Denbighe."[1] Myddelton was most probably deceived by the slaty appearance of the soil into the belief that coal was to be found in the neighbourhood, and after spending a good deal of money in the search, he finally gave it up as a hopeless undertaking.

---

[1] The common story told of Myddelton's subsequent exccution of the New River, is, that he was enabled to carry out the works by means of the large fortune he had realised by the working of a " silver-mine in Wales." This has been repeated by every writer on the subject of Sir Hugh Myddelton's career from Stowe downwards; but it is altogether without foundation, the only mining adventure on which he entered previous to the New River enterprise being that at Denbigh, which proved a total failure.

WHITCHURCH, OR ST. MARCELLUS, DENBIGH.

[By E. M Wimperis, after an original Sketch ]

## CHAPTER III.

In 1603 Hugh Myddelton was returned representative of his native town to the first parliament summoned by James I. In those days the office of representative was not so much coveted as it is now, and boroughs remote from the metropolis were occasionally under the necessity of paying their members to induce them to serve. Thus it was an advantage to the burgesses of Denbigh that they had a man so able to represent them as Hugh Myddelton, resident in London, and who was moreover an alderman and a benefactor of the town. His two brothers—Thomas Myddelton, citizen and grocer, and Robert, citizen and skinner, of London—were members of the same parliament, and we find Hugh and Robert frequently associated on committees of inquiry into matters connected with trade and finance. Among the first committees to which we find the brothers appointed was one on the subject of a bill for explanation of the Statute of Sewers, and another for the bringing of a fresh stream of running water from the river of Lea, or Uxbridge, to the north parts of the city of London.[1] Thus the providing of a better supply of water to the inhabitants of the metropolis came very early under his notice, and doubtless had some influence in directing his future action on the subject.

At the same time the business in Bassishaw-street was not neglected, for, shortly after the arrival of King James in London, we find Myddelton supplying jewelry for

---

[1] 'Commons' Journals,' vol. i. 262.   31st January, 1605.

Queen Anne, whose rage for finery of that sort was excessive. A warrant, in the State Paper-office, orders 250*l.* to be paid to Hugh Myddelton, goldsmith, for a jewel given by James I. to the queen;[1] and it is probable that this connection with the Court introduced him thus early to the notice of the king, and facilitated his approach to him when he afterwards had occasion to solicit His Majesty's assistance in bringing the New River works to completion.

The subject of water supply to the northern parts of the city was still under the consideration of parliamentary committees, of which Myddelton was invariably a member; and at length a bill passed into law, and the necessary powers were conferred. But no steps were taken to carry them into effect. The chief difficulty was not in passing the Act, but in finding the man to execute the work. A proposal made by one Captain Colthurst to bring a running stream from the counties of Hertford and Middlesex, was negatived by the Common Council in 1608. Fever and plague from time to time decimated the population, and the citizens of London seemed as far as ever from being supplied with pure water.

It was at this juncture that Hugh Myddelton stepped forth and declared that if no one else would undertake it, he would, and bring the water from Hertfordshire into London. " The matter," quaintly observes Stowe, " had been well-mentioned though little minded, long debated but never concluded, till courage and resolution lovingly shook hands together, as it appears, in the soule of this

---

[1] " By order, 26th of February, 1604. To Hugh Middleton, Goldsmith, the sum of 250*l.* for a pendant of one diamond bestowed upon the Queen by His Majesty. By writ dated 9th day of January, 1604, 250*l.*"—Extract from the ' Pell Records.' [The sum named would be equivalent to about 1000*l.* of our present money. The Queen's passion for jewels may be inferred from the circumstance stated by Dr. Steven in his 'Memoir of George Heriot,' the King's goldsmith (founder of Heriot's Hospital, Edinburgh), that during the ten years which immediately preceded the accession of King James to the throne of Great Britain, Heriot's bills for the Queen's jewels alone could not amount to less than 50,000*l.* sterling.]

no way to be daunted, well-minded gentleman." When all others held back—lord mayor, corporation, and citizens—Myddelton took courage, and showed what one strong practical man, borne forward by resolute will and purpose, can do.

"The dauntless Welshman," says Pennant, in his excusable admiration for his distinguished kinsman of the Principality, "stept forth and smote the rock, and the waters flowed into the thirsting metropolis." Fuller is no less eulogistic in describing the achievement of this genuine English, or Welsh worthy. "If those," says he, " be recounted amongst David's Worthies, who, breaking through the army of the Philistines, fetcht water from the well of Bethlehem to satisfie the longing of David (founded more in fancy than necessity), how meritorious a work did this worthy man perform, who, to quench the thirst of thousands in the populous city of London, fetcht water on his own cost more than 24 miles, encountering all the way an army of oppositions, grappling with Hills, struggling with Rocks, fighting with Forests, till, in defiance of difficulties, he had brought his project to perfection!"[1]

Myddelton's success in life seems to have been attributable not less to his quick intelligence than to his laborious application and indomitable perseverance. He had, it is true, failed in his project of finding coal at Denbigh; but the practical knowledge which he acquired, during his attempt, of the arts of mining and excavation, had disciplined his skill and given him fertility of resources, as well as cultivated in him that power of grappling with difficulties, which emboldened him to undertake this great work, more like that of a Roman emperor than of a private London citizen.

The corporation were only too glad to transfer to him the powers with which they had been invested by the legislature, together with the labour, the anxiety, the

---

[1] 'Worthies of England,' vol. ii., 590.

expense, and the risk of carrying out the undertaking which they regarded as so gigantic. On the 28th of March, 1609, the corporation accordingly formally agreed to his proposal to bring a supply of water from Amwell and Chadwell, in Hertfordshire, to Islington, as being "a thing of great consequence, worthy of acceptation for the good of the city;" but subject to his beginning the works within two months from the date of their acceptance of his offer, and doing his best to finish the same within four years.[1] A regular indenture was drawn up and executed between the parties on the 21st of April following;[2] and Myddelton began the works and "turned the first sod" in the course of the following month, according to the agreement. The principal spring was at Chadwell, near Ware,[3] and the operations commenced at that point. The second spring was at Amwell, near the same town; both being about twenty miles from London as the crow flies.

The works were no sooner begun than a host of opponents sprang up. The owners and occupiers of lands through which the New River was to be cut, strongly objected to it as destructive of their interests. In the petition presented by them to parliament, they alleged that their meadows would be turned into " bogs and

---

[1] 'Common Council Journals, Corporation of London,' 28 March, 1609.

[2] Curiously enough, there is no record in the Repertories or Journals of the Corporation of this document, which was afterwards cancelled by mutual agreement, and a second (based upon the first, and a letter of attorney) was executed on the 28th March, 1611, and sealed on the 14th April following.

[3] " Ware [in Herts], so named from a sort of dam anciently made there to stop the current [of the river Lea]; commonly called a *Weir* or *Ware*, which, as it is confirmed by an abundance of waters thereabouts, that might put them under a necessity of such contrivances, so particularly from the inundation in the year 1408, when it was almost all drowned; 'since which time,' says Norden, 'and before, there was great provision made by weirs and sluices for the better preservation of the town, and the grounds belonging to the same.' The plenty of waters hereabouts gave occasion to that ingenious and useful project of cutting the channel from hence to London, and conveying thither the New River, to the great convenience and advantage of that city, which river was at first called also 'Myddelton's Waters,' from Sir Hugh Myddelton, a great undertaker in that work." —Camden's 'Brit.,' by Gibson, vol. i., 319, 320.

quagmires," and arable land become "squallid ground;" that their farms would be "mangled" and their fields cut up into quillets and "small peeces;" that the "cut," which was no better than a deep ditch, dangerous to men and cattle, would, upon "soden raines," inundate the adjoining meadows and pastures, to the utter ruin of many poor men; that the church would be wronged in its tithe without remedy; that the highway between London and Ware would be made impassable; and that an infinity of similar evils would be perpetrated, and irretrievable injuries inflicted on themselves and their posterity. The opponents

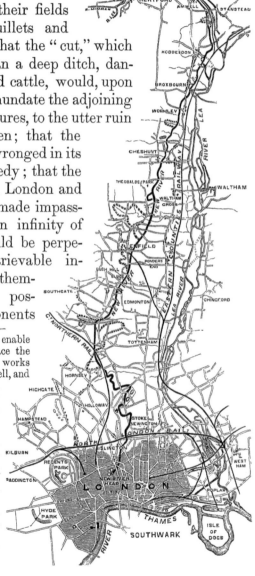

The annexed map will enable the reader easily to trace the line of the New River works between Amwell, Chadwell, and London. The dotted lines indicate those parts of the old course which have since been superseded by more direct cuts, represented by the continuous black line. Where the loops have been detached from the present line of works, they are, in most instances, laid dry, and may be examined and measured correctly as also the soil of which the banks were originally formed.

MAP OF THE NEW RIVER.

[Supplied by W. C. Mylne, Esq., Engineer of the New River Company.]

also pointed out that the Mayor and corporation would have nothing to do with the business, but, by an irrevocable act of the Common Council, had transferred their powers of executing the works to Mr. Myddelton and his heirs, "who doth the same for his own private benefit."[1]

The agitation against the measure was next taken up in parliament. "Much ado there is in the House," writes Mr. Beaulieu, on the 9th of May, 1610, to a friend in the country, "about the work undertaken, and far advanced already by Myddelton, of the cutting of a river and bringing it to London from ten or twelve miles off, through the grounds of many men, who, for their particular interests, so strongly oppose themselves to it, and are like (as it is said) to overthrow it all."[2] On the 20th of June following, a bill was introduced and committed to repeal the Act authorizing the construction of the New River. A committee of ten was appointed a few days after "to view" the river and to certify respecting the progress made with the works, doubtless with the object of ascertaining what damage had actually been done, or was likely to be done, to private property. The committee were directed to make their report in the next session; but as parliament was prorogued in July and did not meet again for four years, the subject is not again mentioned in the Journals of the House.

The corporation of the City did what they could to oppose the bill for the repeal of the New River Acts. On the 25th of May they sent a deputation of aldermen, accompanied by the Town Clerk and Remembrancer, to wait upon the Home Secretary, Chancellor of the Exchequer, and other official men of influence, earnestly

---

[1] These objections and the answers to them are stated in a curious document deposited in the State Paper Office ('Domestic Calendar of State Papers,' vol. 78), entitled, 'The effect of a [proposed] Bill to repeal the Acts of 3 and 4 Jac. I., for bringing the New River into London; stating the objections against those Acts, and answers to the objections.'

[2] Nichols's 'Progresses of James I.,' ii., 313.

entreating them to oppose the Repeal bill, on the ground that the stream of fresh water proposed to be brought into the northern parts of the city " would be a thing very commodious for the preservation of health in the city, and which stream hath been already brought onward about ten miles at the charges of Mr. Hugh Myddelton, the city's deputy, amounting already to the sum of 3000*l.* and above." [1]

Worse than all, was the popular opposition which Myddelton had to encounter. The pastor of Tottenham, writing in 1631, speaks of the New River as " brought with an ill wille from Ware to London." [2] Stowe, who was a contemporary and enthusiastic admirer of Myddelton, says bitterly, " If those enemies of all good endeavours, Danger, Difficulty, Impossibillity, Detraction, Contempt, Scorn, Derision, yea, and Desperate Despight, could have prevailed, by their accursed and malevolent interposition, either before, at the beginning, in the very birth of the proceeding, or in the least stolne advantage of the whole prosecution, this Worke, of so great worth, had never been accomplished." [3] Stowe records that he rode down divers times to see the progress made in cutting and constructing the New River, and " diligently observed that admirable art, pains, and industry were bestowed for the passage of it, by reason that all grounds are not of a like nature, some being oozy and very muddy, others again as stiff, craggy, and stony. The depth of the trench," he adds, " in some places descended full thirty feet, if not more, whereas in other places it required a sprightful art again to mount it over a valley in a trough, between a couple of hills, and the trough all the while borne up by wooden arches, some of them fixed in the ground very deep, and rising in height above twenty-three feet."

---

[1] 'Common Council Journals,' 25th May, 1610. Cambell, Mayor.
[2] 'Briefe Description of the Towne of Tottenham.' By the Rev. William Bedwell, Pastour of the Parish of Tottenham. 4to. London, 1631.
[3] Stowe's ' Survey.' Ed. 1633.

It shortly became apparent to Myddelton that the time originally fixed by the Common Council for the completion of the works had been too short, and we accordingly find him petitioning the Corporation for its extension. This was granted him for five years more, on the ground of the opposition and difficulties which had been thrown in his way by the occupiers and landowners along the line of the proposed stream.[1]   It has usually been alleged that Myddelton fell short of funds, and that the Corporation refused him the necessary pecuniary assistance; but the Corporation records do not bear out this statement, the only application apparently made by Myddelton being for an extension of time.   It has also been stated that he was opposed by the water-carriers, and that they even stirred up the Corporation to oppose the construction of the New River; but this statement seems to be equally without foundation.[2]   The principal obstacle which Myddelton had to encounter was unquestionably the opposition of the landowners and occupiers; and it was so

---

[1] 'Common Council Journals,' 27th February, 1610.   Craven, Mayor.

[2] It is Malcolm who says, in his 'Londinium Redivivum,' that the City refused aid to Myddelton lest, by granting it, they should prejudice the interests of the water-carriers. This, however, is at variance with the fact that they had been instrumental—principally at the instigation of Myddelton himself—in pressing the Corporation to obtain the requisite powers from Parliament to enable fresh water to be brought from Middlesex or Hertfordshire into the city. The Corporation Records contain no facts bearing on the subject; though there is a petition deposited in the Guildhall Library, without date, but, judging from the writing and spelling, probably framed about the close of the 16th or the beginning of the 17th century, entitled: "The humble petition of the whole companie of poore water-tankerd-bearers of the Citie of London, and the suburbs thereof, they and their families being 4000 in num-

ber, living and releeved thereby. Robert Tardy, water-bearer, in the name and behalfe of the reste, followes this petition."   After stating that the water brought to the City "is the most wholesome, purest, and sweetest," which is "not to be doubted or disputed," the petitioners complain that the firemen carry off the water from Newgate, "which was granted for use of that house only;" and that they keep "no hours, but work continually," to the loss of other conduits. The petitioners intimate that the Lady Swinnerton "is allowed but two gallons every hour," and affirm that the conduit, whence she is supplied, "is equal to thirteen gallons and better every hour, as it hath been tried," and that the difference, therefore, is wasted.   The petitioners conclude by praying redress "for this and other grievances."   It is obvious, however, that this petition relates exclusively to the regulations respecting the public conduits, and has no bearing upon the New River project.

obstinate that in his emergency he was driven to apply to the King for assistance.

Though James I. in many respects may have been ridiculous and unkingly, he nevertheless appears throughout his reign to have exhibited a sensible desire to encourage the industry and develop the resources of the kingdom he governed. It was he who made the right royal declaration with reference to the drowned lands in the Fens, that he would not suffer the waters to retain their dominion over the lands which skill and labour might reclaim for human uses. We have seen that he first employed Vermuyden to drain the park at Windsor, and afterwards made over to him the useless swamp of Hatfield Chase to be drained, embanked, and reclaimed. He likewise encouraged the reclamation of Sedgemoor and Malvern Chase; and when the landowners in the Fens took no steps to drain the Great Level, he expressed the determination to become himself the sole undertaker. And now, when Hugh Myddelton's admirable project for supplying the citizens of London with water threatened to break down by reason of the strong local opposition offered to it, and while it was spoken of by many with derision and contempt as an impracticable undertaking, the same monarch came to his help, and while he rescued Myddelton from heavy loss, it might be ruin, he enabled him to prosecute his important enterprise to completion.

James had probably become interested in the works from observing their progress at the point at which they passed through the Royal Park at Theobalds, a little beyond Enfield.[1] Theobalds was the favourite residence of the King, where he frequently indulged in

---

[1] Theobalds, a singularly beautiful place, where Elizabeth held counsel with Burleigh, James often lived, and Charles played with his children. The palace was ordered to be pulled down by the Long Parliament, in spite of the commissioners' report that it was " an excellent building in very good repair;" and, the materials having been sold to the highest bidder, the proceeds were divided amongst the soldiers of Cromwell and Fairfax. The materials alone realised not less than 8275*l.* 11*s.*

the pastime of hunting; and on passing the labourers occupied in cutting the New River, he would naturally make inquiries as to their progress. The undertaking was of a character so unusual, and so much of it passed directly through the King's domains, that he could not but be curious about it. Myddelton, having had dealings with His Majesty as a jeweller, seized the opportunity of making known his need of immediate help, otherwise the project must fall through. Several interviews took place between them at Theobalds and on the ground; and the result was, that James determined to support the engineer with his effective help as King, and also with the help of the State purse, to enable the work to be carried out.[1]

An agreement was accordingly entered into between the King and Myddelton, the original of which is deposited in the Rolls-office, and is a highly interesting document. It is contained on seven skins, and is very lengthy; but the following abstract will sufficiently show the nature of the arrangement between the parties. The Grant, as it is described, is under the Great Seal, and dated the 2nd of May, 1612. It is based upon certain articles of agreement, made between King James I. and Hugh Myddelton, "citizen and goldsmith of London," on the 5th of November preceding. After reciting that Hugh Myddelton had " begun his new cutt or river," and that it promised to be of great convenience and profit to the several districts through which it passed, and more particularly to the city of London, and His Majesty being desirous of seeing perfected so advantageous an undertaking, stipulates, in the first place, to discharge a moiety of all necessary expenses for

---

[1] Salmon, in his ' History of Hertfordshire,' attributes great merit to the King in the following passage, where he says: " King James residing at Theobalds, through whose park the New River runs, was heartily concerned for the success of the endeavour, and promoted it with so great zeal, as perhaps he may be reckoned chief in the work."—Folio Ed. London, 1728. P. 20.

bringing the stream of water within "one mile of the city;" secondly, to pay a moiety of the disbursements "already made" by Hugh Myddelton, upon the latter surrendering an account, and swearing to the truth of the same; thirdly, provision is made for the appointment of an "expenditor," who is to be made "acquainted with and privy to" each item of expenditure, for which purpose he is to keep proper books of account, to be duly subscribed by himself and Myddelton,—the King agreeing, upon the oath of the last mentioned, to discharge a moiety of such account or accounts within twenty-one days next ensuing; fourthly, His Majesty grants an exclusive right to Hugh Myddelton and his assigns to bring water from the springs of Hertfordshire, and the King further stipulates for himself and successors to assent to any Act or Acts of Parliament which may be necessary for confirming and enlarging the powers, &c., originally granted to the Mayor and citizens of London for bringing water to their city, and by them assigned to Myddelton; fifthly, the King grants to Hugh Myddelton a right of way, &c., through his manors, parks, lands, and premises, through which it may be necessary to carry the New River, without charge for loss or damage to the same; sixthly, His Majesty contracts to provide for a moiety of the expenses to be incurred for cisterns and ponds for holding the water, as well as for pipes for distributing the same into small houses, &c. In consideration of these aids and concessions on the part of His Majesty, Myddelton assigns to him a moiety of the interest in, and profits to arise from, the New River "for ever," with the exception of a small quill or pipe of water which the said Myddelton had granted, at the time of his agreement with the City, to the poor people inhabiting St. John-street and Aldersgate-street,—which exception His Majesty allows.

In consideration of these great advantages, Myddelton

executed a conveyance of one-half of the whole undertaking to the King on the 2nd of August following, and the conditions seem to have been faithfully adhered to on both sides. One of the first benefits Myddelton derived from the arrangement was the repayment to him of one-half the expenditure which up to that time had been incurred. It appears from the first certificate delivered to the Lord Treasurer, that the total expenditure to the end of the year 1612 had been 4485*l*. 18*s*. 11*d*., as attested by Hugh Myddelton acting on his own behalf, and Miles Whitacres acting on behalf of the King. The following is a copy of the entry in the Pell Records, in the State Paper-office :—

" Hugh Middleton.[1] 30*th of January.* — By order, dated the last of December, 1612. To Hugh Middleton, of London, goldsmith, the sum of 2242*l*. 19*s*. 5½*d*., being the moiety of 4485*l*. 18*s*. 11*d*. for charges by him disbursed and expended since the 24th August, 1611, until the 1st of December, 1612, inclusive, about the bringing and conveying the New River from the springs of Chadwell and Amwell, in the county of Hertford, unto the north parts of the city of London, which is to be borne by His Majesty, the said charges appearing by a book of the particular expenses thereof, subscribed by the said Hugh Middleton, and Miles Whitacres, gentleman, according to the tenor of the letters patent for warrant hereof, dated 2nd May, 1612 . . £2242 19*s*. 5½*d*."

Further payments were made out of the Treasury to Myddelton, in like manner, for charges disbursed by him in executing the works done to the end of November, 1614, amounting to 4104*l*. 5*s*. 6*d*. ; and in the Domestic State Papers, under date April, 1616, we find a further and final payment from the Exchequer of 2262*l*. 9*s*. 6½*d*. ; making the total payments out of the royal Treasury on

---

[1] We may observe that the name is variously spelled in different documents, as Middleton, Mydelton, Middelton, &c.; but he himself usually signed his name " Myddelton."

account of the New River works amount to 8609*l*. 14*s*. 6*d*. As the books of the New River Company were accidentally destroyed by a fire many years ago, we are unable to test the accuracy of these figures by comparison with the financial records of the Company; but, taken in conjunction with other circumstances hereafter to be mentioned, the amount stated represents, with as near an approach to accuracy as can now be reached, the half of the original cost of constructing the New River works.[1]

---

[1] It appears from the "pageant" which took place on the day of opening, that as many as 600 labourers were employed upon the works at one time. As the pay of labourers was not then more than 6*d*. a day, and of artificers 1*s*. a day, the amount expended on labour during the period the works were under construction—allowing for their suspension for a time through local opposition and bad weather—and on land, materials, inspection, &c., could not have amounted to more than about seventeen thousand pounds. The statements heretofore printed as to the original cost of constructing the New River are, for the most part, gross exaggerations. The assertion that 500,000*l*. of the money of the period, or equal to about two millions of our present money, was expended on the works, has been repeated by various writers, but without any data, excepting the loose statement made by Pennant in his 'Tour in Wales,' to the effect that "2000*l*. a month, which Sir Hugh gained from the Cardiganshire mines, were swallowed up in this river." Whereas the fact was, that Myddelton lost heavily by his first mining enterprise near Denbigh, which proved a complete failure, and he did not enter upon his mining operations in Cardigan until long after the New River had been completed. The fact that so large a sum as 17,000*l*. was expended in the construction of a public work at the beginning of the 17th century is quite strong enough, and stands in no need of exaggeration. It was a very large sum to be expended at that time, when London was comparatively small, and England comparatively poor. It was a larger sum to raise at that period—taking trade, commerce, and public wealth into account—than as many millions would be at this day. It must also be added that Stowe, Maitland, Fuller, Pennant, Morant (who have been generally followed by subsequent encyclopedists and biographic compilers), evidently drew largely upon their imaginations when describing the achievements of Sir Hugh. They all repeat the same story of the "silver-mine in Wales," of his having died in obscurity and poverty, and other like groundless fables. Maitland even magnifies the few and unimportant bridges over the New River to the number of eight hundred, and the fiction is copied by most subsequent writers on the subject.

## CHAPTER IV.

Sir Hugh Myddelton, M.P.—The New River completed.

Although the cutting of the New River may now be regarded as a work of comparatively small account, by men familiar with the canals and railways of this day, it was very different at the time of its projection.  It will be remembered that though successive Acts of Parliament had been obtained to enable such a work to be carried out, none dared venture upon its execution until Hugh Myddelton declared himself willing to be its undertaker.  It was the greatest enterprise of the kind that had yet been attempted in this country.  It was both much more costly and more difficult of execution than the Leet at Plymouth, which consisted mainly of the diversion of one stream into another, whereas this was an entire new river from end to end.  Myddelton had no past experience to serve as his guide, nothing but strong good sense and sound practical judgment, whilst in the earlier period of the enterprise he had to encounter the lot of the bold, original man in all times, —sneers, derision, and ridicule,—and at the same time to battle at every step against the harassing opposition of the tenantry and landowners.  But all these obstructions he eventually overcame, with his own and the King's help.

The general plan adopted by Myddelton was to follow a contour line, as far as practicable, from the then level of the Chadwell spring to the circular pond at Islington, subsequently called the New River Head.[1]  The stream

---

[1] The site of the New River Head had always been a pond, "an open, idell poole," says Hawes, "commonly called the Ducking-pond; being now by the master of this work reduced into a comley pleasant shape, and

originally presented a fall of about 2 inches in the mile, and its City end was at the level of about 82 feet above what is now known as Trinity high water mark. Where the fall of the ground was found inconveniently rapid, a stop-gate was introduced at such places across the stream, penning from 3 to 4 feet perpendicularly, the water flowing over such weirs down to the next level.

By the charter granted by King James, power was given to Myddelton to negotiate and arrange for sufficient land to form a watercourse of 10 feet in width, together with a right or easement to pass with carts and horses along either side of the stream, for the purpose of cleansing and repairs. The bridges over the stream were about a hundred and sixty in number, mostly of timber, and they were invariably executed with a water way under them not exceeding 10 feet. Taking the width of the original river at 10 feet, the probability is that it could not have been more than 4 feet in depth.

To accommodate the cut to the level of the ground as much as possible, numerous deviations were made, and the river was led along the sides of the hills, from which sufficient soil was excavated to form the lower bank of the intended stream. Each valley was traversed on one side until it reached a point where it could be crossed; and there 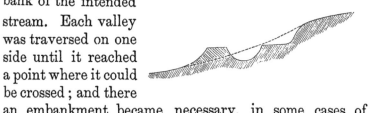 an embankment became necessary, in some cases of from 8 to 10 feet in height, along the top of which the water was conducted in a channel of the dimensions above stated. In those places where embank-

---

many wayes adorned with buildings," &c. The basin is now thrice its original size. The house adjoining it, belonging to the Company, was erected in 1613, but having been new-fronted in 1782, and more recently enlarged, it has lost all external appearance of antiquity. A view of it is to be found in Lewis's ' Islington.'

ments were run out, provision had of course to be made
for the passage of the surface waters from the west
of the line of works into the river Lea, which forms
the natural drain of the district. In some cases these
drainage waters were conveyed under the New River in
culverts, and in others over it by what were termed
flashes.[1] Openings were also left in the banks for the
passage of roads under the stream, the continuity of
which was in such cases maintained either by arches or
timber troughs lined with lead. One of these troughs,
at Bush Hill, near Edmonton, was about 660 feet long,
and 5 feet deep.[2] A brick arch also formed part of this

THE BOARDED RIVER FORMERLY AT BUSH HILL.

[After a Drawing in the 'Gentleman's Magazine,' Vol. LIV.]

---

[1] At each of these "flashes" there
were extensive swamps, where the
flood-waters were upheld to such a
level as to enable them to pass over
the flash, which consisted of a wooden
trough, about twelve feet wide and
three deep, extending across the river;
and from these swamps, as well as from
every other running stream, such apparatus was introduced as enabled the
Company to avail themselves of the supply of water which they afforded, when
required. Mr. Mylne is of opinion that
the river, as originally constructed by
Myddelton, obtained quite as large a
supply from the grass lands along the
hill-sides as was obtained from the
Hertfordshire springs.

[2] The trough was tied together by
imposts seven inches by three; the
uprights were eight inches by four,
with a height of six feet, like but-
tresses; the uprights resting on eighty
brick piers two-and-a-half feet high,

aqueduct, under which flowed a stream which had its
source in Enfield Chase; the arch sustaining the trough

BRICK ARCH UNDER THE NEW RIVER, FORMERLY NEAR BUSH HILL.

[After a Drawing in the 'Gentleman's Magazine,' Vol. LIV.]

and the road along its side.[1]  Another strong timber
aqueduct, 460 feet long and 17 feet high, conducted the
New River over the valley near where it entered the
parish of Islington.  This was long known in the neighbourhood as " Myddelton's Boarded River." [2]  At Islington also there was a brick tunnel executed for a
considerable distance, and another at Newington.  That
at Islington averaged in section about 3 feet by 5, and
appears to have been executed at different periods,
in short lengths.  Such were the principal works along

though these were not erected at equal
distances.   Between  every  two  of
these piers an equal number of imposts and uprights were fixed, resting
on piles of similar dimensions on the
basement timbers of the frame.  Robinson, in his 'History of Edmonton,'
says that in the year 1780, during
the  riots  in  London,  the  rioters
threatened to destroy this aqueduct,
but, on application to Government, a
party of cavalry was sent down to
guard it,  and  they remained there
some days.

[1] This  was  considered  one  of  the
most important structures belonging
to the original New River works.
Salmon ('History of Hertfordshire,'
p. 20), speaking of the " great expense " at which the river was made,
observes of this brick arch and aqueduct that " it is said to have cost
500l." !  It was evidently regarded as
a great work in its time.

[2] These troughs were both removed
by the late Mr. Mylne, Engineer to
the New River Company, and clay
embankments substituted.

the New River. Its original extent was much greater
than it is at present, caused by its frequent windings
along the high grounds for the purpose of avoiding
heavy cuttings and embankments. Although the distance between London and Ware is only about 20 miles,
the New River, as originally constructed, was not less
than 38¾ miles in length.

As the works proceeded, the voice of derision became
hushed, and congratulations began to rise up on all sides
upon the probable early completion of Myddelton's bold
enterprise. The scheme had ceased to be visionary,
for the water was already brought within a mile of
Islington; all that was wanted to admit it to the reservoir being the construction of the tunnel near that place.
At length that too was finished, and now King, corporation, and citizens, vied with each other in doing honour
to the brave and patriotic Hugh Myddelton. The corporation elected his brother, Thomas Myddelton, Grocer
and Citizen, Lord Mayor for the ensuing year; and on
Michaelmas-day, 1613, the citizens assembled in great
numbers to celebrate by a public pageant the entrance of
the New River waters to the metropolis. The ceremony
took place at the new cistern at Islington, in the presence of the Lord Mayor, Aldermen, and Common
Council, amidst a great concourse of spectators. A troop
of some three score labourers, in green Monmouth caps,
bearing spades and mattocks, or such other implements
as they had used in the construction of the work, marched
round about the cistern to the martial music of drums
and trumpets, after which a metrical speech, composed
by one Thomas Middleton,[1] was read aloud as expressive

---

[1] This Thomas Middleton is supposed to have been the dramatist, the
author, amongst other plays, of 'A
Mad World, my Masters,' and 'The
Roaring Girl.' He occasionally wrote
in conjunction with Beaumont and
Fletcher, and other poets of the time.
  A large print was afterwards published by G. Bickham, in comme-
moration of the event, entitled 'Sir
Hugh Myddelton's Glory;' it represents the scene of the ceremony, the
reservoir, with the stream rushing
into it; the Lord Mayor (Sir John
Swinnerton) on a white palfrey, point-
ing exultingly to Sir Hugh; the Re-

of the sentiments of the rest of the workmen.[1] At the conclusion of the recitation the flood-gates were thrown open, and the stream of pure water rushed into the cistern amidst loud huzzas, the firing of mortars, the pealing of bells, and the triumphant welcome of drums and trumpets.

The King signified his gratification at the completion of the undertaking by conferring on Hugh Myddelton the honour of knighthood; but he reserved for him further favours and dignities, in recognition of his valuable public services, as we shall hereafter point out. It is curious to relate that James I. was afterwards nearly drowned in the New River which he had enabled Sir Hugh Myddelton to complete. He had gone out one winter's day after dinner to ride in the park at Theobalds accompanied by his son Prince Charles; when, about three miles from the palace, his horse stumbled and fell, and the King was thrown into the River. It was slightly frozen over at the time, and the King's body disappeared under the ice, nothing but his boots remaining visible. Sir Richard Young rushed in to his rescue, and dragged him out, when " there came much water out of his mouth and body."[2] He was able to ride back to Theobalds, where he got to bed and was soon well again. The King,

---

corder, Sir Henry Montague, afterwards Lord Keeper and Earl of Manchester, and by his side the Lord Mayor elect, the projector's brother, Maister Thomas Myddelton. Various figures gesticulating their admiration occupy the foreground, whilst the foot of the print is garnished with little "chambers," or miniature mortars, spontaneously exploding. There is a copy of the original print in the British Museum.

[1] The following extract may be given, as showing the character of the workmen employed in the undertaking :—

First, here's the Overseer, this try'd man,
An antient souldier and an artizan;
The Clearke; next him the Mathematian;

The Maister of the Timber-worke takes place
Next after these; the Measurer in like case;
Bricklayer, and Enginer; and after those
The Borer, and the Pavier; then it showes
The Labourers next; Keeper of Amwell Head;
The Walkers last;—so all their names are read.
Yet these but parcels of six hundred more,
That, at one time, have been imploy'd before;
Yet these in sight and all the rest will say
That all the weeke they had their Royall pay!

[2] The accident, which occurred on the 9th of January, 1621-2, is related by Nichols in his 'Progresses of James I.,' and by other contemporary writers. -

however, did not forget the circumstance, attributing his
accident to the neglect of Sir Hugh and the Corporation
of London in not taking measures to properly fence
the river; for when the Lord Mayor, Sir Edward Bark-
ham, accompanied by the Recorder, Sir Heneage Finch,
attended the King at Greenwich, in June, 1622, to be
knighted, James took occasion, in rather strong terms,
to remind the Lord Mayor and his brethren of his recent
mischance in "Myddelton's Water."

It is scarcely necessary to point out the great benefits
conferred upon the inhabitants of London by the construction of the New River, through the provision by its
means of an abundant and unremitting supply of pure
water for domestic and other purposes.[1] Along this new
channel were poured into the city several millions of
gallons daily; and the reservoirs at New River Head
being, as before stated, at an elevation of 82 feet above
the level of high water in the Thames, they were thus
capable of supplying through pipes the basement stories
of the greater number of houses then in the metropolis.[2]

The pipes which were laid down in the first instance
to convey the water to the inhabitants were made of
*wood*, principally elm; and at one time the New River
Company had wooden pipes laid down through the

---

[1] In the 'Gentleman's Magazine,'
vol. xxiii., pp. 114-6, is to be found a
curious paper by Sir Christopher Wren
on the subject of the distribution of
the water of the New River—"that
noble acqueduct," as he designates it.

[2] The distribution of the water was,
in the opinion of Mr. Mylne, the engineer, by far the most expensive part
of the original undertaking. The
powers granted for the formation of
the New River did not extend to the
purchase of lands across which it became necessary to lay the pipes to
connect the water in the ponds with
the pipes and the public ways leading
to the city. About a hundred acres
of grass-land then surrounded the
New River Head, about one-half of
which belonged to a Welsh family

named Lloyd, and the other half to
the Northampton family. A grant,
for a term of years only, was obtained
to lay the pipes across those lands so
as to reach the public thoroughfares,
at a fixed rent for each line of pipes;
and as the requirements of the public
rapidly increased, so did the heavy
charge for the easement; and ultimately the lands became covered with
a network of wooden pipes traversing
them in all directions. The annual
burden became so heavy that it seriously affected the profits of the New
River Company, until it became extinguished by purchase; but this portion of the expenditure did not fall on
Sir Hugh Myddelton, who at an early
period sold the greater part of his shares
in the concern.

streets to the extent of about 400 miles! But the
leakage was so great through the porousness of the
material,—about one-fourth of the whole quantity of
water supplied passing away by filtration,—the decay of
the pipes in ordinary weather was so rapid, and they
were so liable to burst during frosts, that they were
ultimately abandoned when mechanical skill was sufficiently advanced to enable pipes of cast iron to be substituted for them.   For a long time, however, a strong
prejudice existed against the use of water conveyed
through pipes of any kind,
and the cry of the water
carriers long continued to
be familiar to London ears,
of " Any New River water
here ! "   " Fresh and fair
New River water ! none of
your pipe sludge ! "

"NEW RIVER WATER!"
[After an Ancient Print.]

Many persons also conti-
nued to support the water
carriers because they were
poor people, with nothing
but water-carrying to turn
to for a living.  The alleged
dearness of the water supplied through the pipes was
another ground of popular objection to them ; though
there really seems to have been but little reason for
complaint on this score.   The following fact will convey
an idea of the mode of charging for a supply of water
adopted at an early period.   In 1614 the Court of Common Council ordered Mr. Chamberlain to pay Mr. Hugh
Myddelton twenty shillings for a fine, and five shillings
a quarter yearly, in consideration of a quill of water of
half an inch bore, taken from the pipes of the said Hugh
Myddelton, to serve the Sessions House at Justice Hall,
and Richard Weaver to see that no waste be made thereof.[1]

[1] ' City of London Corporation Records.'

In another case we find a sum of twenty-six shillings and eight pence paid yearly for a similar pipe, to supply the yard and kitchen of a householder at Islington.[1]

Among the many important uses to which the plentiful supply of New River water was put, was the extinction of fires, then both frequent and destructive, in consequence of the greater part of the old houses in London being built of wood. Stowe[2] particularly mentions the case of a fire which broke out in Broad-street, on the 12th November, 1623, in the house of Sir William Cockaigne, which speedily extended itself to several of the adjoining buildings. We are told by the chronicler, that " Sir Hugh Myddelton, upon the first knowledge thereof, caused all the sluices of the water-cisterne in the field to be left open, whereby there was plenty of water to quench the fire. The water "[ of the New River], he continues, " hath done many like benefits in sundrie like former distresses." [3]

We now proceed to follow the fortunes of Sir Hugh in connexion with the New River Company. The year after the public opening of the cistern at Islington, we find him a petitioner to the Corporation for a loan of 3000*l.*, for three years, at six per cent., which was granted him " in consideration of the benefit likely to accrue to the city from his New River ;" his sureties being the Lord Mayor (Hayes), Mr. Robert Myddelton (his brother), and Mr. Robert Bateman.[4] There is every reason to believe that Myddelton had involved himself in difficulties by locking up his capital in this costly undertaking; and that he was driven to solicit the loan to carry him through until he had been enabled to dispose of the greater part of his

---

[1] Nelson's ' History of Islington.'

[2] Stowe's ' Chronicle.'

[3] The main of the New River at Islington was, it is said, shut down at the time of the Great Fire of London in 1666; and it was believed by some, who pretended to the means of knowing, that the supply of water had been stopped by Captain John Graunt, a papist, under whose name Sir William Petty published his observations on the bills of mortality. —Burnet's ' Own Times,' ed. 1823. Vol. i., p. 401.

[4] ' City of London Corporation Records,' 6th Sept., 1614.

interest in the undertaking to other capitalists.  This he seems to have done very shortly after the completion of the works.  The capital was divided into seventy-two shares, one half of which belonged to Sir Hugh and the other half to the King, in consideration of the latter having borne one half of the cost.  Of the thirty-six shares owned by Sir Hugh, as many as twenty-eight were sold by him to other persons, most probably for considerable sums.[1]  The enterprise was now regarded as successful; its projector had been crowned with public honours, and there could be no doubt as to the large consumption of the New River water.  In that stage of affairs, before the actual paying qualities of the concern could be ascertained, but while the most sanguine anticipations were formed as to the large profits to be derived from the sale of the water, it seems probable that Sir Hugh would have little difficulty in inducing other capitalists to embark in the undertaking.  At all events he did sell twenty-eight of his shares, and that he must have realized a considerable sum is countenanced by the circumstance that we find him shortly after embarked in an undertaking hereafter to be described, requiring the command of a very large capital.

The shareholders were incorporated by letters patent on the 21st of June, 1619, under the title of "The Governors and Company of the New River brought from Chadwell and Amwell to London."[2]  The government of the corporation was vested in the twenty-nine adventurers who held amongst them the thirty-six shares originally belonging to Sir Hugh, who had by that time

---

[1] In Pennant's 'London' it is stated that the original shares in the concern were 100*l.* each, whereas Entick makes them to have amounted to not less than 7000*l.* each! This is only another illustration of the hap-hazard statements put forth respecting Sir Hugh and his works. We have seen that the original cost of the New River probably did not amount to more than 18,000*l.*, in which case the capital represented by each original share would be about 250*l.*

[2] On the occasion of its subsequent confirmation by parliament, Sir Edward Coke said: "This is a very good bill, and prevents one great mischief that hangs over the city. *Nimis potatio: frequens incendium.*"

reduced his holding to only two shares. From the first Minute of the Court of the Company, held on the 2nd of November, 1619, it appears that his son Hugh also held one share, and his brother, Sir Thomas, another. Several persons of distinction held shares, but none more than two each. Amongst these original proprietors we find the names of Sir Henry Montague the Lord Chief Justice, Sir William Borlase, Sir Lawrence Hide, and Sir Henry Nevill. There were four other shareholders of the name of Middleton, but it does not appear that they were relatives of the goldsmith. At this first Court Sir Hugh was appointed Governor, and Robert Bateman Deputy-Governor of the Company.

Under the arrangement made by Sir Hugh with the King, the latter was precluded from taking any part in the management, in order to prevent what, it was thought, might lead to undue influence; and he was only allowed a representative at the meetings of the Company, to prevent injustice to the royal interests. On the 9th of November, 1619, we find a grant made to Sir Giles Mompesson[1] of the Surveyorship of the profits of the New River, with authority to attend their meetings, inspect their accounts, &c., with a grant to him for such service of 200*l.* per annum of the King's moiety of the profits of the said river. It was long, however, before there were any profits to be divided; for the expenses of making repairs and improvements, and laying down wooden pipes, continued to be very great for many years; and the ingenious method of paying dividends out of capital, to keep up the price of shares and invite further speculation, had not yet been invented. In fact, no dividend whatever was paid until after the lapse of twenty years from the date of opening the New River at Islington; and the

---

[1] 'Domestic Calendar of State Papers.' This Sir Giles Mompesson is understood to have been the original of Sir Giles Overreach in Massinger's 'New Way to Pay Old Debts.'

first dividend only amounted to 15*l*. 3*s*. 3*d*. a share. The next dividend of 3*l*. 4*s*. 2*d*. was paid three years later, in 1636; and as the concern seemed to offer no great prospect of improvement, and a further call on the proprietors was expected, Charles I., who required all his available means for other purposes, finally re-granted his thirty-six "King's shares" to the Company, under his great seal, in consideration of a fee farm rent of 500*l*., which is to this day paid yearly into the King's exchequer.

Notwithstanding this untoward commencement of the New River Company, it made great and rapid progress when its early commercial difficulties had been overcome; and after the year 1640 its prosperity steadily kept pace with the population and wealth of the metropolis.[1]  By the end of the seventeenth century the dividend paid was at the rate of about 200*l*. per share; at the end of the eighteenth century the dividend was above 500*l*. per share; and at the present date each share produces

---

[1] The New River Company has from time to time enlarged its works, widening the stream to about 25 feet, and adding to its supplies of water from various sources. About forty-two square feet of water flows from the river into London, at the rate of two miles an hour, all the year round. Some 70,000 houses are supplied from this source, besides large breweries and manufactories.  The charge to each house is on the average less than a penny a day for 241 gallons of water.  The principal additional supply has been obtained from the River Lea.  The first water taken from that source was permitted without opposition, but the Lea Trustees found it necessary to check the diversion of their river, and protracted litigation was the consequence.  Mutual arrangements were then entered into, and the New River supply was strictly regulated by fixing the diameter of the pipe through which the Lea water was to flow, and subsequently by the provision of a carefully-regulated balance-engine erected near Hertford. On one occasion the New River Company contrived to make a very good bargain through the dexterity of their surveyor.  Finding the quantity of Lea River water first agreed upon to be too small for their purpose, the Company offered to give *double* the price it then paid, for a pipe of *double* the diameter.  To this the agent of the Lea River trustees weakly assented, being so ignorant of the business as not to be aware that the orifices are to each other as the *squares* of their diameters.  In consequence of this want of a very slight amount of mathematical knowledge, the Lea River Company was obliged to furnish *four* times the quantity of water which it originally supplied, for only *twice* the sum which it at first received.  Well might their agent exclaim, "O, I see mathematics may be good for something, after all!"

about 850*l*. a year. At only twenty years' purchase, the capital value of a single share at this day would be about 17,000*l*. But most of the shares have in course of time, by alienation and bequeathment, become very much subdivided; the possessors of two or more fractional parts of a share being enabled, under a decree of Lord Chancellor Cowper, in 1711, to depute a person to represent them in the government of the Company.

SOURCE OF NEW RIVER AT AMWELL, NEAR WARE.[1]

[By Percival Skelton, after his Original Drawing.]

---

[1] The monumental pedestal represented in the foreground of the engraving was erected by the late Robert Mylne, Esq., the Company's engineer, to the genius and patriotism of Sir Hugh Myddelton. Its inscription runs thus:—

Sacred to the Memory of
Sir Hugh Mydelton, Baronet;
whose successful care,
assisted by the patronage of his King,
conveyed this stream to London;
an immortal work,
since man cannot more nearly
imitate the Deity
than in bestowing health.

# CHAPTER V.

## SIR HUGH MYDDELTON'S EMBANKMENT OF BRADING HAVEN.

No sooner had the New River Company been formed and its operations organized, than we find Sir Hugh Myddelton engaged in the new and bold enterprise of enclosing a large tract of drowned land from the sea. The scene of his operations on this occasion was the eastern extremity of the Isle of Wight, at a place now marked on the maps as Brading Harbour. This harbour or haven consists of a tract of about eight hundred acres in extent.[1]  At low water it lies a wide mud flat, through the middle of which a small stream, called the Yar, winds its way from near the village of Brading, at the head of the haven, to the sea at its eastern extremity ;[2] whilst at high tide it forms a beautiful and apparently inland lake, embayed between hills of moderate elevation covered with trees, in many places down to the water's edge.  At its seaward margin Bembridge Point stretches out as if to meet the promontory on the opposite shore, where stands the old tower of St. Helen's Church, now used as a sea-mark ; and, as seen from most points, the bay seems to be completely land-locked.

The reclamation of so large a tract of land, apparently so conveniently situated for the purpose, had long been matter of speculation.  It is not improbable that at some early period neither swamp nor lake existed at Brading Haven, but a green and fertile valley ; for

---

[1] The extent of land reclaimed by Myddelton at Brading Haven has, with the inaccuracy that characterises almost everything heretofore published relating to him, been stated at 2000 acres.  Sir Richard Worsley, in his 'History of the Isle of Wight,' gives the whole area of the haven as only 856 acres.

[2] See the engraving at p. 84.

in the course of the works undertaken by Sir Hugh
Myddelton for its recovery from the sea, a well, strongly

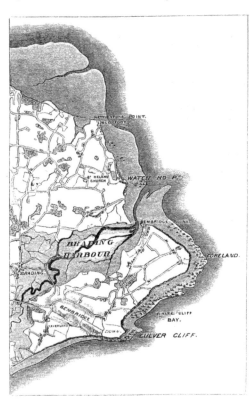

MAP OF BRADING HAVEN.

[Ordnance Survey ]

cased with stone,
was discovered
near the middle
of the haven, in-
dicating the exist-
ence of a former
settled population
on the soil. The
sea must, how-
ever, have burst
in and destroyed
the settlements,
laying the whole
valley under wa-
ter.[1] In King
James's reign,
when the inning
of drowned lands
began to receive
an unusual degree
of attention, the
project of reclaim-
ing Brading Ha-
ven was again re-
vived ; and in the

year 1616 a grant was made of the drowned district[2] to
one John Gibb, the King reserving to himself a rental
of 20l. per annum. The owners of the adjoining lands
contested the grant, claiming a prior right to the property in the haven, whatever its worth might be. But
the verdict of the Exchequer went against the landowners, and the right of the King to grant the area of

---

[1] Sundry traditions are extant in
the neighbourhood as to the circumstances connected with the inundation.

See 'Land We Live in,' vol. i., 262.
[2] Grant by Privy Seal, 18 July,
14 Jac. I.

the haven for the purpose of reclamation was maintained. It appears that Gibb sold his grant to one Sir
Bevis Thelwall, a page of the King's bedchamber, who
at once invited Sir Hugh Myddelton to join him in
undertaking the work; but Thelwall would not agree
to pay Gibb anything until the enterprise had been
found practicable. In 1620 we find that a correspondence was in progress as to "the composition to be made
by the Solicitor-General with Sir Hugh touching the
draining of certain lands in the Isle of Wight, and
the bargain having been made according to such directions as His Majesty hath given, then to prepare the
surrender, and thereupon such, other assurance for His
Majesty as shall be requisite." [1]

A satisfactory arrangement having been made with
the King, Myddelton began the work of reclaiming the
haven in the course of the same year. He sent to Holland for Dutch workmen familiar with such undertakings;
and from the manner in which he carried out his embankment, it is obvious that he mainly followed the
Dutch method of reclamation, which, as we have already
seen in the case of the drainage of the Fens by Vermuyden, was in many respects but ill adapted for English
practice. But it would also appear, from a patent for
draining land which he took out in 1621,[2] that he employed some invention of his own for the purpose of
facilitating the work. The introduction to the grant of
the patent runs as follows :—

" WHEREAS wee are given to vnderstand that our welbeloved
subiect Hugh Middleton, Citizen and Goldsmith of London, hath
to his very great charge maynteyned many strangers and others,
and bestowed much of his tyme to invent a new way, and by his
industrie, greate charge, paynes, and long experience, hath devised
and found out ' A NEW INVENÇON, SKILL, OR WAY FOR THE
WYNNING AND DRAYNING OF MANY GROUNDꝰ WHICH ARE DAYLIE

<hr>

[1] 'Domestic Calendar of State Pa-
pers.' Docquet, 13th August, 1620.     [2] 'Record of Patents,' No. 19.
Sealed 2nd July, 1621.

AND DESPERATELIE SURROUNDED WITHIN OUR KINGDOME OF ENG-
LAND AND DOMINION OF WALES,' and is now in very great hope to
bringe the same to good effect, the same not being heretofore
knowne, experimented, or vsed within our said realme or dominion,
whereby much benefitt, which as yet is lost, will certenly be
brought both to vs in particular and to our comon wealth in
generall, and hath offered to publish and practise his skill
amongest our loving subiectℇ. . . . . . . . . ., KNOWE YEE, that
wee, tendring the weale of this our kingdom and the benefitt of
our subiectℇ, and out of our princely care to nourish all artℇ,
invencions, and studdies whereof there may be any necessary
or pffitable vse within our dominions, and out of our desire to
cherish and encourage the industries and paynes of all other our
loving subiectℇ in the like laudable indeavors, and to recompence
the labors and expences of the said Hugh Middleton disbursed and
to be susteyned as aforesaid, and for the good opinion wee have conceived of the said Hugh Middleton, for that worthy worke of his in
bringing the New River to our cittie of London, and his care and
industrie in busines of like nature tending to the publicke good
. . . . . . doe give and graunt full, free, and absolute licence,
libertie, power, and authoritie vnto the said Hughe Middleton, his
deputies," &c. to use and practise the same during the terme of
fowerteene years next ensuing the date hereof.

No description is given of the particular method
adopted by Sir Hugh in forming his embankments; but
it would appear that he proceeded by driving piles into
the bottom of the Haven near Bembridge Point where it
is about the narrowest, and thus formed a strong embankment at its junction with the sea, but, as would afterwards appear, without making adequate provision for the
egress of the inland waters.

A curious contemporary manuscript by Sir John
Oglander is still extant, preserved amongst the archives
of the Oglander family, who have held the adjoining
lands from a period antecedent to the date of the Conquest, which we cannot do better than quote, as giving
the most authentic account extant of the circumstances
connected with the inning of Brading Haven by Sir
Hugh Myddelton. This manuscript says :—

"Brading Haven was begged first of all of King James by one Mr. John Gibb, being a groom of his bedchamber, and the man that King James trusted to carry the reprieve to Winchester for my Lord George Cobham and Sir Walter Rawleigh, when some of them were on the scaffold to be executed. This man was put on to beg it of King James by one Sir Bevis Thelwall, who was then one of the pages of the bedchamber. After he had begged it, Sir Bevis would give him nothing for it until the haven were cleared; for the gentlemen of the island whose lands joined to the haven challenged it as belonging unto them. King James was wonderful earnest in the business, both because it concerned his old servant, and also because it would be a leading case for the fens in Lincolnshire. After the verdict went in the Chequer against the gentlemen, then Sir Bevis Thelwall would give nothing for it till he could see that it was feasible to be inned from the sea; whereupon one Sir Hugh Myddelton was called in to assist and undertake the work, and Dutchmen were brought out of the Low Countries, and they began to inn the haven about the 20th of December, 1620. Then, when it was taken in, King James compelled Thelwall and Myddelton to give John Gibb (who the King called 'Father') 2000*l*.[1] Afterwards Sir Hugh Myddelton, like a crafty fox and subtle citizen, put it off wholly to Sir Bevis Thelwall, betwixt whom afterwards there was a great suit in the Chancery; but Sir Bevis did enjoy it some eight years, and bestowed much money in building of a barnhouse, mill, fencing of it, and in many other necessary works.

"But now let me tell you somewhat of Sir Bevis Thelwall and Sir Hugh Myddelton, and of the nature of the ground after it was inned, and the cause of the last breach. Sir Bevis was a gentleman's son in Wales, bound apprentice to a mercer in Cheapside, and afterwards executed that trade till King James came into England: then he gave up, and purchased to be one of the pages of the bedchamber, where, being an understanding man, and knowing how to handle the Scots, did in that infancy gain a fair estate by getting the Scots to beg for themselves that which he first found out for them, and then himself buying of them with ready money under half the value. He was a very bold fellow, and one that King James very

---

[1] On the 30th June, 1622, the haven was granted by the King (the original grant to Gibb having been cancelled) to Hugh Myddelton, Esquire, Robert Bateman, Citizen and Skynner, of London, and Richard Middleton, Citizen and Grocer, in consideration of 2000*l*. paid to the King by Hugh Myddelton, viz., 1000*l*. down, and the remaining 1000*l*. by two half-yearly payments at Lady-day and Michaelmas, 1622; the King passing the 1000*l*. and the bonds for the two sums of 500*l*. to John Gibb.

well affected.  Sir Hugh Myddelton was a goldsmith in London.
This and other famous works brought him into the world, viz., his
London waterwork, Brading Haven, and his mine in Wales.

" The nature of the ground, after it was inned, was not answerable to what was expected, for almost the moiety of it next to the
sea was a light running sand, and of little worth.  The best of it
was down at the farther end next to Brading, my Marsh, and
Knight's Tenement, in Bembridge.  I account that there was
200 acres that might be worth 6s. 8d. the acre, and all the rest
2s. 6d. the acre.  The total of the haven was 706 acres.  Sir Hugh
Myddelton, before he sold, tried all experiments in it : he sowed
wheat, barley, oats, cabbage seed, and last of all rape seed, which
proved best; but all the others came to nothing.  The only inconvenience was in it that the sea brought in so much sand and ooze
and seaweed that choked up the passage of the water to go out,
insomuch as I am of opinion that if the sea had not broke in Sir
Bevis could hardly have kept it, for there would have been no current for the water to go out ; for the eastern tide brought so much
sand as the water was not of force to drive it away, so that in time
it would have laid to the sea, or else the sea would have drowned the
whole country.  Therefore, in my opinion, it is not good meddling
with a haven so near the main ocean.

" The country (I mean the common people) was very much
against the inning of it, as out of their slender capacity thinking
by a little fishing and fowling there would accrue more benefit than
by pasturage ; but this I am sure of, it caused, after the first three
years, a great deal of more health in these parts than was ever
before ; and another thing is remarkable, that whereas we thought
it would have improved our marshes, certainly they were the worse
for it, and rotted sheep which before fatted there.

" The cause of the last breach was by reason of a wet time when
the haven was full of water, and then a high spring tide, when both
the waters met underneath in the loose sand.  On the 8th of March,
1630, one Andrew Ripley that was put in earnest to look to Brading
Haven by Sir Bevis Thelwall, came in post to my house in Newport to inform me that the sea had made a breach in the said haven
near the easternmost end.  I demanded of him what the charge
might be to stop it out ; he told me he thought 40s., whereupon I
bid him go thither and get workmen against the next day morning,
and some carts, and I would pay them their wages ; but the sea the
next day came so forcibly in that there was no meddling of it, for
Ripley went up presently to London to Sir Bevis Thelwall himself,

to have him come down and take some further course; but within four days after the sea had won so much on the haven, and made the breach so wide and deep, that on the 15th of March when I came thither to see it I knew not well what to judge of it, for whereas at the first 5*l.* would have stopped it out, now I think 200*l.* will not do it, and what will be the event of it time will tell. Sir Bevis on news of this breach came into the island on the 17th of March, 1630, and brought with him a letter from my Lord Conway to me and Sir Edward Dennies, desiring us to cause my Lady Worsley, on behalf of her son, to make up the breach which happened in her ground through their neglect. She returned us an answer that she

ENTRANCE TO BRADING HARBOUR, FROM ST. HELEN'S OLD TOWER.[1]

[By R. P. Leitch, after a Sketch by the Author.]

[1] The above view represents the present state of the entrance to Brading Haven. A wide ridge of drifted sand lies across it, in front of the old bank raised by Sir Hugh, which extended from a point below the hill under "Mrs. Grant's house," a little to the westward of the village of Bembridge (seen on the opposite shore) to what are now called "The Boat Houses," situated towards the northern side of the haven, and behind the sand-ridge extending across the view. The black piles driven into the bottom of the haven in the process of embankment are still to be seen sticking up at low-water; and only a few years since the old gates which served for a sluice were dug up near the Boat Houses. At the extremity of the sand-ridge there is a ferry across to the village of Bembridge, in front of which is the narrow entrance into the haven. There have been serious encroachments of the sea on that side of late years, and the channel has become much impeded; so much so that it has been feared that the navigation would be lost. The old church-tower of St. Helen's, faced with brick and whitewashed, on the right of the view, is still used as a sea-mark.

thought that the law would not compel her unto it, and therefore desired to be excused, which answer we returned to my lord. What the event will be I know not, but it seemeth to me not reasonable that she should suffer for not complying with his request. If he had not inned the haven this accident could never have happened; therefore he giving the cause, that she should apply the cure I understand not. But this I am sure, that Sir Bevis thinketh to recover of her and her son all his charges, which he now sweareth every way to be 2000*l*. For my part, I would wish no friend of mine to have any hand in the second inning of it. Truly all the better sort of the island were very sorry for Sir Bevis Thelwall, and the commoner sort were as glad as to say truly of Sir Bevis that he did the country many good offices, and was ready at all times to do his best for the public and for everyone.

" Sir Hugh Myddelton took it first in, and it was proper for none but him, because he had a mine of silver in Wales to maintain it. It cost at the first taking of it in 4000*l*., then they gave 2000*l*. to Mr. John Gibb for it, who had begged it of King James; afterwards, in building the barn and dwelling-house, and water-mill, with the ditching and quick-setting, and making all the partitions, it could not have cost less than 200*l*. more: so in the total it stood them, from the time they began to take it in, until the 8th of March, a loss of 7000*l*."

It will thus be observed that the loss of this undertaking fell upon Thelwall, and not upon Myddelton; Sir Hugh having sold out of the adventure long before the sea burst through his embankment. The date of his conveyance to Sir Bevis Thelwall was the 4th September, 1624, nearly six years before the final ruin of the work. He had, therefore, got his capital out of the adventure, most probably with his profit as contractor, and was thus free to embark in the important mining enterprise in Wales, on which we find him next engaged.

# CHAPTER VI.

## MINING ENTERPRISE IN WALES—MYDDELTON'S DEATH.

SIR HUGH continued to maintain his Parliamentary connection with his native town of Denbigh, of which he was still the representative. We do not find that he took an active part in political questions. The name of his brother, Sir Thomas, frequently appears in the parliamentary debates of the time, and he was throughout a strong opponent of the Court party; but that of Sir Hugh only occurs in connection with commercial topics or schemes of internal improvement, on which he seems to have been consulted as an authority.

Sir Hugh's occasional visits to his constituents brought him into connection with Welsh families, and made him acquainted with the mining enterprises then on foot in different parts of Wales—so rich in ores of copper, lead, and iron. It appears that the Governor and Company of Mines Royal in Cardiganshire [1] were incorporated in the year 1604, for the purpose of working the lead and silver mines of that county. The principal were those at Cwmsymlog and the Darren Hills, situated about midway, as the crow flies, between Aberystwith and the mountain of Plinlimmon, and at Tallybout, about midway between Aberystwith and the estuary at the mouth of the River Dovey. They were all situated in the township of Skibery Coed, in the northern part of the county of Cardigan. For many years these mines (which were

---

[1] Meyrick, in his 'History of Cardiganshire,' gives the names of the twenty-two members who constituted the original Company, under their Governor, Philip Earl of Pembroke; but the name of Myddelton does not appear amongst them.

first opened out by the Romans) were worked by the
Corporation of Mines Royal; but it does not appear that

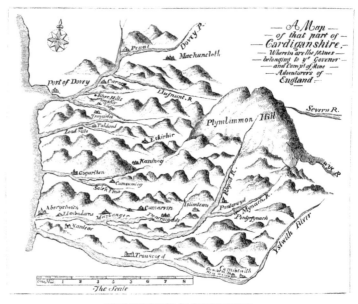

CHART OF MINES IN NORTH WALES.

[From an old Print in the British Museum.]

much success attended their operations. Mining was
little understood then, and all kinds of pumping and
lifting machinery were clumsy and inefficient. Although
there was no want of ore, the mines were so drowned
by water that the metal could not well be got at and
worked out.

Myddelton's spirit of enterprise was excited by the
prospect of battling with the water and getting at the
rich ore, and he had confidence that his mechanical
ability would enable him to overcome the difficulty.
The Company of Mines Royal were only too glad to get
rid of their unprofitable undertaking, and they agreed
to farm their mines to Sir Hugh at the rental of 400*l.*
per annum.  This was in the year 1617, some time after
he had completed his New River works, but before he

had commenced the embankment of Brading Haven,—
and Sir Bevis Thelwall was also a partner with him in
this new venture. It took him some time to clear the
mines of water, which he did by pumping-machines of
his own contrivance; but at length sufficient ore was
raised to enable it to be tested, and it was then found
to contain a considerable proportion of silver. His operations seem to have been attended with success, for we
shortly find him sending considerable quantities of silver
to the Royal Mint to be coined.

King James was so much gratified by these further
proofs of Myddelton's skill and enterprise, as displayed
in his embankment of Brading Harbour and his successful mining operations in Wales, that he raised him
to the dignity of a Baronet on the 19th of October,
1622; and the compliment was all the more marked
by His Majesty directing that Sir Hugh should be discharged from the payment of the customary fees,
amounting to 1095*l.*, and that the dignity should be
conferred upon him without any charge whatever.[1] The
patent of baronetcy granted on the occasion sets forth
the " reasons and considerations " which induced the
King to confer the honour; and it may not be out of
place to remark, that though more eminent industrial
services have been rendered to the public by succeeding
engineers, there has been no such cordial or graceful
recognition of them by any succeeding monarch. The
patent states that King James had made a baronet of
Hugh Myddelton, of London, goldsmith, for the following
reasons and considerations :—

" 1. For bringing to the city of London, with excessive charge
and greater difficulty, a new cutt or river of fresh water, to the
great benefit and inestimable preservation thereof. 2. For gaining

---

[1] Sloane MS. (British Museum), vol. ii., 4177. Also 'Calendar of Domestic State Papers,' Oct. 19th, 1622. Grant to Hugh Myddelton of the rank of Baronet, granting discharge of 1095*l.* due on being made a Baronet.

a very great and spacious quantity of land in Brading Haven, in the Isle of Wight, out of the bowells of the sea, and with bankes and pyles and most strange defensible and chargeable mountains, fortifying the same against the violence and fury of the waves. 3. For finding out, with a fortunate and prosperous skill, exceeding industry, and noe small charge, in the county of Cardigan, a royal and rych myne, from whence he hath extracted many silver plates which have been coyned in the Tower of London for current money of England." [1]

The King, however, did more than confer the title— he added to it a solid benefit in confirming the lease made to Sir Hugh by the Governor and Company of Mines Royal, " as a recompense for his industry in bringing a new river into London," [2] waiving all claim to royalty upon the silver produced, although the Crown was entitled, according to the then interpretation of the law, [3] to a royalty on all gold and silver found in the lands of a subject; and it is certain that the lessee [4] who succeeded Sir Hugh did pay such royalty into the State Exchequer. It also appears from documents preserved amongst the State Papers, that large offers of royalty were actually made to the King at the very time that this handsome concession was granted to Sir Hugh. [5]

The discovery of silver in the Welsh mountains doubt- less caused no small talk at the time, and, as in Australia and California now, there were many attempts made by lawless persons to encroach upon the diggings. On this, a royal proclamation was published, warning such persons

---

[1] Harleian MS., No. 1507, Art. 40. (British Museum.)

[2] 'Domestic Calendar of State Papers,' Feb. 21, 1625. Vol. clxxxiv. 15.

[3] Sir S. R. Meyricke. Introduction to the 'History of Cardiganshire,' pp. ccx.; 12-14.

[4] Subsequent to 1636, Thomas Bushell (who purchased the lease) paid 1000*l*. per annum to the King; and some years after, in 1647, we find him agreeing to pay 2500*l*. per annum to the Parliament. As a

curious fact, we may here add that, under date Die Sabbati, 14 August, 1641, Parliament granted an order or license to Thomas Bushell to dig turf on the King's wastes within the limits of Cardiganshire, for the purpose of smelting and refining the lead ores, &c., his predecessor (Myddelton) hav- ing used up almost all the wood grow- ing in the neighbourhood of the mines.

[5] Proposals by William Gomeldon, Feb., 1625.

against the consequences of their trespass, and orders were issued that summary proceedings should be taken against them.[1] It appears that Sir Hugh and his partners continued to work the mines with profit for a period of about sixteen years, although it is stated that during that time, in consequence of the large quantity of water met with, little more than the superficies of the mine could be worked.[2] The water must, however, have been sufficiently got under to enable so much ore eventually to be raised. Waller says an engine was employed at Cwmsymlog; and a tradition long existed among the neighbouring miners that there were two engines placed about the middle of the work. There were also several " levels" at Cwmsymlog, one of which is called to this day " Sir Hugh's Level." The following rude cut, from Pettus' ' Fodinæ Regales,' may give an idea of the

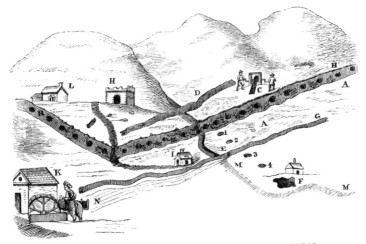

PLAN OF MYDDELTON'S SILVER MINING WORKS AT CWMSYMLOG.

A. The old works of Myddelton and Bushell.
B. The round holes are the shafts of the mine.
C. Windlace to wind up ore from the shafts.
D. A new vein.
E. Sir H. Myddelton's adit.
F. A new adit.
G. Adits to drain works.

H. Myddelton's decayed chapel.
I. Old stamping-house.
K. The smelting mills, supposed six miles from the hill.
L. Unwrought ground.
M. The brook that divides the hill.
N. The stream which drives the mills.

---

[1] ' State Papers,' Jas. I., vol. clii., 22.

[2] So stated in the certificate from

the miners presented to the Right Hon. the Lords and others of H. M. Privy Council (1642).

manner in which the works of Cwmsymlog (facetiously
styled by the author or his printer " Come-some-luck! ")
were laid out.

From a statement made by Bushell to parliament of
the results of the working subsequent to 1636, it appears
that the lead alone amounted to the value of above
5000*l*. a year, to which there was to be added the value of
the silver—Bushell alleging, in his petition to Charles I.,
deposited in the State Paper-office,[1] that Sir Hugh had
brought " to the Minte theis 16 yeares of puer silver
100 poundes weekly." A ton of the lead ore is said to
have yielded about a hundred ounces of silver, and the
yield at one time was such that Myddelton's profits
were alleged by Bushell to have amounted to at least
two thousand pounds a month. There is no doubt, therefore, that Myddelton realized considerable profits by the
working of his Welsh mines, and that towards the close
of his useful life he was an eminently prosperous man.

Successful as he had been in his enterprise, he was
ready to acknowledge the Giver of all Good in the
matter. He took an early opportunity of presenting a
votive cup, manufactured by himself out of the Welsh
silver, to the corporation of Denbigh, and another to
the head of his family at Gwaenynog in its immediate
neighbourhood, both of which are still preserved. On
the latter is inscribed " Mentem non munus—Omnia a
Deo—Hugh Myddelton."

While conducting the mining operations, Sir Hugh
resided at Lodge, now called Lodge Park, in the immediate neighbourhood of the mines. The house was the
property of Sir John Pryse, of Gogerddan, whose son
Richard, afterwards created a baronet, was married to
Myddelton's daughter Hester. The house stood on the
top of a beautifully wooded hill, overlooking the estuary

---

[1] Dated 22nd October, 1636. The prayer of Bushell's petition to Charles I. is, that His Majesty will ratify his agreement with Lady Myddelton (by that time a widow) for the purchase of the residue of her lease.

of the Dovey and the great bog of Gorsfochno, the view being bounded by picturesque hills on the one hand and by the sea on the other. Whilst residing here, on one of his visits to the mines, a letter reached him from his cousin, Sir John Wynn, of Gwydir, dated the 1st September, 1625, asking his assistance in an engineering project in which he was interested. This was the reclamation of the large sandy marshes, called Traeth-Mawr and Traeth-Bach, situated at the junction of the counties of Caernarvon and Merioneth, at the northern extremity of the bay of Cardigan. Sir John, after hailing his good cousin as " one of the great honours of the nation," congratulated him on the great work which he had performed in the Isle of Wight, and added, " I may say to you what the Jews said to Christ, We have heard of thy greate workes done abroade, doe now somewhat in thine own country." After describing the nature of the land proposed to be reclaimed, Sir John declares his willingness " to adventure a brace of hundred pounds to joyne with Sir Hugh in the worke," and concludes by urging him to take a ride to Traeth-Mawr, which was not above a day's journey from where Sir Hugh was residing, and afterwards to come on and see him at Gwydir House, which was at most only another day's journey or about twenty-five miles further to the north-west of Traeth-Mawr. The following was Sir Hugh's reply :—

" HONOURABLE SIR,

" I have received your kind letter. Few are the things done by me ; for which I give God the glory. It may please you to understand my first undertaking of public works was amongst my owne kindred, within less than a myle of the place where I hadd my first being, 24 or 25 years since, in seekinge of coales for the town of Denbighe.

" Touching the drowned lands near your lyvinge, there are many things considerable therein. Iff to be gayned, which will hardlie be performed without great stones, which was plentiful at the Weight [Isle of Wight], as well as wood, and great sums of money to be spent, not hundreds, but thousands ;[1] and first of all his Majesty's interest must be got. As for myself, I am grown into years, and full of business here at the mynes, the river at London, and other places, my weeklie charge being above 200*l.* ; which maketh me verie unwillinge to undertake any other worke ; and the least of theis, whether the drowned lands or mynes, requireth a whole man, with a large purse. Noble sir, my desire is great to see you, which should draw me a farr longer waie ; yet such are my occasions at this tyme here, for the settlinge of this great worke, that I can hardlie be spared one howr in a daie. My wieff being also here, I cannot leave her in a strange place. Yet my love to publique works, and desire to see you (if God permit), maie another tyme drawe me into those parts. Soe with my heartie comendations I comit you and all your good desires to God.

" Your assured lovinge couzin to command,

" HUGH MYDDELTON.

Lodge, Sept. 2nd, 1625."

At the date of this letter Sir Hugh was an old man of seventy, yet he still continued industriously to apply himself to business affairs. Like most men with whom work has become a habit, he could not be idle, and an active pursuit seems to have become necessary to his happiness. To the close of his life we find him engaged in correspondence on various subjects — on mining, draining, and general affairs. When in London he continued to occupy his house in Bassishaw-street, where the goldsmith business was carried on in his absence by his son William. Thus we find him addressing a letter to Lord Secretary Conway, dated Bassishaw-street, 15th

---

[1] A long time passed before the attempt was made to reclaim the large tract of land at Traeth-Mawr; but after the lapse of two centuries, it was undertaken by William Alexander Madocks, Esq., and accomplished in spite of many formidable difficulties. Two thousand acres of Penmorfa Marsh were first enclosed on the western side of the river, after which an embankment was constructed across the estuary, about a mile in length, by which 6000 additional acres were secured. The sums expended on the works are said to have exceeded 100,000*l.* ; but the expenditure has proved productive, and the principal part of the reclaimed land is now under cultivation. Tremadoc, or Madock's Town, and Port Madoc, are two thriving towns, built by the proprietor on the estate thus won from the sea.

January, 1627,[1] on behalf of a young man "of civill life and honest conversation," whom he desired to serve by commending him to the notice of his lordship. He also continued to maintain his pleasant country house at Bush Hill, near Edmonton, which he occupied when engaged on the engineering business of the New River, near to which it was conveniently situated.

At length all correspondence ceases, and the busy hand and head of the old man find rest in death. Sir Hugh died on the 10th of December, 1631, at the advanced age of seventy-six. In his will, which he made on the 21st November, three weeks before his death, when he was "sick in bodie" but "strong in mind," for which he praised God, he directed that he should be buried in the church of St. Matthew, Friday-street, where he had officiated as churchwarden and where six of his sons and five of his daughters had been baptized. It had been his parish church, and was hallowed in his memory by many associations of family griefs as well as joys; for there he had buried several of his children in early life, amongst others his two eldest-born sons. The church of St. Matthew, however, has long since ceased to exist, though its registers have been preserved: it was destroyed in the great fire of 1666, and the monumental record of Sir Hugh's last resting-place perished in the common ruin.

· The popular and oft-repeated story of Sir Hugh Myddelton having died in poverty and obscurity is only one of the numerous fables which have accumulated about his memory.[2] He left fair portions to all the

[1] 'State Paper Office: Domestic Papers,' Charles I., vol. xlix., Art. 41.

[2] The tradition still survives that Sir Hugh retired in his old age to the village of Kemberton, near Shiffnal, Salop, where he lived in great indigence under the assumed name of Raymond, and that he was there occasionally employed as a street paviour! The parish register is said to contain an entry of his burial on the 11th of March, 1702; by which date Hugh Myddelton, had he lived until then, would have been about 150 years old! The entry in the register was communicated by the rector of the parish in 1809 to the 'Gentleman's Magazine' (vol. lxxix., p. 795), but it is scarcely

children who survived him, and an ample provision to his widow.[1] His eldest son and heir, William, who succeeded to the baronetcy, inherited the estate at Ruthin, and afterwards married the daughter of Sir Thomas Harris, Baronet, of Shrewsbury. Elizabeth, the daughter of Sir William, married John Grene, of Enfield, clerk to the New River Company, and from her is lineally descended the Rev. Henry Thomas Ellacombe, M.A., rector of Clyst St. George, Devon, who still holds two shares in the New River Company, as trustee for the surviving descendants of Myddelton in his family. Sir Hugh left to his two other sons, Henry and Simon,[2] besides what he had already given them, one share each in the New River Company (after the death of his wife) and 400*l.* a-piece. His five daughters seem to have been equally well provided for. Hester was left 900*l.*, the remainder of her portion of 1900*l.* ; Jane having already had the same portion on her marriage to Dr. Chamberlain, of London. Elizabeth and Ann, like Henry and Simon, were left a share each in the New River Company and 500*l.* a-piece. He bequeathed to his wife, Lady Myddelton, the house at Bush Hill, Edmonton, and the furniture in it, for use during her life, with remainder to his youngest son Simon and his heirs. He also left her all the "chains, rings, jewels, pearls, bracelets, and

---

necessary to point out that it can have no reference whatever to the subject of this memoir.

[1] On the 24th June, 1632, Lady Myddelton memorialised the Common Council of London with reference to the loan of 3000*l.* advanced to Sir Hugh, which does not seem to have been repaid; and more than two years later, on the 10th Oct., 1634, we find the Corporation allowed 1000*l.* of the amount, in consideration of the public benefit conferred on the city by Sir Hugh through the formation of the New River, and for the losses alleged to have been sustained by him through breaches in the water-pipes on the occasion of

divers great fires, as well as for the " present comfort " of Lady Myddelton. It is to be inferred that the balance of the loan of 3000*l.* was then repaid. Lady Myddelton died at Bush Hill on the 19th July, 1643, aged sixty-three, and was interred in the chancel of Edmonton Church, Middlesex. On her monumental tablet it is stated that she was " the mother of fifteen children."

[2] Simon's son Hugh was created a Baronet, of Hackney, Middlesex, in 1681. He married Dorothy, the daughter of Sir William Oglander, of Nunwell, Baronet.

gold buttons, which she hath in her custody and useth to wear at festivals, and the deep silver basin, spout pot, maudlin cup, and small bowl;" as well as "the keeping and wearing of the great jewel given to him by the Lord Mayor and Aldermen of London, and after her decease to such one of his sons as she may think most worthy to wear and enjoy it." By the same will Lady Myddelton was authorised to dispose of her interest in the Cardiganshire mines for her own benefit; and it afterwards appears, from documents in the State Paper Office, that Thomas Bushell, "the great chymist," as he was called, purchased it for 400*l.* cash down, and 400*l.* per annum during the continuance of her grant, which had still twenty-five years to run after her husband's death.[1] Besides these bequeathments, and the gifts of land, money, and New River shares, which he had made to his other children during his lifetime, Sir Hugh left numerous other sums to relatives, friends, and clerks; for instance, to Richard Newell and Howell Jones 30*l.* each, "to the end that the former may continue his care in the works in the Mines Royal, and the latter in the New River water-works," where they were then respectively employed. He also left an annuity of 20*l.* to William Lewyn, who had been engaged in the New

---

[1] Bushell is said to have made a large fortune out of the mines after Sir Hugh Myddelton's death. He was authorised under an indenture with Charles I., dated the 30th July, 1637, to erect a mint in the Castle of Aberystwith, where he coined the bullion drawn from the mines into half-crowns, shillings, sixpences, half-groats, and halfpence. When the civil wars broke out, Bushell was not ungrateful to the King—presenting him with a loan, or rather gratuity, of 40,000*l.*, and raising a regiment for the royal service amongst his miners, which he continued to maintain until a late period in the contest between the King and the Parliament. On the defeat of the former, he took re-fuge in the Isle of Lundy. Numerous wild traditions are still related of Bushell by the country people in the neighbourhood of Lodge, where he resided. There is a curious old well in Lodge Park, known as "Bushell's Well," where he is said to have killed and thrown in his wife; and the people still believe that her headless corpse haunts the wood round the well. Fifty or sixty years after Sir Hugh Myddelton's time the mines were worked by Lewis Morris, the well-known Welsh antiquarian writer. Most of them are now abandoned. An advertisement of a new company to reopen those which had ceased to be worked recently appeared, but the design seems to have been abandoned.

River undertaking from its commencement. Nor were his men and women servants neglected, for he bequeathed to each of them a gift of money, not forgetting "the boy in the kitchen," to whom he left forty shillings. He remembered also the poor of Henllan, near Denbigh, "the parish in which he was born," leaving to them 20*l.*; a similar sum to the poor of Denbigh, which he had represented in several successive parliaments; and 5*l.* to the parish of Amwell, in Hertfordshire. To the Goldsmiths' Company, of which he had so long been a member, he bequeathed a share in the New River Company, for the benefit of the more necessitous brethren of that guild, "especially to such as shall be of his name, kindred, and county." [1]

Such was the life and such the end of Sir Hugh Myddelton, a man full of enterprise and resources, an energetic and untiring worker, a great conqueror of obstacles and difficulties, an honest and truly noble man, and one of the most distinguished benefactors the city of London has ever known.

---

[1] Several of the descendants of Sir Hugh Myddelton, when reduced in circumstances, obtained assistance from this fund. It has been stated, and often repeated, that Lady Myddelton, after her husband's death, became a pensioner of the Goldsmiths' Company, receiving from them 20*l.* a year. But this annuity was paid, not to the widow of the first Sir Hugh, but to the mother of the last Sir Hugh, more than a century later. The last who bore the title was an unworthy scion of this distinguished family. He could raise his mind no higher than the enjoyment of a rummer of ale; and towards the end of his life existed upon a pension granted him by the New River Company. The statements so often published (and which, on more than one occasion, have brought poor persons up to town from Wales to make inquiries) as to an annuity of 100*l.* said to have been left by Sir Hugh and unclaimed for a century, and of an advertisement calling upon his descendants to apply for the sum of 10,000*l.* alleged to be lying for them at the Bank of England, are altogether unfounded. No such annuity has been left, no such sum has accrued, and no such advertisement has appeared.

ANCIENT CAUSEWAY IN COCK MILL WOOD, NEAR WHITBY, YORKSHIRE.

[By Percival Skelton, after an original Sketch by Miss Simpson.]

# EARLY ROADS

### AND

# MODES OF TRAVELLING.

## CHAPTER I.

### OLD ROADS.

ROADS have in all times been among the most influential agencies of society; and the makers of roads, by enabling men readily to communicate with each other, have properly been regarded as among the most effective pioneers of civilization. Roads are literally the pathways not only of industry, but of social and national intercourse. Wherever a line of communication between men is formed, it renders commerce practicable; and, where commerce penetrates, it invariably creates a civilization and leaves a history. Roads place the city and the town in connection with the village and the farm, open up markets for field produce, and provide outlets for manufactures. They enable the natural resources of a country to be developed, facilitate travelling and intercourse, break down local jealousies, and in all ways tend to bind together society and bring out fully that healthy spirit of industry which is the life and soul of every great nation.

The road is so necessary an instrument of social well-being that in every new colony it is one of the first things thought of. First roads, then institutions, followed by schools, churches, and newspapers. The new country, as well as the old, can only be effectually "opened up," as the common phrase is, by roads, and until these are made it is virtually closed. Freedom itself cannot exist without free communication, every limitation of movement on the part of the members of society amounting to a positive abridgment of their personal liberty. Hence roads, canals, and railways, by providing the greatest possible facilities for locomotion and information, are essential for the freedom of all classes, of the poorest as well as the richest. By bringing the ends of a kingdom together, they reduce the inequalities of fortune and station, and, by equalizing the price of commodities, to that extent they render them accessible to all. Without their assistance the concentrated populations of our large towns could neither be clothed nor fed; but by their instrumentality an immense range of country is brought as it were to their very doors, and the sustenance and employment of our large masses of people become comparatively easy. In the raw materials required for food, for manufacturing, and for domestic purposes, the cost of transport necessarily forms a considerable item ; and it is clear that the more this cost can be reduced by facilities of communication, the cheaper do these articles become, the more they are multiplied, and so enter into the consumption of the community at large. Let any one imagine what would be the effect of closing the roads, railways, and canals of England. The country would be brought to a dead lock, employment would be restricted in all directions, and a large proportion of the inhabitants concentrated in the large towns must at certain seasons perish of cold and hunger.

In the earlier periods of English colonization roads were of comparatively less consequence. While the population was thin and scattered, and men lived by hunting and pastoral pursuits, the track across the down, the heath, and the moor, sufficiently answered their purpose. Yet even in those districts unencumbered with wood, where the first settlements were made—

as on the downs of Wiltshire, the moors of Devonshire, and the wolds of Yorkshire—stone tracks were laid down by the tribes between one village and another. We have given, at the beginning of this chapter, a representation of one of those ancient trackways, as its remains still exist, in the neighbourhood of Whitby, in Yorkshire; and there are many of the same description of old roads to be met with in other parts of England. In some districts they are called trackways or ridgeways, being narrow causeways usually following the natural ridge of the country, and probably serving in early times as local boundaries. On Dartmoor they are constructed of stone blocks, irregularly laid down on the surface of the ground, forming a rude causeway of about five or six feet wide.[1]

The Romans, with many other arts, first brought into England the art of road-making. They thoroughly understood the value of good roads, regarding them as the essential means for the maintenance of their empire in the first instance, and of social prosperity in the next. It was the road, not less than the legion, which made them masters of the world. Wherever they went they opened up the communications of the countries they subdued, and the roads which they made were certainly among the very best of their kind.[2] For centuries after they had left England the Roman roads continued to be the main highways of internal communication, and their remains are even to this day to be traced in many parts of the country. Settlements were made and towns sprang up along these old "streets;" and the numerous Stretfords and Stratfords, and towns

---

[1] An interesting description of these old trackways in Devonshire is given by the Rev. Samuel Rowe, M.A., Vicar of Crediton, in his ' Perambulations of the Ancient and Royal Forest of Dartmoor.' London, 1848.

[2] The *Curator Viarum* of the Romans was an official of distinction,

wielding great authority. Plutarch says of Caius Gracchus, when appointed supreme director for making roads, &c., that the people were charmed to see him go forth on his tours of road-making, followed by such numbers of architects, artificers, ambassadors, and magistrates.

ending in " le-street"—as Ardwick-le-street, in Yorkshire, and Chester-le-street, in Durham—mostly mark
the direction of these ancient lines of road.[1] There is
one peculiarity in the roads constructed by the Romans
which must have struck many observers—their remarkable straightness. Level does not seem to have been
of consequence, compared with directness. This peculiarity is supposed to have originated in an imperfect
knowledge of mechanics, for the Romans do not appear
to have been acquainted with the moveable joint in
wheeled carriages. The carriage-body rested solidly
upon the axles, which in four-wheeled vehicles were
rigidly parallel with each other. Being unable readily
to turn a bend in the road, it has been concluded that for
this reason all their great highways were constructed in
as straight lines as possible.

But most of these old roads having been neglected
and allowed·to fall into decay, the forest and the waste
gradually resumed their dominion over them in most
places, until the roads of England became about the
worst in Europe. We find, however, that numerous
attempts were made in early times to preserve the
ancient ways and enable a communication to be maintained between the metropolis and the rest of the
country, as well as between one market town and
another. The state of the highways may be inferred
from the character of the legislation applying to them.
One of the first laws on the subject was passed in
1285, directing that all bushes and trees along the roads
leading from one market to another should be cut down
for two hundred feet on either side, to prevent robbers
lurking therein; but nothing was proposed for amend-
ing the condition of the ways themselves. In 1346,
Edward III. authorised the first toll to be levied for the
repair of the roads leading from St. Giles's-in-the-Fields

---

[1] In Yorkshire, the old Roman road | in part by the name of " Street-
from Tadcaster to York is still known | Houses."

to the village of Charing (now Charing Cross), and from the same quarter to near Temple Bar (down Drury-Lane), as well as the highway then called Perpoole (now Gray's Inn Lane). The footway at the entrance of Temple Bar was interrupted by thickets and bushes, and in wet weather was almost impassable. The roads further west were so bad that when the sovereign went to Parliament faggots were thrown into the ruts in King-street, Westminster, to enable the royal cavalcade to pass along. In Henry VIII.'s reign several remarkable statutes were passed relating to certain worn-out and impracticable roads in Sussex and the Weald of Kent. From these it would appear that when the old roads were found too deep and miry to be passed, they were merely abandoned and a new track was selected. This is apparent from the Act 14 Henry VIII., chap. 6, which giveth liberty to every man having highway that is worn deep and incommodious for passage to lay out another way in some such other place of his land as shall be thought meet by the view of two justices of the peace and twelve other men of wisdom and discretion. Another Act passed in the same reign related to the repairs of bridges and of the highways at the ends of bridges. But as these Acts were for the most part merely permissive, they could have had but little practical effect in improving the communications of the kingdom. In the reign of Philip and Mary (in 1555), the Act was passed providing that each parish should elect two surveyors of highways to see to the maintenance of their repairs by compulsory labour; and to this day parish and cross roads are maintained on the principle of Mary's Act, though the compulsory labour has since been commuted into a compulsory tax. In Elizabeth and James's reigns other Acts were passed, but, from the statements of contemporary writers, it would appear that very little substantial progress was made in consequence, and travelling continued to be

encompassed with difficulties. Even in the neighbourhood of the metropolis the roads were in certain seasons
scarcely passable. The great Western road into London
was especially bad, and about Knightsbridge, in winter,
the traveller had to wade through deep mud. Wyatt's
men entered the city by this approach in the rebellion
of 1554, and were called the " draggle-tails," because of
their wretched plight.

At a greater distance from the metropolis the roads
were still worse. They were in many cases but rude
tracks across heaths and commons, as furrowed with
deep ruts as ploughed fields, and in winter to pass along
one of them was like travelling in a ditch. The attempts
made by the adjoining occupiers to mend them were for
the most part confined to throwing large stones into the
bigger holes to fill them up. It was easier to allow
new tracks to be made than to mend the old ones. The
lands of the country were still mostly unenclosed, and it
was thus possible, in fine weather, to get from place to
place, in one way or another, with the help of a guide.
In the absence of bridges, guides were necessary to point
out the safest fords as well as to pick out the least miry
tracks. The most frequented lines of road were struck
out from time to time by the drivers of pack-horses,
who, to avoid the bogs and the sloughs, were usually
careful to keep along the higher grounds; but, to prevent those horsemen who departed from the beaten track
being swallowed up in quagmires, beacons were erected
to warn them against the more dangerous places.[1]

In some of the older-settled districts of England the
old roads are still to be traced in the hollow Ways or

---

[1] See Ogilvy's ' Britannia Depicta,'
the traveller's ordinary guide-book
between 1675 and 1717, as Bradshaw's Railway Time-book is now.
The Grand Duke Cosmo, in his ' Travels in England in 1669,' speaks of
the country between Northampton
and Oxford as for the most part unenclosed and uncultivated, abounding
in weeds. From Ogilby's fourth edition, published in 1749, it appears
that the roads in the midland and
northern districts of England were
still, for the most part, entirely unenclosed.

Lanes, which are met with, in some places, eight and ten feet deep.    Horse-tracks in summer, and rivulets in winter, the earth became gradually worn into these deep furrows, many of which, in Wilts, Somerset, and Devon, represent the tracks of roads as old as, if not older than, the Conquest.    When the ridgeways of the earliest settlers on Dartmoor, above alluded to, were abandoned, the tracks were formed through the valleys, but the new roads were no better than the old ones.    They were narrow and deep, fitted only for a horse passing along laden with its crooks, as so capitally described in the ballad of " The Devonshire Lane." [1]

---

[1] This ballad is so descriptive of the old roads of the south-west of England that we are tempted to quote it at length.    It was written by the Rev. John Marriott, sometime vicar of Broadclist, Devon;  and Mr. Rowe, vicar of Crediton, says, in his ' Perambulation of Dartmoor,' that he can readily imagine the identical lane near Broadclist, leading towards Poltemore, which might have *sat* for the portrait.

In a Devonshire lane, as I trotted along
T'other day, much in want of a subject for song,
Thinks I to myself, I have hit on a strain,
Sure marriage is much like a Devonshire lane.

In the first place 'tis long, and when once you are in it,
It holds you as fast as a cage does a linnet;
For howe'er rough and dirty the road may be found,
Drive forward you must, there is no turning round.

But tho' 'tis so long, it is not very wide,
For two are the most that together can ride;
And e'en then, 'tis a chance but they get in a pother,
And jostle and cross and run foul of each other.

Oft poverty meets them with mendicant looks,
And care pushes by them, o'erladen with crooks;
And strife's grazing wheels try between them to pass,
And stubbornness blocks up the way on her ass.

Then the banks are so high, to the left hand and right,
That they shut up the beauties around them from sight;
And hence, you'll allow, 'tis an inference plain,
That marriage is just like a Devonshire lane.

But thinks I, too, these banks, within which we are pent,
With bud, blossom, and berry, are richly besprent;
And the conjugal fence, which forbids us to roam,
Looks lovely, when deck'd with the comforts of home.

In the rock's gloomy crevice the bright holly grows;
The ivy waves fresh o'er the withering rose,
And the ever-green love of a virtuous wife
Soothes the roughness of care, cheers the winter of life.

Then long be the journey, and narrow the way,
I'll rejoice that I've seldom a turnpike to pay;
And whate'er others say, be the last to complain,
Though marriage is just like a Devonshire lane.

Similar roads existed until recently in the immediate
neighbourhood of Birmingham, long the centre of considerable traffic.  The sandy soil was sawn through, as
it were, by generation after generation of human feet,
and by pack-horses, helped by the rains, until in some
places the tracks were as much as from twelve to fourteen yards deep;[1] one of these, partly filled up, retaining to this day the name of Holloway Head.  In the
neighbourhood of London there was also a Hollow way,
which now gives its name to a populous metropolitan
parish.  Hagbush Lane was another of such roads;
before the formation of the Great North Road it was one
of the principal bridle-paths leading from London to
the northern parts of England, but it was so narrow as
barely to afford passage for more than a single horseman, and so deep that the rider's head was beneath the
level of the ground on either side.

The roads of Sussex long preserved an infamous noto-
riety.  Chancellor Cowper, when a barrister on circuit,
wrote to his wife in 1690, that "the Sussex ways are
bad and ruinous beyond imagination.  I vow 'tis melancholy consideration that mankind will inhabit such a
heap of dirt for a poor livelihood.  The country is a
sink of about fourteen miles broad, which receives all
the water that falls from two long ranges of hills on
both sides of it, and not being furnished with convenient
draining, is kept moist and soft by the water till the
middle of a dry summer, which is only able to make it
tolerable to ride for a short time."  It was almost as
difficult for old persons to get to church in Sussex during
winter as it was in the Lincoln Fens, where they rowed
there in boats.  Fuller saw an old lady being drawn
to church in her own coach by the aid of six oxen.  The
Sussex roads were indeed so bad as to pass into a bye-
word.  A contemporary writer says, that in travelling

---

[1] Hutton's ' History of Birmingham.'  Ed. 1836, p. 21.

a slough of extraordinary miryness, it used to be called " the Sussex bit of the road ;" and he satirically alleged that the reason why the Sussex girls were so long-limbed was because of the tenacity of the mud in that county ; the practice of pulling the foot out of it " by the strength of the ancle " tending to stretch the muscle and lengthen the bone ! [1]

But the roads in the immediate neighbourhood of London seem to have been almost as bad as those in Sussex. Thus, when the poet Cowley retired to Chertsey, in 1665, he wrote to his friend Sprat to visit him, and, by way of encouragement, told him that he might sleep the first night at Hampton town ; thus occupying two days in the performance of a journey of twenty-two miles in the immediate neighbourhood of the metropolis. As late as 1736 we find Lord Hervey, writing from Kensington, complaining that " the road between this place and London is grown so infamously bad that we live here in the same solitude as we would do if cast on a rock in the middle of the ocean ; and all the Londoners tell us that there is between them and us an impassable gulf of mud." The mud was no respecter of persons, either ; for we are informed that the carriage of Queen Caroline could not, in bad weather, be dragged from St. James's Palace to Kensington in less than two hours, and occasionally the royal coach stuck fast in a rut, or was even overthrown into the mud. The streets of London themselves were no better at that time, the kennel being still permitted to flow in the middle of the street, which was paved with round stones,—flagstones for the pedestrians being as yet unknown.

---

[1] ' *Iter Sussexiense*.' By Dr. John Burton.

## CHAPTER II.

### EARLY MODES OF CONVEYANCE.

SUCH being the early state of the roads, the only practicable modes of travelling were on foot and on horseback. The poor walked and the rich rode. Kings rode and Queens rode. Gentlemen rode and robbers rode. Judges rode circuit in jack-boots, and the Bar sometimes walked and sometimes rode. Chaucer's ride to Canterbury will be remembered as long as the English language lasts. Hooker rode to London on a hard-paced nag, that he might be in time to preach his first sermon at St. Paul's. Ladies rode on pillions, holding on by the gentleman or the serving-man mounted before. Shakespeare incidentally describes the same style of travelling among the humbler classes in his 'Henry IV.' [1] The party, afterwards set upon by Falstaff and his companions, bound from Rochester to London, were up by two in the morning, expecting to perform the journey of thirty miles by close of day, and to get to town "in time to go to bed with a candle." Two are carriers, one of whom has "a gammon of bacon and two razes of ginger, to be delivered as far as Charing Cross;" the other has his panniers full of turkeys. There is also a franklin of Kent, and another, "a kind of auditor," probably a tax-collector, with several more, forming in all a company of eight or ten, who travel together for mutual protection. Their robbery on Gad's Hill, as painted by Shakespeare, is but a picture, by no means exaggerated, of the adventures and dangers of the road at the time of which he wrote.

---

[1] King Henry the Fourth (Part I.), Act II. Scene 1.

Distinguished personages sometimes rode in horse-litters, but riding on horseback was generally preferred. Queen Elizabeth made most of her journeys in this way,[1] and when she went into the City she rode on a pillion behind her Lord Chancellor.    The Queen, however, was at length provided with a coach, which must have been a very remarkable machine.    This royal vehicle is said to have been one of the first coaches used in England, and it was introduced by the Queen's own coachman, one Boomen, a Dutchman.[2]    It was little better than a cart without springs, the body resting solid upon the axles. Taking the bad roads and ill-paved streets into account, it must have been an excessively painful mode of conveyance.    Indeed, at one of the first audiences which the Queen gave to the French ambassador in 1568, she recounted to him the nature of the jolting she had received in it a few days before, and described "la douleur qu'elle sentoit à son couste, pour s'y estre heurtrée quelques jours auparavant, en ung coche où elle allait ung peu trop viste."[3]

Such coaches were in the first place used only for state processions.  The roads, even in the neighbourhood of London, were so bad and so narrow that the vehicles could not well be taken into the country.   But as the fashion of using them spread, the aristocracy removed to the western parts of the metropolis, where they could be used, and in course of time they even extended into the country.    They were still, however, neither more nor less than waggons, and, indeed, were called by that name; but wherever they went they

---

[1] Part of the riding road along which the Queen was accustomed to pass on horseback between her palaces at· Greenwich and Eltham is still in existence, a little to the south of Morden College, Blackheath. It winds irregularly through the fields, broad in some places and narrow in others. Probably it is very little different from what it was when used as a royal road. It is now very appropriately termed "Muddy Lane."

[2] For much curious information on this subject see a paper by J. H. Markland, F.R.S., entitled "Remarks on the Early Use of Carriages," in 'Archæologia,' vol. xx. (London, 1824), pp. 443-76.

[3] 'Dépêches de La Mothe Fénélon,' 8vo., 1838. Vol. i., p. 27.

excited much wonder. It is related of "that valyant knyght Sir Harry Sidney," that on a certain day in the year.1583 he entered Shrewsbury in his waggon, "with his Trompeter blowynge, verey joyfull to behold and see." [1]

From this time the use of coaches gradually spread, more particularly amongst the nobility, superseding the horse-litters which had till then been used for the conveyance of ladies and others unable to bear the fatigue of riding on horseback. The first carriages were heavy and lumbering, and upon the execrable roads of the time they went pitching over the stones and into the ruts, with the pole dipping and rising like a ship in a rolling sea. That they had no springs, is clear enough from the statement of Taylor, the water-poet, that in the paved streets of London men and women were so "tost, tumbled, rumbled, and jumbled about in them." Although the road from London to Dover, along the old Roman Watling-street, was then one of the best in England, the journey of the French household of Queen Henrietta, when they were sent forth from the palace of Charles I., occupied four tedious days before they reached Dover.

But it was only a few of the main roads leading from the metropolis that were practicable for coaches; and on the occasion of a royal progress, or the visit of a lord-lieutenant, there was a general turn out of labourers and masons to mend the ways and render the bridges at least temporarily secure. When the judges, usually old men and bad riders, took to going the circuit in their coaches, juries were often kept waiting until their lordships could be dug out of a bog or hauled out of a slough by the aid of plough-horses. In the seventeenth century scarcely a Quarter Sessions passed without presentments from the grand jury against certain districts on account of the bad state of the roads, and many were the fines

---

[1] Nichols's ' Progresses,' vol. ii., 309.

which the judges imposed upon them as a set-off against their bruises and other damages while on circuit.

For a long time the roads continued barely practicable for wheeled vehicles of the rudest sort, though Fynes Morison (writing of the time of James I.) gives an account of " carryers, who have long covered waggons, in which they carry passengers from place to

THE OLD STAGE WAGGON.

[By Louis Huard, after a Print by Rowlandson.]

place; but this kind of journeying," he says, " is so tedious, by reason they must take waggon very early and come very late to their innes, that none but women and people of inferior condition travel in this sort." The waggons of which Morison wrote, made only from ten to fifteen miles in a long summer's day : that is, supposing them not to have broken down by pitching over the boulders laid along the road, or stuck fast in a quagmire, when they had to wait for the arrival of the next team of horses to help to drag them out. The waggon, however, continued to be adopted as a popular mode of travelling until late in the eighteenth century ; and Hogarth's picture illustrating the practice will be remembered, of the cassocked parson on his lean horse, attending his daughter newly alighted from the York waggon.

The introduction of stage-coaches about the middle of the seventeenth century formed a new era in the history of travelling by road. At first they were only a better sort of waggon, and confined to the more practicable highways near London. Their pace did not exceed four miles an hour, and the jolting of the unfortunate passengers conveyed in them must have been very hard to bear. It used to be said of their drivers that they were "seldom sober, never civil, and always late." The first mention of coaches for public accommodation is made by Sir William Dugdale in his Diary, from which it appears that a Coventry coach was on the road in 1659. But probably the first coaches, or rather waggons, were run between London and Dover, as one of the most practicable routes for the purpose. M. Sobrière, a French man of letters, who landed at Dover on his way to London in the time of Charles II., alludes to the existence of a stage-coach, but it seems to have had no charms for him, as the following passage will show : " That I might not," he says, " take post or be obliged to use the stage-coach, I went from Dover to London in a waggon. I was drawn by six horses, one before another, and driven by a waggoner, who walked by the side of it. He was clothed in black, and appointed in all things like another St. George. He had a brave monteror on his head and was a merry fellow, fancied he made a figure, and seemed mightily pleased with himself."

Shortly after, coaches seem to have been running as far north as Preston in Lancashire, as appears by a letter from one Edward Parker to his father, dated November, 1663,[1] in which he says, " I got to London

---

[1] Paper in ' Archæologia,' quoted above.

on Saturday last; but my journey was noe ways pleasant, being forced to ride in the boote all the waye.
Ye company yt came up with mee were persons of greate
quality, as knights and ladyes. My journey's expense
was 30s. This traval hath soe indisposed mee, yt I am
resolved never to ride up againe in ye coatch." These
vehicles must, however, have considerably increased, as
we find a popular agitation was got up against them; the
Londoners nicknamed them "hell-carts;" pamphlets were
written recommending their abolition; and attempts were
even made to have them suppressed by Act of Parliament.

Thoresby occasionally alludes to stage-coaches in his
Diary, speaking of one that ran between Hull and York
in 1679, from which place he had to proceed by Leeds
in the usual way on horseback. This Hull coach did
not run in winter, because of the state of the roads;
stage-coaches being usually laid up in that season like
ships during Arctic frosts.[1] Afterwards, when a coach
was put on between York and Leeds, it performed
the journey of twenty-four miles in eight hours;[2] but
the road was so bad and dangerous that the travellers
were accustomed to get out and walk the greater part
of the way. Thoresby often waxes eloquent upon the
subject of his manifold deliverances from the dangers of
travelling by coach. He was especially thankful when
he had passed the ferry over the Trent in journeying
between Leeds and London, having on several occasions narrowly escaped drowning there. Once, on his

---

[1] "4th May, 1714. Morning: we
dined at Grantham, had the annual
solemnity (this being the first time
the coach passed the road in May),
and the coachman and horses being
decked with ribbons and flowers, the
town music and young people in
couples before us; we lodged at Stamford, a scurvy, dear town. 5th May:
had other passengers, which, though
females, were more chargeable with
wine and brandy than the former

part of the journey, wherein we had
neither; but the next day we gave
them leave to treat themselves."—
Thoresby's 'Diary,' vol. ii., 207.
[2] "May 22, 1708. At York. Rose
between three and four, the coach
being hasted by Captain Crome (whose
company we had) upon the Queen's
business, that we got to Leeds by
noon; blessed be God for mercies to
me and my poor family."—Thoresby's
'Diary,' vol. ii., 7.

journey to London, some showers fell, which "raised the washes upon the road near Ware to that height that passengers from London that were upon that road swam, and a poor higgler was drowned, which prevented me travelling for many hours; yet towards evening we adventured with some country people, who conducted us over the meadows, whereby we missed the deepest of the Wash at Cheshunt, though we rode to the saddle-skirts for a considerable way, but got safe to Waltham Cross, where we lodged."[1] On another occasion Thoresby was detained four days at Stamford by the state of the roads, and was only extricated from his position by a company of fourteen members of the House of Commons travelling towards London, who took him into their convoy, and set out on their way southward attended by competent guides. When the "waters were out," as the saying went, the country became closed, the roads being simply impassable. During the Civil Wars eight hundred horse were taken prisoners while sticking in the mud.[2] When rain fell, pedestrians, horsemen, and coaches alike came to a standstill until the roads dried again and enabled the wayfarers to proceed. Thus we read of two travellers stopped by the rains within a few miles of Oxford, who found it impossible to accomplish their journey in consequence of the waters that covered the country thereabout.

The introduction of stage-coaches, like every other public improvement, was at first regarded with prejudice, and had considerable obloquy to encounter. In a curious book published in 1673, entitled 'The Grand Concern of England Explained in several Proposals to Parliament,'[3] stage-coaches and caravans were denounced as one of the greatest evils that had happened of late years to the kingdom, mischievous to the public, destructive to trade, and prejudicial to lands. It was alleged that travelling

---

[1] Thoresby's 'Diary,' vol. i., 295.
[2] Waylen's 'Marlborough.'
[3] Reprinted in the 'Harleian Miscellany,' vol. viii., p. 547. Supposed to have been written by one John Gressot, of the Charterhouse.

by coach was calculated to destroy the breed of horses, and make men careless of good horsemanship,—that it hindered the training of watermen and seamen, and interfered with the public resources. The reasons given are curious. It was said that those who were accustomed to travel in coaches became weary and listless when they rode a few miles, and were unwilling to get on horseback—"not able to endure frost, snow, or rain, or to *lodge in the fields*;" that to save their clothes and keep themselves clean and dry, people rode in coaches, and thus contracted an idle habit of body; that this was ruinous to trade, for that "most gentlemen, before they travelled in coaches, used to ride with swords, belts, pistols, holsters, portmanteaus, and hat-cases, which, in these coaches, they have little or no occasion for: for, when they rode on horseback, they rode in one suit and carried another to wear when they came to their journey's end, or lay by the way; but in coaches a silk suit and an Indian gown, with a sash, silk stockings, and beaver-hats, men ride in, and carry no other with them, because they escape the wet and dirt, which on horseback they cannot avoid; whereas, in two or three journeys on horseback, these clothes and hats were wont to be spoiled; which done, they were forced to have new very often, and that increased the consumption of the manufactures and the employment of the manufacturers; which travelling in coaches doth no way do." The writer of the same protest against coaches gives some idea of the extent of travelling by them in those days; for to show the gigantic nature of the evil he is contending against, he avers that between London and the three principal towns of York, Chester, and Exeter, not fewer than eighteen persons, making the journey in five days, travel by them weekly (the coaches running thrice in the week), and a like number back; "which come, in the whole, to eighteen hundred and seventy-two in the year." Another great nuisance, the writer alleged, which flowed from

the establishment of the stage-coaches, was, that not only
did the gentlemen from the country come to London in
them oftener than they need, but their ladies either came
with them or quickly followed them. "And when they
are there they must ·be in the mode, have all the new
fashions, buy all their clothes there, and go to plays,
balls, and treats, where they get such a habit of jollity
and a love to gaiety and pleasure, that nothing afterwards in the country will serve them, if ever they should
fix their minds to live there again ; but they must have
all from London, whatever it costs." Then there were
the grievous discomforts of stage-coach travelling to be
set against the more noble method of travelling by horseback, as of yore. "What advantage is it to men's health,"
says the writer, waxing wroth, "to be called out of their
beds into these coaches, an hour before day in the morning ; to be hurried in them from place to place, till one
hour, two, or three within night ; insomuch that, after
sitting all day in the summer-time stifled with heat and
choked with dust, or in the winter-time starving and
freezing with cold or choked with filthy fogs, they are
often brought into their inns by torchlight, when it is
too late to sit up to get a supper ; and next morning
they are forced into the coach so early that they can get
no breakfast ? What addition is this to men's health or
business to ride all day with strangers, oftentimes sick,
antient, diseased persons, or young children crying ; to
whose humours they are obliged to be subject, forced to
bear with, and many times are poisoned with their nasty
scents and crippled by the crowd of the boxes and
bundles ? Is it for a man's health to travel with tired
jades, to be laid fast in the foul ways and forced to wade
up to the knees in mire ; afterwards sit in the cold till
teams of horses can be sent to pull the coach out ? Is it
for their health to travel in rotten coaches and to have
their tackle, perch, or axle-tree broken, and then to wait
three or four hours (sometimes half a day) to have them

mended, and then to travel all night to make good their
stage? Is it for a man's pleasure, or advantageous to
his health and business, to travel with a mixed company
that he knows not how to converse with; to be affronted
by the rudeness of a surly, dogged, cursing, ill-natured
coachman; necessitated to lodge or bait at the worst inn
on the road, where there is no accommodation fit for
gentlemen; and this merely because the owners of the
inns and the coachmen are agreed together to cheat the
guests?" Hence the writer loudly calls for the suppression of stage-coaches forthwith as a great nuisance and
crying evil.

Travelling by coach was in early times a very delibe-
rate affair. Time was of less consequence than safety,
and coaches were advertised to start "God willing," and
"about" such and such an hour "as shall seem good"
to the majority of the passengers. The difference of a
day in the journey from London to York was a small
matter, and Thoresby was even accustomed to leave the
coach and go in search of fossil shells in the fields on
either side the road while making the journey between
the two places. The long coach "put up" at sun-down,
and "slept on the road." Whether the coach was to
proceed or stop short at some favourite inn was determined by the vote of the passengers, who usually
appointed a chairman at the beginning of the journey.
In 1700 York was a week distant from London, and
Tunbridge Wells, now reached in an hour, was two days.
Salisbury and Oxford were also two days' journeys, and
Exeter five. The Fly coach from London to Exeter *slept*
at the latter place the fifth night from town; the coach
proceeded next morning to Axminster, where it breakfasted, and there a woman barber "*shaved* the coach."[1]
Between London and Edinburgh, as late as 1763, a fortnight was consumed, the coach only starting once a

---

[1] Roberts's 'Social History of the Southern Counties,' p. 494.

month.[1] The risk of break-downs in driving over the execrable roads may be inferred from the circumstance that every coach carried with it a box of carpenter's tools, and the hatchets were occasionally used in lopping off the branches of trees overhanging the road and obstructing the travellers' progress.

No wonder, therefore, that a great deal of the travelling of the country continued to be performed on horseback, this being by far the pleasantest as well as most expeditious mode of journeying. Even Dr. Johnson rode from Birmingham to Derby with his Tetty on the day of their marriage, the Doctor taking the opportunity of the journey to give his bride her first lesson in marital discipline. At a later period James Watt rode from Glasgow to London, when proceeding thither to learn the art of mathematical instrument making. Nearly all the commercial gentlemen rode, carrying their samples and luggage in two bags at their saddle-bow, and hence their appellation of Riders or Bagmen. For safety's sake, these usually journeyed in company; for the dangers of travelling were by no means confined to the ruggedness of the roads. The highways were infested by troops of

---

[1] Mr. Pennant has left us the following account of his journey in the Chester stage to London in 1739-40: "The first day," says he, "with much labour, we got from Chester to Whitchurch, twenty miles; the second day to the 'Welsh Harp;' the third, to Coventry; the fourth, to Northampton; the fifth, to Dunstable; and, as a wondrous effort, on the last, to London, before the commencement of night. The strain and labour of six good horses, sometimes eight, drew us through the sloughs of Mireden and many other places. We were constantly out two hours before day, and as late at night, and in the depth of winter proportionally later. The single gentlemen, then a hardy race, equipped in jack-boots and trowsers, up to their middle, rode post through thick and thin, and, guarded against the mire, defied the frequent stumble and fall, arose and pursued their journey with alacrity; while, in these days, their enervated posterity sleep away their rapid journeys in easy chaises, fitted for the conveyance of the soft inhabitants of Sybaris." In 1710 a Manchester manufacturer, taking his family up to London, hired a coach for the whole way, which, in the then state of the roads, must have made it a journey of probably eight or ten days. And, in 1742, the system of travelling had so little improved that a lady, wanting to come with her niece from Worcester to Manchester, wrote to a friend in the latter place to send her a hired coach, because the man *knew the road*, having brought from thence a family some time before."—Aikin's 'Manchester.'

robbers and vagabonds who lived by plunder.  Turpin and Bradshaw beset the Great North Road; Duval, Macheath, Maclean, and hundreds more notorious highwaymen infested Hounslow Heath, Finchley Common, Shooter's Hill, and all the approaches to the metropolis.  A sight common to be seen then, was a gibbet erected by the roadside, with the skeleton of some former malefactor hanging from it in chains; and "Hangman's-lanes" were very numerous in the neighbourhood of London.[1]  It was considered most unsafe to travel after dark, and when the first "night coach" was started, the risk was thought too great, and it was not patronised. Travellers armed themselves on setting out on a journey as if they were going to battle, and a blunderbuss was considered as indispensable for a coachman as a whip. Dorsetshire and Hampshire, like most other counties, were beset with gangs of highwaymen; and when the Grand Duke Cosmo set out from Dorchester to travel to London in 1669, he was "convoyed by a great many horse-soldiers belonging to the militia of the county, to secure him from robbers."[2]  Thoresby, in his Diary, frequently alludes with awe to his having passed safely "the great common where Sir Ralph Wharton slew the highwayman," and he also makes special mention of Stonegate Hole, "a notorious robbing place" near Grantham. Like every other traveller, the pious man carried loaded pistols in his bags, and on one occasion he was thrown

---

[1] Lord Campbell mentions the remarkable circumstance that Popham, afterwards Lord Chief Justice in the reign of Elizabeth, took to the road in early life, and robbed travellers on Gad's Hill.  Highway robbery could not, however, have been considered a very ignominious pursuit at that time, as during Popham's youth a statute was made by which, on a first conviction for robbery, a peer of the realm or lord of parliament was entitled to have benefit of clergy, "though he cannot read!"  What is still more extraordinary is, that Popham is supposed to have continued in his course as a highwayman even after he was called to the Bar.  This seems to have been quite notorious, for when he was made Serjeant the wags reported that he served up some wine destined for an Alderman of London, which he had intercepted on its way from Southampton.—Aubrey, iii., 492.— Campbell's 'Chief Justices,' i., 210.

[2] 'Travels of Cosmo the Third, Grand Duke of Tuscany,' p. 147.

into great consternation near Topcliffe, in Yorkshire, on missing them, believing that they had been abstracted by some designing rogues at the inn where he had last slept.[1]  No wonder that, before setting out on a journey in those days, men were accustomed to make their wills.

When Mrs. Calderwood, of Coltness, travelled from Edinburgh to London in 1756, she relates in her Diary that she travelled in her own postchaise, attended by John Rattray, her stout serving man, on horseback, with pistols at his holsters and a good broad sword by his side.  The lady had also with her in the carriage a case of pistols, for use upon an emergency.  Robberies were then of frequent occurrence in the neighbourhood of Bawtry, in Yorkshire, and one day a suspicious-looking character, whom they took to be a highwayman, made his appearance; but " John Rattray, talking about powder and ball to the postboy, and showing his whanger, the fellow made off."  Mrs. Calderwood started from Edinburgh on the 3rd of June, when the roads were dry and the weather was fine, and she reached London on the evening of the 10th, which was considered a rapid journey in those days.

The danger, however, from footpads and highwaymen was not greatest in remote country places, but in and about the metropolis itself.  The proprietors of Bellsize House and gardens, in the Hampstead-road, then one of the principal places of amusement, had the way to London patrolled during the season by twelve " lusty fellows;" and Sadler's Wells, Vauxhall, and Ranelagh

---

[1] " It is as common a custom, as a cunning policie in thieves, to place chamberlains in such great inns where cloathiers and graziers do lye; and by their large bribes to infect others, who were not of their own preferring; who noting your purses when you draw them, they'l gripe your cloak-bags, and feel the weight, and so inform the master thieves of what they think, and not those alone, but the Host himself is oft as base as they, if it be left in charge with them all night; he to his roaring guests either gives item, or shews the purse itself, who spend liberally, in hope of a speedie recruit." See 'A Brief yet Notable Discovery of Housebreakers,' &c., 1659.  See also 'Street Robberies Considered; a Warning for Housekeepers,' 1676; 'Hanging not Punishment Enough,' 1701, &c.

advertised similar advantages. Foot passengers proceeding towards Kensington and Paddington in the evening, would wait until a sufficiently numerous band had collected to set footpads at defiance, and then they started in company. Carriages were stopped in broad daylight in Hyde Park, and even in Piccadilly itself, and pistols presented at the breasts of fashionable people, who were called upon to deliver up their purses. Horace Walpole relates a number of curious instances of this sort, he himself having been robbed in broad day, with Lord Eglinton, Sir Thomas Robinson, Lady Albemarle, and many more. A curious robbery of the Portsmouth mail, in 1757, illustrates the imperfect postal communication of the period. The boy who carried the post had dismounted at Hammersmith, about three miles from Hyde Park Corner, and called for beer, when some thieves took the opportunity of cutting the mailbag from off the horse's crupper and got away undiscovered!

The means adopted for the transport of merchandise were as tedious and difficult as those ordinarily employed for the conveyance of passengers. Corn and wool were sent to market on horses' backs,[1] manure was carried to the fields in panniers, and fuel was conveyed from the moss or the forest in the same way. The little coal used in the southern counties was principally seaborne, though pack-horses occasionally carried coal inland for the supply of the blacksmiths' forges. When Wollaton Hall was built by John of Padua for Sir Francis Willoughby in 1580, the stone was all brought on horses'

---

[1] The food of London was then principally brought to town in panniers. The population being comparatively small, the feeding of London was still practicable in this way; besides, the city always possessed the great advantage of the Thames, which secured a supply of food by sea. In 'The Grand Concern of England Explained,' it is stated that the hay, straw, beans, peas, and oats, used in London, were principally raised within a circuit of twenty miles of the metropolis; but large quantities were also brought from Henley-on-Thames and other western parts, as well as from below Gravesend, by water; and many ships laden with beans came from Hull, and with oats from Lynn and Boston.

backs from Ancaster, in Lincolnshire, thirty-five miles distant, and they loaded back with coal, which was taken in exchange for the stone.

The roads being almost impassable in certain seasons and difficult at all times, there was necessarily very little trade between one part of the kingdom and another. For without ready communication either by land or water, the commercial exchange of bulky articles—raw produce or manufactured commodities — is simply impossible. Hence England was not, and as yet could not be, very much of a commercial country. It was cheaper to bring foreign wares to London by sea than to bring them by tedious journeys on horses' backs from the interior of the country. Two centuries ago the inland carriage of goods from Norwich was as much as the sea freight from Lisbon. From London to Birmingham the charge was from 5l. to 7l. a ton, and from London to Exeter 12l. A century later the charge between Birmingham and London was reduced to between 8s. and 9s. a ton for every ten miles, or an average of about 5l. a ton;[1] but at the same time the rate of carriage between Leeds and London was 13l. a ton. This rate, it will readily be imagined, was prohibitory as regarded the large mass of manufactured articles in general consumption. But many articles now in common domestic use even amongst the poorest classes were then comparatively little known. No manufacture of pottery but of the very coarsest kind existed; vessels of wood, of pewter, and even of leather, formed the principal part of the household and table utensils of genteel and opulent families; and we long continued to import our cloths, our linen, our glass, our " Delph " ware, our cutlery, our paper, and even our hats, from France, Germany, and Holland.

The little trade which existed between one part of the kingdom and the other was carried on by means of pack-

---

[1] 'A History of Inland Navigation.' London, 1769, p. 73.

horses, along roads little better than bridle-paths. These horses travelled in lines, with the bales or panniers strapped across their backs. The foremost horse bore a bell or a collar of bells, and was hence called the " bell-horse." He was selected because of his sagacity ; and by

THE PACK-HORSE CONVOY.
[By Louis Huard after his original Drawing.]

the tinklings of the bells he carried, the movements of his followers were regulated. The bells also gave notice of the approach of the convoy to those who might be advancing from the opposite direction. This was a matter of some importance, as in many parts of the path there was not room for two loaded horses to pass each other, and quarrels and fights between the drivers of the pack-horse trains were frequent as to which of the meeting convoys was to pass down into the dirt and allow the other to pass along the bridleway. The pack-horses not only carried merchandise but passengers, and at certain times scholars proceeding to and from Oxford and Cambridge. When Smollett travelled from

Glasgow to London, he rode partly on pack-horses, partly by waggon, and partly on foot; and the adventures which he described as having befallen Roderick Random are supposed to have been drawn in a great measure from his own experiences during the journey.

A cross-country merchandise traffic gradually sprang up between the northern counties, since become pre-eminently the manufacturing districts of England; and long lines of pack-horses laden with bales of wool and cotton traversed the hill ranges which divide Yorkshire from Lancashire. Whitaker says that as late as 1753 the roads near Leeds consisted of a narrow hollow way little wider than a ditch, barely allowing of the passage of a vehicle drawn by horses in a single line; this deep narrow road being flanked by an elevated causeway covered with flags or boulder stones. When travellers encountered each other on this narrow track, they often tried to wear out each other's patience rather than descend into the dirt alongside. The raw wool and bale goods of the district were nearly all carried along these flagged ways on the backs of single horses; and it is difficult to imagine the delay, the toil, and the perils by which the conduct of the traffic was attended. On horseback before daybreak and long after nightfall, these hardy sons of trade pursued their object with the spirit and intrepidity of foxhunters; and the boldest of their country neighbours had no reason to despise either their horsemanship or their courage.[1] The Manchester trade was carried on in the same way. The chapmen there used to keep their gangs of pack-horses, which accompanied them to all the principal towns, bearing their goods in packs, which they sold to their customers,

---

[1] 'Loides and Elmete,' by T. D. Whitaker, LL.D., 1816, p. 81. Notwithstanding its dangers, Dr. Whitaker seems to have been of opinion that the old mode of travelling was even safer than that which immediately followed it : " Under the old state of roads and manners," he says, " it was impossible that more than one death could happen at once; what, by any possibility, could take place analogous to a race betwixt two stage-coaches, in which the lives of thirty or forty distressed and helpless

bringing back sheep's wool and other raw materials of manufacture.[1]

The only records of this long-superseded mode of communication are now to be traced on the signboards of wayside public-houses. Many of the old roads still exist in Yorkshire and Lancashire; but all that remains of the former traffic is the pack-horse painted on these village signs—things as retentive of odd bygone facts as the picture-writing of the ancient Mexicans.[2]

---

individuals are at the mercy of two intoxicated brutes?"

[1] The author of ·the ' Original ' says :—" I have by tradition the following particulars of the mode of carrying on the home trade by one of the principal merchants of Manchester, who was born at the commencement of the last century, and who realised a sufficient fortune to keep a carriage when not half-a-dozen were kept in the town by persons connected with business. He sent the manufactures of the place into Nottinghamshire, Lincolnshire, Cambridgeshire, and the intervening counties, and principally took in exchange feathers from Lincolnshire and malt from Cambridgeshire and Nottinghamshire. All his commodities were conveyed on pack-horses, and he was from home the greater part of every year, performing his journeys entirely on horseback. His balances were received in guineas, and were carried with him in his saddle-bags. He was exposed to the vicissitudes of the weather, to great labour and fatigue, and to constant danger. In Lincolnshire he travelled chiefly along bridle-ways, through fields where frequent gibbets warned him of his perils, and where flocks of wild-fowl continually darkened the air. Business carried on in this manner required a combination of personal attention, courage, and physical strength not to be hoped for in a deputy ; and a merchant then led a much more severe and irksome life than a bagman afterwards, and still more than a ' traveller' of the present day. In the earlier days of the merchant above-mentioned the wine-merchant who supplied Manchester resided at Preston, then always called Proud Preston, because exclusively inhabited by gentry. The wine was carried on horses, and a gallon was considered a large order. Men in business confined themselves generally to punch and ale, using wine only as a medicine, or on extraordinary occasions ; so that a considerable tradesman somewhat injured his credit amongst his neighbours by being so extravagant as to send to a tavern for wine to entertain a London customer."

[2] Earl of Ellesmere's ' Essays,' p. 244. In the curious collection of old coins at the Guildhall there are several halfpenny tokens issued by the proprietors of inns bearing the sign of the pack-horse. Some of these would indicate that pack-horses were kept for hire. We append a couple of illustrations of these curious old coins.

PACK-HORSE HALFPENNY TOKENS.
[From the Guildhall Collection.]

## CHAPTER III.

### MANNERS AND CUSTOMS INFLUENCED BY THE STATE OF THE ROADS.

WHILST the road communications of the country remained thus imperfect, the people of one part of England knew next to nothing of the people of the other parts. When a shower of rain had the effect of rendering the highways impassable, even horsemen were cautious in venturing far from home, and it was only a limited number who could afford to travel on horseback. The labouring people journeyed a-foot, and the limited middle class used the waggon or the coach. But the amount of intercourse between the people of different districts—then exceedingly limited at all times—was, in a country so wet as England, necessarily suspended during the greater part of the year. This slight degree of communication consequently produced numerous distinct and strongly-marked local dialects, local prejudices, and local customs, which survive to this day, though they are rapidly disappearing, to the regret of many, under the influence of our improved facilities for travelling. Every village had its witches, sometimes of different sorts, and there was scarcely an old house but had its white lady or moaning old man with the long beard. There were ghosts in the fens which walked on stilts, whilst the sprites of the hill country rode on flashes of fire. But those village witches and local ghosts have long since disappeared, excepting perhaps in a few of the less penetrable districts, where they still survive.

It is curious to find that down even to the beginning of the seventeenth century, the inhabitants of the southern districts of the island regarded those of the north as a kind of ogres.   Lancashire was supposed to be almost impenetrable—as indeed it was to a considerable extent, —and inhabited by a half-savage race.   Camden vaguely described it, previous to his visit in 1607, as that part of the country lying "beyond the mountains towards the Western Ocean."  He acknowledged that he approached the Lancashire people "with a kind of dread," but determined at length "to run the hazard of the attempt," trusting in the Divine assistance.   Camden was favoured in his northern visit even beyond his expectations, and after making his survey of the county, he succeeded in returning within the bounds of civilization in safety.

About a century later, in 1700, the Rev. Mr. Brome, rector of Cheriton in Kent, entered upon a series of travels in England as if it had been a newly-discovered country.   He set out in spring, so soon as the roads had become passable.   His friends convoyed him on the first stage of his journey, and left him, commending him to the care of Providence.   He was, however, careful to employ guides to conduct him from one place to another, and in the course of his three years' travels he saw many new and wonderful things; but when the winter and wet weather set in, he was compelled to suspend his travelling and lay up, like an arctic voyager, for several months, until spring came round again.   He passed through Northumberland into Scotland, down the western side of the island towards Devonshire, where he found the farmers gathering in their corn upon horseback, the roads being so narrow that it was impossible for them to use waggons.   He desired to travel into Cornwall, the boundaries of which he reached, but was prevented proceeding farther by the rains, and accordingly made the best of his way homewards.[1]

The vicar of Cheriton was considered a wonderful

---

[1] 'Three Years' Travels in England, Scotland, and Wales.' By James | Brome, M.A., Rector of Cheriton, Kent.  London, 1726.

man in his day,—almost as adventurous as we should
now regard a traveller in Central Africa. Twenty
miles of sloughs, or an unbridged river between two
parishes, were greater impediments to intercourse than
the Atlantic Ocean now is between England and
America. There were towns situated even in the same
county, more widely separated, for all practical purposes,
than London and Glasgow are at the present day.
There were many districts which travellers never visited,
and where the appearance of a stranger produced as
great an excitement as the arrival of a white man in an
African village.[1] Although this comparative seclusion
of most districts produced a picturesqueness and variety
of manners throughout England, it also produced a considerable amount of brutality, of which the local amusements of bull-running, cock-fighting, cock-throwing,
Plough-Monday, and such like, were the fitting ex-
ponents. People knew little except of their own
narrow district. The world beyond was as good as
closed against them. Almost the only intelligence of
general affairs which reached them was communicated
by pedlars and packmen, who were accustomed to retail
to their customers the news of the day with their wares;
or, at most, a news-letter from London, after it had been
read nearly to pieces at the great house of the district,
would find its way to the village, and its driblets of

---

[1] The treatment he received was
occasionally even less polite. When
William Hutton, of Birmingham, accompanied by another gentleman,
went to view the field of Bosworth, in
1770, "the inhabitants," he says,
"set their dogs at us in the street,
merely because we were strangers.
Human figures not their own are seldom seen in these inhospitable re-
gions. Surrounded with impassable
roads, no intercourse with man to hu-
manise the mind, nor commerce to
smooth their rugged manners, they
continue the boors of Nature." In
certain villages in Lancashire and
Yorkshire, not very remote from large
towns, the appearance of a stranger,
down to a comparatively recent period,
excited a similar commotion amongst
the villagers, and the word would
pass from door to door, "Dost knaw
'im?" "Naya." "Is 'e straunger?"
"Ey, for sewer." "Then paus' 'im
—Eave a duck at 'im—Fettle 'im!"
And the "straunger" would straightway find the "ducks" flying about
his head, and be glad to make his
escape from the village with his life.

information would thus become diffused amongst the little community.   Matters of public interest were long in becoming known in the remoter districts of the country. Macaulay relates that the death of Queen Elizabeth was not heard of in some parts of Devon until the courtiers of her successor had ceased to wear mourning for her. The news of Cromwell's being made Protector only reached Bridgewater nineteen days after the event, when the bells were set a-ringing; and the churches in the Orkneys continued to put up the usual prayers for James II. three months after he had taken up his abode at St. Germains.

There were then no shops in the smaller towns or villages, and comparatively few in the larger; even these being badly furnished with articles in general use.   The country people were irregularly supplied by hawkers, who sometimes bore their stocks upon their backs, and occasionally on pack-horses.   Pots, pans, and household utensils were thus sold from door to door; and until a comparatively recent period the whole of the pottery-ware manufactured in Staffordshire was hawked about and disposed of in this way.   The pedlars carried frames resembling camp-stools, on which they were accustomed to display their wares when the opportunity occurred for showing them to advantage.   The articles which they sold were chiefly of a fanciful kind—ribbons, laces, and female finery; the housewives' great reliance for the supply of general clothing in those days being on domestic industry.   In autumn the mistress of the household was accustomed to lay in a store of articles sufficient to serve for the entire winter.   It was like laying in a stock of provisions and clothing for a siege during the time that the roads were closed.   The greater part of the meat required for winter's use was killed and salted down at Martinmas, whilst stockfish and baconed herrings were provided for Lent.   Scatcherd says that in his district the clothiers united in groups of three or four,

and at the Leeds winter fair they would purchase an ox,
which, having divided, they salted and hung the pieces
for their winter's food.[1] There was also the winter's
stock of firewood to be provided, and the rushes with
which to strew the floors—carpets being a comparatively
modern invention ; besides, there was the store of wheat
and barley for bread, the malt for ale, the honey for
sweetening (then used for sugar), the salt, the spiceries,
and the savoury herbs so much employed in the ancient
cookery.    When the stores were laid in, the housewife
was in a position to bid defiance to bad roads for six
months to come.   This was the case of the well-to-do ;
but the poorer classes, who could not lay in a store for
winter, were often very badly off both for food and
firing, and in many hard seasons they literally starved.
But charity was active in those days, and many a poor
man's store was eked out by his wealthier neighbour.

When the household stores were thus laid in, the
mistress, with her daughters and servants, sat down to
their distaffs and spinning-wheels ; for the manufacture
of the family clothing was usually the work of the
winter months.   The fabrics then worn were almost
entirely of wool, silk and cotton being scarcely known.
The wool, when not grown on the farm, was purchased
in a raw state, and was carded, spun, dyed, and in many
cases woven at home : so also with the linen clothing,
which, until quite a recent date, was entirely the produce
of female fingers and household spinning-wheels.   This
kind of work occupied the winter months, occasionally
alternated with knitting, embroidery, and tapestry work.
Many of our old country houses to this day bear witness
to the steady industry of the ladies of even the highest
ranks in those times, in the fine tapestry hangings with
which the walls of many of the older rooms in such
mansions are covered.   Amongst the humbler classes
the same winter's work went on.   The women sat round

---

[1] Scatcherd, 'History of Morley.'

log fires knitting, plaiting, and spinning by their light, even in the daytime. Glass had not yet come into general use, and the openings in the wall which served for windows had necessarily to be shut up close to keep out the cold, though at the same time this shut out the light. The chimney, usually of lath and plaster ending overhead in a cone and funnel for the smoke, was so roomy in the old cottages as to accommodate almost the whole family sitting around the fire of logs piled in the reredosse in the middle, and there they carried on their winter's work. Such was the domestic occupation of women in the rural districts in olden times; and it may perhaps be questioned whether the revolution in our social system, which has taken out of their hands so many branches of household manufacture and useful domestic employment, be an altogether unmixed blessing.

Winter at an end, and the roads once more available for travelling, the Fair of the district was looked forward to with interest. Fairs were among the most important institutions of past times, rendered necessary by the then imperfect communications. Every town had its fair, which was also its festival, held under the protection of some patron saint; and the business as well as the gaiety of the neighbourhood usually centred on the occasion. High courts were held by the Bishop or Lord of the Manor, to accommodate which special buildings were erected, used only at fair time. Royal charters were granted to certain towns authorising them to hold fairs, and granting to them peculiar privileges. Amongst those of the first class were Winchester, St. Botolph's Town (Boston), and St. Ives. We find the great London

---

¹ The charter of Portsdown Fair, held on Portsdown Hill, near Portsmouth, is said to have been granted on condition that a loaf of bread was presented to the lord of the manor from wheat that had been raised, ground, and baked on the ground on which the fair was held. We are told that the ancient ceremony has been recently performed; but Portsdown Fair, which formerly was a great cloth-market, has now degenerated into a saturnalia of Waterloo-flys, merry-go-rounds, gin, and gingerbread-nuts.

merchants travelling thither in caravans, bearing with
them all manner of goods, and bringing back the wool
purchased by them in exchange.

Winchester Great Fair attracted merchants from all
parts of Europe.  It was held on the hill of St. Giles,
and was divided into streets of booths, named after the
merchants of the different countries who exposed their
wares in them.  " The passes through the great woody
districts, which English merchants coming from London
and the West would be compelled to traverse, were on
this occasion carefully guarded by mounted ' serjeants-
at-arms,' since the wealth which was being conveyed to
St. Giles's-hill attracted bands of outlaws from all parts
of the country." [1]   Weyhill Fair, near Andover, was
another of the great fairs in the same district, which was
to the West country agriculturists and clothiers what
Winchester St. Giles's Fair was to the general merchant.
That of St. Botolph's Town was one of the principal fairs
for the northern districts, to which people resorted from
great distances to buy and sell.   Thus we find, from
the ' Compotus ' of Bolton Priory,[2] that the monks of
that house sent their wool to St. Botolph's Fair to be sold,
though it was a good hundred miles distant, and there
they bought their groceries, spiceries, and other necessary
articles,   That fair, too, was often beset by robbers, and
on one occasion a strong party of them, under the disguise
of monks, attacked and robbed certain booths, setting
fire to the rest; and such was the amount of destroyed
wealth, that it is said the veins of molten gold and
silver ran along the streets.

The concourse of persons attending these fairs was
immense.  The nobility and gentry, the heads of the
religious houses, the yeomanry and the commons, resorted
to them to buy and sell all manner of agricultural pro-

---

[1] Murray's ' Handbook of Surrey, Hants, and Isle of Wight,' 168.

[2] Whitaker's ' History of Craven.'

duce. The farmers sold their wool and cattle, and hired
their servants there. The housewives sold the surplus
produce of their winter's industry, and bought their
cutlery, bijouterie, and more tasteful articles of apparel.
There were caterers for all customers—stuffs and wares
offered for sale from all countries. And in the wake of
this business part of the fair there invariably followed a
crowd of ministers to the popular tastes—quack doctors
and merry andrews, jugglers and minstrels, singlestick
players, grinners through horse-collars, and sportmakers
of every kind.

Smaller fairs were held in all districts for similar
purposes of exchange. At these the staples of the district
were sold and servants usually hired. Many were for
special purposes—cattle fairs, leather fairs, cloth fairs,
bonnet fairs, fruit fairs. Scatcherd says that less than a
century ago a large fair was held between Huddersfield
and Leeds, in a field still called Fairstead, near Birstal,
which used to be a great mart for fruit, onions, and such
like ; and that the clothiers resorted thither from all the
country round to purchase the articles, which were stowed
away in barns, and sold at booths by lamplight in the
morning.[1]

Even Dartmoor had its fair, on the site of an ancient
British village or temple near Merivale Bridge, testify-
ing to its great antiquity; for it is surprising how an
ancient fair lingers about the place on which it has been
accustomed to be held, long after the necessity for it has
ceased. The site of this old fair at Merivale Bridge
is the more curious, as in its immediate neighbourhood, on the road between Two Bridges and Tavistock,
is found the singular-looking granite rock, bearing so
remarkable a resemblance to the Egyptian sphynx, in a
mutilated state : it is of similarly colossal proportions,
and stands in a district almost as lonely as that in which

---

[1] Scatcherd's ' History of Morley,' 226.

the Egyptian sphynx looks forth over the sands of the Memphean Desert.[1]

SITE OF AN ANCIENT BRITISH VILLAGE AND FAIR ON DARTMOOR.

[By Percival Skelton, after his original Drawing.]

The last occasion on which the fair was held in this secluded spot was in the year 1625, when the plague raged at Tavistock ; and there is a part of the ground, situated amidst a line of pillars marking a stone avenue— a characteristic feature of the ancient aboriginal worship

---

[1] Vixen Tor is the name of this singular-looking rock. Mr. Rowe thus describes it : " Fronting the river [Walkham] the huge masses of which the tor is composed are piled up tier after tier, in a rude but noble façade, divided into three compartments by fissures, through which an ascent to the summit can be effected, whereon appearances of rock-basins will be observed. The river-front faces directly south, and this lofty Vixen-rock is traditionally reported to have been resorted to in past times for astronomical purposes. Vixen Tor, whether considered in itself or with reference to the striking scenery of which it forms the central object, is one of the most interesting in the moorland district."—' Perambulations of Dartmoor,' 186. It is proper, however, to add, that the appearance of the rock is most probably accidental, and that the head of the Sphynx is produced by the three angular blocks of rock being seen in profile. But Mr. Borlase, in his ' Antiquities of Cornwall,' expresses the opinion that the rock-basins on the summit of the rock were used by the Druids for purposes connected with their religious ceremonies.

—which is to this day pointed out and called by the name of the "Potatoe market."

But the glory of the great fairs has long since departed. They declined with the extension of turnpikes, and railroads gave them their deathblow. Shops now exist in every little town and village, drawing their supplies regularly by road and canal from the most distant parts. St. Bartholomew, the great fair of London,[1] and Donnybrook, the great fair of Dublin, have been suppressed as nuisances; and nearly all that remains of the dead but long potent institution of the Fair, is the occasional exhibition, at periodic times in country places, of pig-faced ladies, dwarfs, giants, double-bodied calves, and such-like wonders, amidst a blatant clangour of drums, gongs, and

---

[1] The provisioning of London, now grown so populous, would be almost impossible but for the perfect system of roads now converging on it from all parts. In early times, London, like country places, had to lay in its stock of salt-provisions against winter, drawing its supplies of vegetables from the country within easy reach of the capital. Hence the London market-gardeners petitioned against the extension of turnpike-roads about a century ago, as they afterwards petitioned against the extension of railways, fearing lest their trade should be destroyed by the competition of country-grown cabbages. But the extension of the roads had become a matter of absolute necessity, in order to feed the huge and ever-increasing mouth of the Great Metropolis, the population of which has grown in about two centuries from four hundred thousand to three millions. This enormous population has, perhaps, never at any time more than a fortnight's supply of food in stock, and most families not more than a few days; yet no one ever entertains the slightest apprehension of a failure in the supply, or even of a variation in the price from day to day in consequence of any possible shortcoming. That this should be so would be one of the most surprising things in the history of modern London, but that it is sufficiently accounted for by the magnificent system of roads, canals, and railways, which connect it with the remotest corners of the kingdom. Modern London is mainly fed by steam. The Express Meat-Train, which runs nightly from Aberdeen to London, drawn by two engines, and makes the journey in twenty-four hours, is but a single illustration of the rapid and certain method by which modern London is fed. The north Highlands of Scotland have thus, by means of railways, become grazing-grounds for the metropolis. Express fish-trains from Dunbar and Eyemouth (Smeaton's harbours), augmented by fish-trucks from Cullercoats and Tynemouth on the Northumberland coast, and from Redcar, Whitby, and Scarborough on the Yorkshire coast, also arrive in London every morning. And what with steam-vessels bearing cattle, meat, and fish, arriving by sea, and canal-boats laden with potatoes from inland, and railway-vans laden with butter and milk drawn from a wide circuit of country, and road-vans piled high with vegetables within easy drive of Covent Garden, the Great Mouth is thus from day to day regularly, satisfactorily, and expeditiously filled.

cymbals. Like the sign of the Pack-Horse' over the village inn door, the modern village fair, of which the principal article of merchandise is gingerbread-nuts, is but the vestige of a state of things that has long since passed away.

There were, however, remote and almost impenetrable districts which long resisted modern inroads. Of such was Dartmoor, which we have already more than once referred to. The difficulties of road-engineering in that quarter, as well as the sterility of a large proportion of the moor, had the effect of preventing its becoming opened up to modern traffic; and it is accordingly curious to find how much of its old manners, customs, traditions, and language has been preserved. It looks like a piece of England of the Middle Ages, left behind on the march. Witches still hold their sway on Dartmoor, where there exist no less than three distinct kinds—white, black, and grey,[1]—and there are still professors of the craft, male as well as female, in most of the villages. As might be expected, the pack-horses held their ground in Dartmoor the longest, and in some parts of North Devon they are not even yet extinct. When our artist was in the neighbourhood, sketching the ancient bridge on the moor[2] and the site of the old fair, a farmer said to him, "I well remember the train of pack-horses and the effect of their jingling bells on the silence of Dartmoor. My grandfather, a respectable farmer in the north of Devon, was the first to use a 'butt' (a square box without wheels, dragged by a horse) to carry manure to field; he was also the first man in the district to use an umbrella, which on Sundays he hung in the church-porch, an object of curiosity to the villagers." We are also informed by a gentleman who resided for some time at South Brent,

---

[1] The white witches are kindly disposed, the black cast the "evil eye," and the grey are consulted for the discovery of theft, &c.

[2] See Part IV.—Frontispiece to 'Bridges, Harbours, and Ferries.'

on the borders of the Moor, that the introduction of the
first cart in that district is remembered by many now
living, and the bridges were shortly afterwards widened
to accommodate the wheeled vehicles.

But the primitive features of the district are perhaps
best represented by the interesting little town of Chagford, situated in the valley of the North Teign, an ancient
stannary and market town, backed by a wide stretch of
moor. The houses of the place are built of moor stone
—grey, ancient-looking, and substantial—some with
projecting porch and parvise room over, and granite-
mullioned windows; the ancient church, built of granite,
with a stout old steeple of the same material, its em-
battled porch and granite-groined vault springing from
low columns with Norman-looking capitals, forming
the sturdy centre of this ancient town clump. A post-
chaise is still a phenomenon in the place, the roads and
lanes leading to it being so steep and rugged as to be
but ill adapted for springed vehicles of any sort. The
upland road or track to Tavistock scales an almost precipitous hill, and though well enough adapted for the
pack-horse of the last century, is quite unfitted for the
cart and waggon traffic of this. Hence the horse with
panniers maintains its ground in the Chagford district,
and the double-horse, furnished with a pillion for the
lady riding behind, is still to be met with in the country
roads. Among the patriarchs of the hills the straight-
breasted blue coat may yet be seen, with the shoe
fastened with buckle and strap, as in the days when
George III. was king; and old women are still found
retaining the cloak and hood of their youth. Old agricultural implements continue in use. The slide or
sledge is seen in the fields; the flail, with its mono-
tonous strokes, resounds from the barn-floors; the corn
is sifted by the windstow—the wind merely blowing
away the chaff from the grain when shaken out of sieves
by the motion of the hand on some elevated spot; the

old wooden plough is still at work, and the goad is still
used to urge the yoke of oxen in dragging it along.

" In such a place as Chagford," says Mr. Rowe, " the cooper or
rough carpenter will still find a demand for the pack-saddle, with
its accompanying furniture of *crooks, crubs,* or *dung-pots.*   Before

THE DEVONSHIRE CROOKS.
[By Louis Huard, after an original Sketch.]

the general introduction of carts, these rough and ready contrivances
were found of great utility in the various operations of husbandry,
and still prove exceedingly convenient in situations almost, or
altogether, inaccessible to wheel-carriages.   The *long crooks* are
used for the carriage of corn in sheaf from the harvest-field to the
mowstead or barn, for the removal of furze, browse, faggot-wood,
and other light materials.   The writer of one of the happiest effu-
sions of the local muse,[1] with fidelity to nature equal to Cowper or

---

[1] See ' The Devonshire Lane,' above quoted, note to p. 161.

Crabbe, has introduced the figure of a Devonshire pack-horse bending under the 'swagging load' of the high-piled *crooks* as an emblem of care toiling along the narrow and rugged path of life. The force and point of the imagery must be lost to those who have never seen (and, as in an instance which came under my own knowledge, never heard of) this unique specimen of provincial agricultural machinery. The crooks are formed of two poles,[1] about ten feet long, bent, when green, into the required curve, and when dried in that shape are connected by horizontal bars. A pair of crooks, thus completed, is slung over the pack-saddle—one 'swinging on each side to make the balance true.' The short crooks, or *crubs*, are slung in a similar manner. These are of stouter fabric, and angular shape, and are used for carrying logs of wood and other heavy materials. The dung-pots, as the name implies, were also much in use in past times, for the removal of dung and other manure from the farmyard to the fallow or plough lands. The *slide*, or sledge, may also still occasionally be seen in the hay or corn fields, sometimes without, and in other cases mounted on low wheels, rudely but substantially formed of thick plank, such as might have brought the ancient Roman's harvest load to the barn some twenty centuries ago." [2]

---

[1] Willow saplings, crooked and dried in the required form. Mrs. Bray says the crooks are called by the country people " Devil's tooth-picks." A correspondent informs us that the queer old crook-packs represented in our illustration are still in use in North Devon. He adds : " The pack-horses were so accustomed to their position when travelling in line (when going in double file) and so jealous of their respective places, that if one got wrong and took another's place, the animal interfered with would strike at the offender with his crooks."

[2] Rowe's 'Perambulation of Dartmoor,' pp. 87, 8, 9. The primitive contrivance (Mr. Rowe further observes) for hanging the gates of the moorland crofts and commons may also be seen in this neighbourhood. No iron hinge of any kind, nor gate-post, is employed. An oblong moor-stone block, in which a socket is drilled, is built into the wall, from which it projects sufficiently to receive the back-stanchion of the gate, while a corresponding socket is sunk in a similar stone fixed in the ground below, unless a natural rock should be found *in situ*, suitable for the purpose, which is frequently the case. The gate, thus secured, swings freely, swivel-like, in these sockets ; and thus, from materials on the spot, without the assistance of iron, a simple, durable, and efficient hinge is formed by the rural engineer.—P. 89.

# CHAPTER IV.

## ROADS AND TRAVELLING TOWARDS THE END OF LAST CENTURY.

THE progress made in the improvement of the roads throughout the kingdom was exceedingly slow. Though some of the main thoroughfares were mended up so as to admit of stage-coach travelling at the rate of from four to six miles an hour, the lesser-frequented roads continued to be all but impassable. Travelling was still difficult, tedious, and dangerous. Only those who could not well avoid it ever thought of undertaking a journey, and travelling for pleasure was out of the question. A writer in the 'Gentleman's Magazine,' in 1752, says that a Londoner at that time would no more think of travelling into the west of England for pleasure than of going to Nubia.

But signs of progress were not awanting. In 1754 some enterprising Manchester men advertised a "flying coach" for the conveyance of passengers between that town and the metropolis; and, lest they should be classed with projectors of the Munchausen kind, they heralded their enterprise with this statement: "However incredible it may appear, this coach will actually (barring accidents) arrive in London in four days and a half after leaving Manchester!" Fast coaches were also established on other of the northern roads, though not with very extraordinary results as to speed. When John Scott, afterwards Lord Chancellor Eldon, travelled from Newcastle to Oxford in 1766, he mentions that he journeyed in what was denominated "a fly," because of its rapid travelling; yet he was three or four days and nights on the road. There was no such velocity, however, as to endanger overturning or other mischief.  On the panels of the coach were painted the appropriate motto of *Sat cito si sat bene*—quick enough if well enough —a motto which the future Lord Chancellor made his own.[1]

The journey by coach between London and Edinburgh still occupied a week or more, according to the state of the weather, as late as 1763.  Between Bath or Birmingham and London occupied between two and three days.  The road across Hounslow Heath was so bad, that it was stated before a Parliamentary Committee that it was frequently known about that time to be two feet deep in mud.  The rate of travelling was about six and a half miles an hour; but the work was so heavy that it "tore the horses' hearts out," as the common saying went, so that they only lasted for two or three years.  When the Bath road became improved, Burke was enabled, in the summer of 1774, to travel from London to Bristol, to meet the electors there, in little more than four and twenty hours; but his biographer takes care to relate that he "travelled with incredible speed." Glasgow was still a fortnight's distance from the metropolis, and the arrival of the mail there was so important an event that a gun was fired to announce its coming in. Sheffield set up a "flying machine on steel springs" to London in 1760 : it "slept" the first night at the Black Man's Head Inn, Nottingham ; the second at the Angel, Northampton ; and arrived at the Swan with Two Necks, Lad-lane, on the evening of the third day.  The fare was 1*l.* 17*s.*, and 14 lbs. of luggage was allowed.  But the

---

[1] We may incidentally mention three other journeys south by future Lords Chancellors.  Mansfield rode up from Scotland to London when a boy, taking two months to make the journey on his pony.  Wedderburn's journey by coach from Edinburgh to London, in 1757, occupied him six days. "When I first reached London," said the late Lord Campbell, "I performed the same journey in three nights and two days, Mr. Palmer's mail-coaches being then established ; but this swift travelling was considered dangerous as well as wonderful, and I was gravely advised to stay a day at York, as several passengers who had gone through without stopping had died of apoplexy from the rapidity of the motion !"

heaviest part of the charge was for living and lodging
on the road, not to mention the fees to guards and drivers.
The style in which the journey was performed may be
inferred from the circumstance that on one occasion,
when a quarrel took place between the guard and a pas-
senger, the coach stopped to see them fight it out on the
road.   Though the Dover road was still one of the best
in the kingdom, the Dover flying-machine, carrying only
four passengers, took a long summer's day to perform
the journey.   It set out from Dover at four o'clock in
the morning, breakfasted at the Red Lion, Canterbury,
and the passengers ate their way up to town at various
inns on the road, arriving in London in time for supper.
Smollett complained of the innkeepers along this route
as the greatest set of extortioners in England.

What a ride by coach was in those days has been so
well described by a Prussian clergyman, Mr. Charles H.
Moritz, that we cannot do better than give his account of
one from Leicester to Northampton, and from thence to
London.   The journey was made in the year 1782 :—

"Being obliged," he says, "to bestir myself to get back to
London, as the time grew near when the Hamburgh captain with
whom I intended to return had fixed his departure, I determined to
take a place as far as Northampton on the outside.   But this ride
from Leicester to Northampton I shall remember as long as I live.

"The coach drove from the yard through a part of the house.
The inside passengers got in from the yard, but we on the outside
were obliged to clamber up in the street, because we should have
had no room for our heads to pass under the gateway.   My companions on the top of the coach were a farmer, a young man very
decently dressed, and a black-a-moor.   The getting up alone was
at the risk of one's life, and when I was up I was obliged to sit
just at the corner of the coach, with nothing to hold by but a sort
of little handle fastened on the side.   I sat nearest the wheel, and
the moment that we set off I fancied that I saw certain death before
me.   All I could do was to take still tighter hold of the handle,
and to be strictly careful to preserve my balance.   The machine
rolled along with prodigious rapidity over the stones through the
town, and every moment we seemed to fly into the air, so much so

that it appeared to me a complete miracle that we stuck to the coach at all.    But we were completely on the wing as often as we passed through a village or went down a hill.

" This continual fear of death at last became insupportable to me, and, therefore, no sooner were we crawling up a rather steep hill, and consequently proceeding slower than usual, than I carefully crept from the top of the coach, and was lucky enough to get myself snugly ensconced in the basket behind.

" ' O, Sir, you will be shaken to death !' said the black-a-moor ; but I heeded him not, trusting that he was exaggerating the unpleasantness of my new situation.    And truly, as long as we went on slowly up the hill it was easy and pleasant enough ; and I was just on the point of falling asleep among the surrounding trunks and packages, having had no rest the night before, when on a sudden the coach proceeded at a rapid rate down the hill.    Then all the boxes, iron-nailed and copper-fastened, began, as it were, to dance around me ; everything in the basket appeared to be alive, and every moment I received such violent blows that I thought my last hour had come.    The black-a-moor had been right, I now saw clearly ; but repentance was useless, and I was obliged to suffer horrible torture for nearly an hour, which seemed to me an eternity. At last we came to another hill, when, quite shaken to pieces, bleeding, and sore, I ruefully crept back to the top of the coach to my former seat.    ' Ah, did I not tell you that you would be shaken to death ? ' inquired the black man, when I was creeping along on my stomach.    But I gave him no reply.    Indeed, I was ashamed ; and I now write this as a warning to all strangers who are inclined to ride in English stage-coaches, and take an outside seat, or, worse still, horror of horrors, a seat in the basket.

" From Harborough to Northampton I had a most dreadful journey.    It rained incessantly, and as before we had been covered with dust, we now were soaked with rain.    My neighbour, the young man who sat next me in the middle, every now and then fell asleep ; and when in this state he perpetually bolted and rolled against me, with the whole weight of his body, more than once nearly pushing me from my seat, to which I clung with the last strength of despair.    My forces were nearly giving way, when at last, happily, we reached Northampton, on the evening of the 14th July, 1782, an ever-memorable day to me.

" On the next morning I took an *inside* place for London.    We started early in the morning.    This journey from Northampton to the metropolis, however, I can scarcely call a ride, for it was a per-

petual motion, or endless jolt from one place to another, in a close
wooden box, over what appeared to be a heap of unhewn stones and
trunks of trees scattered by a hurricane.   To make my happiness
complete, I had three travelling companions, all farmers, who slept
so soundly that even the hearty knocks with which they hammered
their heads against each other and against mine did not awake
them.   Their faces, bloated and discoloured by ale and brandy and
the knocks aforesaid, looked, as they lay before me, like so many
lumps of dead flesh.

" I looked, and certainly felt, like a crazy fool when we arrived
at London in the afternoon." [1]

THE BASKET COACH, 1780.

[By Louis Huard, after a Print by Rowlandson.]

Arthur Young, in his books, inveighs strongly against
the execrable state of the roads in all parts of England
towards the end of last century.   In Essex he found the
ruts " of an incredible depth," and he almost swears at
one near Tilbury.   " Of all the cursed roads," he says,

---

[1] C. H. Moritz : 'Reise eines Deutschen in England im Jahre 1782.
Berlin, 1783.

"that ever disgraced this kingdom in the very ages of barbarism, none ever equalled that from Billericay to the King's Head at Tilbury.  It is for near twelve miles so narrow that a mouse cannot pass by any carriage.  I saw a fellow creep under his waggon to assist me to lift, if possible, my chaise over a hedge.  To add to all the infamous circumstances which concur to plague a traveller, I must not forget the eternally meeting with chalk waggons, themselves frequently stuck fast, till a collection of them are in the same situation, and twenty or thirty horses may be tacked to each to draw them out one by one!"[1]  Yet, will it be believed the proposal to form a turnpike-road from Chelmsford to Tilbury was resisted, as Arthur Young says, " by the Bruins of the country, whose horses were worried to death with bringing chalk through those vile roads!"  No better did he find the turnpike between Bury and Sudbury, in Suffolk: "I was forced to move as slow in it," he says, " as in any unmended lane in Wales.  For ponds of liquid dirt, and a scattering of loose flints just sufficient to lame every horse that moves near them, with the addition of cutting vile grips across the road under the pretence of letting the water off, but without effect, altogether render at least twelve out of these sixteen miles as infamous a turnpike as ever was beheld."  Between Tetsworth and Oxford he found the so-called turnpike abounding in loose stones as large as one's head, full of holes, deep ruts, and withal so narrow that with great difficulty he got his chaise out of the way of the Witney waggons.  " Barbarous " and " execrable " are the words which he constantly employs in speaking of the roads; parish and turnpike, all seemed to be alike bad.  From Gloucester to Newnham, a distance of twelve miles, he found a " cursed road," " infamously stony," with " ruts all the way."  From Newnham to Chepstow

---

[1] Arthur Young's 'Six Weeks' Tour through the Southern Counties of England and Wales.' 2nd Ed., 1769, pp. 88-9.

he notes another bad feature in the roads, and that is the perpetual hills; "for," he says, "you will form a clear idea of them if you suppose the country to represent the roofs of houses joined, and the road to run across them." Passing still further west, the unfortunate traveller, who seems scarcely able to find words to express his sufferings, continues :—

"But, my dear Sir, what am I to say of the roads in this country! the turnpikes! as they have the assurance to call them and the hardiness to make one pay for? From Chepstow to the half-way house between Newport and Cardiff they continue mere rocky lanes, full of hugeous stones as big as one's horse, and abominable holes. The first six miles from Newport they were so detestable, and without either direction-posts or milestones, that I could not well persuade myself I was on the turnpike, but had mistook the road, and therefore asked every one I met, who answered me, to my astonishment, ' Ya-as! ' Whatever business carries you into this country, avoid it, at least till they have good roads: if they were good, travelling would be very pleasant." [1]

At a subsequent period Arthur Young visited the northern counties; but his account of the roads in that quarter is not more satisfactory. Between Richmond and Darlington he found them like to " dislocate his bones," being broken in many places into deep holes, and almost impassable; " yet," says he, " the people will drink tea! "—a decoction against the use of which the traveller is found constantly declaiming. The roads in Lancashire made him almost frantic, and he gasps for words to express his rage. Of the road between Proud Preston and Wigan he says: " I know not in the whole range of language terms sufficiently expressive to describe this infernal road. Let me most seriously caution all travellers who may accidentally propose to travel this terrible country, to avoid it as they would the devil; for a thousand to one they break their necks or their limbs

---

[1] ' Six Weeks' Tour in the Southern Counties of England and Wales,' pp. 153-5.

by overthrows or breakings-down.   They will here meet with ruts, which I actually measured *four feet deep*, and floating with mud only from a wet summer.   What, therefore, must it be after a winter ?   The only mending it receives is tumbling in some loose stones, which serve no other purpose than jolting a carriage in the most intolerable manner.   These are not merely opinions, but facts ; for I actually passed *three carts broken down* in those eighteen miles of execrable memory." [1]

It would even appear that the bad state of the roads in the Midland counties about the same time had nearly caused the death of the heir to the throne.   On the 2nd of September, 1789, the Prince of Wales left Wentworth Hall, where he had been on a visit to Earl Fitzwilliam, and took the road for London in his carriage. When about two miles from Newark the Prince's coach was overturned by a cart in a narrow part of the road ; it rolled down a slope, turning over three times, and landed at the bottom, shivered to pieces.   Fortunately, the Prince escaped with only a few bruises and a sprain ; but the incident had no effect in stirring up the local authorities to make any improvement in the road, which remained in the same wretched state until a comparatively recent period.

We may briefly refer to the several stages of improvement—if improvement it could be called—in the most frequented highways of the kingdom, and to the action of the legislature with reference to the extension of turnpikes.   The trade and industry of the country had been steadily improving ; but the greatest obstacle to their further progress was always felt to be the disgraceful state of the roads.   As long ago as the year 1663 an Act was passed [2] authorising the first toll-gates or turnpikes to be erected, at which collectors were stationed to levy small sums from those using the road, for the

---

[1] 'A Six Months' Tour through the North of England,' vol. iv., p. 431.
[2] Act 15 Car. II., c. 1.

purpose of defraying the needful expenses of their maintenance. This Act, however, only applied to a portion of the Great North Road between London and York, and authorised the new toll-bars to be erected at Wade's Mill in Hertfordshire, at Caxton in Cambridge-shire, and at Stilton in Huntingdonshire.[1] This Act was not followed by any others for a quarter of a century, and even after that lapse of time the Acts passed of a similar character were few and far between. For nearly a century more, travellers from Edinburgh to London met with no turnpikes until within about 110 miles of the metropolis. North of that point there was only a narrow causeway fit for pack-horses, flanked with clay sloughs on either side. It is, however, stated that the Duke of Cumberland and the Earl of Albemarle, when on their way to Scotland in pursuit of the rebels in 1746, did contrive to reach Durham in a coach and six; but there the roads were found so wretched that they were under the necessity of taking to horse, and Mr. George Bowes, the county member, made His Royal Highness a present of his nag to enable him to proceed on his journey.

The rebellion of 1745 gave a great impulse to the construction of roads for military as well as civil purposes. The nimble Highlanders, without baggage or waggons, had been able to cross the border and pene-trate almost to the centre of England before any definite knowledge of their proceedings had reached the rest of

---

[1] The preamble of the Act recites that " The ancient highway and post-road leading from London to York, and so into Scotland, and likewise from London into Lincolnshire, lieth for many miles in the counties of Hertford, Cambridge, and Hunting-don, in many of which places the road, by reason of the great and many loads which are weekly drawn in waggons through the said places, as well as by reason of the great trade of barley and malt that cometh to Ware, and so is conveyed by water to the city of London, as well as other car-riages, both from the north parts as also from the city of Norwich, St. Edmondsbury, and the town of Cam-bridge, to London, is very ruinous and become almost impassable, inso-much that it is become very dangerous to all his Majesty's liege people that pass that way," &c.

the kingdom. In the metropolis itself little information could be obtained of the movements of the rebel army for several days after they had left Edinburgh. Light of foot, they outstripped the cavalry and artillery of the royal army, which were delayed at all points by impassable roads. No sooner was the rebellion put down than Government directed its attention to the best means of securing the permanent subordination of the Highlands, and with this object the construction of good highways was declared to be indispensable. The expediency of opening up the communication between the capital and the principal towns of Scotland was also generally admitted; and from that time, though slowly, the construction of the main high routes between north and south made steady progress. The extension of the turnpike system, however, encountered violent opposition from the people, being regarded as a grievous tax upon their freedom of movement from place to place. Armed bodies of men assembled to destroy the turnpikes; and they burnt down the toll-houses and blew up the posts with gunpowder. The resistance was the greatest in Yorkshire, along the line of the Great North Road towards Scotland, though riots also took place in Somersetshire and Gloucestershire, and even in the immediate neighbourhood of London. At Selby, in Yorkshire, the public bellman summoned the inhabitants one May morning to assemble with their hatchets and axes at midnight, to cut down the turnpikes erected by Act of Parliament; and they were not slow to accept his invitation. Soldiers were then sent into the district to protect the tollbars and the tolltakers; but this was a difficult matter, for the tollgates were numerous, and wherever a "pike" was left unprotected for a night, it was found destroyed in the morning. The Yeadon and Otley mobs, near Leeds, were especially violent. On the 18th of June, 1753, they made quite a raid upon the turnpikes, burning or destroying about a dozen of them in one week.   A score of the rioters were apprehended, and while on their way to York Castle a rescue was attempted, when the soldiers were under the necessity of firing, and many persons were killed and wounded.

The prejudices entertained against the turnpikes were so strong, that in some places the country people would not even use the improved roads after they were made.[1] For instance, the driver of the Marlborough coach obstinately refused to use the New Bath road, but stuck to the old waggon-track, called " Ramsbury." He was an old man, he said : his grandfather and father had driven the aforesaid way before him, and he would continue in the old track till death.[2]   Petitions were also presented to Parliament against the extension of turnpikes; but here the opposition was of a much less honest character than that of the misguided and prejudiced country folks. The agriculturists in the neighbourhood of the metropolis, having secured the advantages which the turnpike-roads first constructed had conferred upon them, desired to retain a monopoly of their improved means of communication.   They alleged that if turnpike-roads were extended into the remoter counties, the greater cheapness of labour there would enable the distant farmers to sell their grass and corn cheaper in the London market than themselves, and that thus they would be ruined.[3]   This opposition, however, did not prevent the progress of turnpike and highway legislation; and we find that, from 1760 to 1774, no fewer than four hundred and fifty-two Acts were passed for making and repairing highways.   Nevertheless the roads of the kingdom long continued in a very unsatisfactory state, chiefly

---

[1] The Blandford waggoner said, " Roads had but one object—for waggon-driving.   He required but four-foot width in a lane, and all the rest might go to the devil." He added, " The gentry ought to stay at home, and be d——d, and not run gossiping up and down the country."—Roberts's ' Social History of the Southern Counties.'

[2] ' Gentleman's Magazine' for December, 1752.

[3] Adam Smith's ' Wealth of Nations,' book i., chap. xi., part i.

arising from the extremely imperfect manner in which they were made.

Road-making as a profession was as yet unknown. Deviations were made in the old roads to make them more easy and straight; but the deep ruts were merely filled up with any materials that lay nearest at hand, and stones taken from the quarry, instead of being broken and laid on carefully to a proper depth, were tumbled down and roughly spread, the country road-maker trusting to the operation of cart-wheels and waggons to crush them into a proper shape. Men of eminence as engineers—and there were very few such at the time—considered road-making beneath their consideration; and it was even thought singular that, in 1768, the distinguished Smeaton should have condescended to make a road across the valley of the Trent, between Markham and Newark.

The making of the new roads was thus left to such persons as might choose to take up the trade, special skill not being thought at all necessary on the part of a road-maker. It is only in this way that we can account for the remarkable fact, that the first extensive maker of roads who pursued it as a business, was not an engineer, not even a mechanic, but a Blind Man, bred to no trade, and possessing no experience whatever in the arts of surveying or bridge-building, yet a man possessed of extraordinary natural gifts, and unquestionably most successful as a road-maker. We allude to John Metcalf, commonly known as "Blind Jack of Knaresborough," to whose biography, as the constructor of nearly two hundred miles of capital roads—as, indeed, the first great English road-maker—we propose to devote our next chapter.

## CHAPTER V.

### John Metcalf, Road Maker.

JOHN METCALF was born at Knaresborough in 1717, the son of poor working people. When only six years old he was seized with virulent small-pox, which totally destroyed his sight. The blind boy, when sufficiently recovered to go abroad, first learnt to grope from door to door along the walls on either side of his parents' dwelling. In about six months he was able to feel his way to the end of the street and back without a guide, and in three years he could go on a message to any part of the town. He grew strong and healthy, and longed to join in the sports of boys of his age. He went bird-nesting with them, and climbed the trees while the boys below directed him to the nests, receiving his share of the eggs and young birds. Thus he shortly became an expert climber, and could mount with ease any tree that he was able to grasp. He rambled into the lanes and fields alone, and soon knew every foot of the ground for miles round Knaresborough. He next learnt to ride, delighting above all things in a gallop. He contrived to keep a dog and coursed hares: indeed, the boy was the marvel of the neighbourhood. His unrestrainable activity, his acuteness of sense, his shrewdness, and his cleverness, astonished everybody.

The boy's confidence in himself was such, that though blind, he was ready to undertake almost any adventure. Among his other arts he learnt to swim in the Nidd, and became so expert that on one occasion he saved the lives of three of his companions. Once, when two men were drowned in a deep part of the river, Metcalf was

sent for to dive for them, which he did, and brought up
one of the bodies at the fourth diving : the other had
been carried down the stream.  He thus also saved a
manufacturer's yarn, a large quantity of which had been
carried by a sudden flood into a deep hole under the

METCALF'S BIRTHPLACE, KNARESBOROUGH.[1]

[By E. M. Wimperis, after a Sketch by T. Sutcliffe, Leeds.]

[1] We give Mr. Sutcliffe's account
of his search after Blind Jack's birthplace, and of the origin of the above
illustration, in his own words :—" It
was market-day in Knaresborough,"
he says, " and my first act on arriving
was to go into the market and put a
stop to all other business by assem-
bling the old country people and
holding a public conference as to
where Jack was born.  Although
many of them had seen him, none of
them knew where he was born or
where he had died ; but they assured
me I could ' larn all aboot him fra th'
toon-crier, wha was an auld man and

High Bridge. At home, in the evenings, he learnt to play the fiddle, and became so skilled on the instrument, that he was shortly able to earn money by playing dance music at country parties. At Christmas time he played waits, and during the Harrogate season he played to the assemblies at the Queen's Head and the Green Dragon.

On one occasion, towards dusk, he acted as guide to a belated gentleman along the difficult road from York to Harrogate. The road was then full of windings and turnings, and in many places it was no better than a track across unenclosed moors. Metcalf brought the gentleman safe to his inn, the Granby, late at night, and was invited to join in a tankard of negus. On Metcalf leaving the room, the gentleman observed to the landlord—" I think, landlord, my guide must have drunk a great deal of spirits since we came here." " Why so, Sir ? " " Well, I judge so, from the appearance of *his eyes.*" " Eyes ! bless you, Sir," rejoined the landlord, " don't you know that he is *blind ?*" " Blind ! What do you mean by that ? " " I mean, Sir, that he cannot see—he is as blind as a stone." " Well, landlord," said the gentleman, " this is really too much : call him in." Enters Metcalf. " My friend, are you *really* blind ? "

---

knew ivverybody.' To the town-crier I went, but could learn nothing satisfactory from him. I then walked up and down the streets, looking into the doors and windows, and wherever I saw an old head I walked into the house, and asked the possessor if he or she knew Blind Jack. I was at last directed to the Rev. Thomas Collins, an aged gentleman, to whose father's house Jack had been accustomed to go regularly to play the fiddle. Mr. Collins did not know his exact birthplace, but directed me to seek out a certain Mr. Calvert, aged ninety-five, the author of ' The History of Knaresborough,' and from this gentleman I learned that Jack's birthplace was pulled down to make room for the house in which the Rev. Mr. Collins himself resided. He also described to me Jack's house, its character and extent; from which I could form a very accurate idea of it. As Jack's garden and croft came down as far as the churchyard, my next move was to the top of the church-tower, from which place I sketched, designed, and filled up, from the descriptions given me, the view of Blind Jack's birthplace ; and I believe it to be as faithful a representation of the house as it is possible, under the circumstances, to present. The gable-end, at the low right-hand corner, is the grammar-school, which was then only one-story high. A street of houses and cottages must have gone up from thence towards the castle at the top of the hill, though trees might have hid most of them from sight."

"Yes, Sir," said he, "I lost my sight when six years old." "Had I known that, I would not have ventured with you on that road from York for a hundred pounds." "And I, Sir," said Metcalf, "would not have lost my way for a thousand."

Metcalf now thrived and saved money, and he bought and rode a horse of his own! He had a great affection for the animal, and when he called, it would immediately answer him by neighing. The most surprising thing is that he was a good huntsman; and to follow the hounds was one of his greatest pleasures. He was as bold a rider as ever took the field. He trusted much, no doubt, to the sagacity of his horse; but he himself was apparently regardless of danger. The hunting adventures which are related of him, considering his blindness, seem altogether marvellous. He would also run his horse for the petty prizes or plates given at the feasts in the neighbourhood, and he attended the races at York and other places, where he made bets with considerable skill, keeping well in his memory the winning and losing horses. After the races, he would return to Knaresborough late at night, guiding others who but for him could never have made out the way.

On one occasion he rode his horse in a match in Knaresborough Forest. The ground was marked out by posts, including a circle of a mile, and the race was three times round. Great odds were laid against the blind man, because of his supposed inability to keep the course. But his ingenuity was never at fault. He procured a number of dinner-bells from the Harrogate inns and set men to ring them at the several posts. Their sound was enough to direct him during the race, and the blind man came in the winner! After this race was over, a gentleman who owned a notorious runaway horse came up and offered to lay a bet with Metcalf that he could not gallop the horse fifty yards and stop it within two hundred. Metcalf accepted the bet, with the condition that he might choose his ground.   This was agreed to, but there was to be neither hedge nor wall in the distance.   Metcalf forthwith proceeded to the neighbourhood of the large bog near the Harrogate Old Spa, and having placed a person on the line in which he proposed to ride, who was to sing a song to guide him by its sound, he mounted and rode straight into the bog, where he had the horse effectually stopped within the stipulated two hundred yards, stuck up to his saddle-girths in the mire. Metcalf scrambled out and claimed his wager; but it was with the greatest difficulty that the horse could be extricated.

The blind man also played at bowls very successfully, receiving the odds of a bowl extra for the deficiency of each eye.   He had thus three bowls for the other's one; and he took care to place one friend at the jack and another midway, who, keeping up a constant discourse with him, enabled him readily to judge of the distance. In athletic sports, such as wrestling and boxing, he was also a great adept; and being now a full-grown man, of great strength and robustness, about six feet two in height, few durst try upon him the practical jokes which cowardly persons are sometimes disposed to play upon the blind.

Notwithstanding his mischievous tricks and youthful wildness, there must have been something exceedingly winning about the man, possessed of a strong, daring, manly, and affectionate nature; and we are not, therefore, surprised to learn that the daughter of the landlord of the Granby fairly fell in love with Blind Jack and married him, much to the disgust of her relatives.   When asked how it was that she could marry such a man, her woman-like reply was, "Because I could not be happy without him : his actions are so singular, and his spirit so manly and enterprising, that I could not help loving him."   But, after all, Dolly was not so far wrong in her choice as her parents thought her.   As the result proved,

Metcalf had in him elements of success in life, which, even according to the world's estimate, made him eventually a very "good match," and the woman's clear sight in this case stood her in good stead.

But before this marriage was consummated, Metcalf had wandered far and "seen" a good deal of the world, as he termed it. He travelled on horseback to Whitby, and from thence he sailed for London, taking with him his fiddle, by the aid of which he continued to earn enough to maintain himself for several weeks in the metropolis. Returning to Whitby, he sailed from thence to Newcastle to "see" some friends there, whom he had known at Harrogate while visiting that watering-place. He was welcomed by many families and spent an agreeable month, afterwards visiting Sunderland, still supporting himself by his violin playing. Then he returned to Whitby for his horse, and rode homeward alone to Knaresborough by Pickering, Malton, and York, over very bad roads, the greater part of which he had never before travelled upon, yet without once missing his way. When he arrived at York it was the dead of night, and he found the city gates at Middlethorp shut. They were of strong planks, with iron spikes fixed on the top; but throwing his horse's bridle-rein over one of the spikes, he climbed up, and by the help of a corner of the wall that joined the gates, he got safely over; then opening them from the inside, he led his horse through.

After another season at Harrogate he made a second visit to London, in the company of a North countryman who played the small pipes. He was kindly entertained by Colonel Liddell, of Ravensworth Castle, who gave him a general invitation to his house. During this visit, which was in 1730-1, Metcalf ranged freely over the metropolis and visited Maidenhead and Reading, returning by Windsor and Hampton Court. The Harrogate season being at hand, he prepared to proceed thither—Colonel Liddell, who was also about setting out for

Harrogate, offering him a seat behind his coach. Metcalf thanked him, but declined the offer, observing that he could, with great ease, walk as far in a day as he, the Colonel, was likely to travel in his carriage; besides, he preferred the walking. That a blind man should undertake to walk a distance of two hundred miles over an unknown road, in the same time that it took a gentleman to perform the same distance in his coach, dragged by post-horses, seems almost incredible; yet Metcalf actually arrived at Harrogate before the Colonel, and that without hurrying by the way. The circumstance is at once accounted for by the deplorable state of the roads, which made travelling by foot on the whole considerably more expeditious than travelling by coach. The story is even extant of a man with a wooden leg being once offered a lift upon a stage coach; but he declined, with "Thank'ee, I can't wait; I'm in a hurry." And he stumped on, ahead of the stage-coach.

The account of Metcalf's journey on foot from London to Harrogate is not without a special bearing on our subject, as illustrative of the state of the roads at that time. He started on a Monday morning, about an hour before the Colonel in his carriage, and his suite, which consisted of sixteen servants on horseback. It was arranged that they should sleep that night at Welwyn, in Hertfordshire. Metcalf made his way to Barnet; but a little north of that town, where the road branches off to St. Albans, he took the wrong way, and thus made a considerable détour. Nevertheless he arrived at Welwyn first, to the surprise of the Colonel. Next morning he set off as before, and reached Biggleswade; but there he found the river swollen and no bridge provided to enable travellers to cross to the further side. He made a considerable circuit, in the hope of finding some method of crossing the stream, and was so fortunate as to fall in with a fellow wayfarer, who led the way across some planks, Metcalf following the sound of his feet. Arrived at the other side, Metcalf, taking some pence from his pocket, said, "Here, my good fellow, take that and get a pint of beer." The stranger declined, saying he was welcome to his services. Metcalf, however, pressed upon his guide the small reward, when the other asked, "Pray, can you see very well?" "Not remarkably well," said Metcalf. "My friend," said the stranger, "I do not mean to tithe you : I am the rector of this parish; so God bless you, and I wish you a good journey." Metcalf set forward again with the blessing, and reached his journey's end safely, again before the Colonel. On the Saturday after thus setting out from London, the travellers reached Wetherby, where Colonel Liddell desired to rest until the Monday; but Metcalf proceeded on to Harrogate, thus completing the journey in six days—the other arriving two days later.

He now renewed his musical performances at Harrogate, and was also in considerable request at the Ripon assemblies, which were attended by most of the families of distinction in that neighbourhood. When the season at Harrogate was over, he retired to Knaresborough with his young wife, and having purchased an old house, he had it pulled down and another built on its site—he himself getting the requisite stones for the masonry out of the bed of the adjoining river. The uncertainty of the income derived from musical performances led him to think of following some more settled pursuit, now that he had a wife to maintain as well as himself. He accordingly set up a four-wheeled and a one-horse chaise for the public accommodation—Harrogate up to that time having been without any vehicle for hire. The innkeepers of the town soon followed his example, on which he gave up the trade and took to fish-dealing. He bought fish at the coast, which he conveyed on horseback to Leeds and other towns for sale. He continued very indefatigable at this business for some time, being on the road often for nights together; but he was at length forced to abandon it in consequence of the inadequacy of the returns. He was therefore under the necessity of again taking up his violin; and he was employed as a musician, in the Long Room at Harrogate, at the time of the outbreak of the Rebellion of 1745.

The news of the rout of the Royal army at Preston-pans, and the intended march of the Highlanders south-wards, put a stop to business as well as pleasure, and caused a general consternation throughout the northern counties. The great bulk of the people were, however, comparatively indifferent to the measures of defence which were adopted; and but for the energy displayed by the country gentlemen in raising forces in support of the established government, the Stuarts might again have been seated on the throne of Britain. Among the county gentlemen of York who distinguished themselves on the occasion was William Thornton, Esq., of Thorn-ville Royal. The county having voted ninety thousand pounds for raising, clothing, and maintaining a body of four thousand men, Mr. Thornton proposed, at a public meeting held at York, that they should be embodied with the regulars and march with the King's forces to meet the Pretender in the field. This proposal was, however, over-ruled, the majority of the meeting resolving that the men should be retained at home for purposes merely of local defence. On this decision being come to, Mr. Thornton determined to raise a company of volunteers at his own expense, and to join the Royal army with such force as he could muster. He then went abroad among his tenantry and servants, and endeavoured to induce them to follow him, but altogether without success.

Still determined on raising his company, Mr. Thornton next cast about him for other means; and who should he think of in his emergency but Blind Jack! Metcalf had often played to his family at Christmas time, and Thornton knew him to be one of the most popular men in the neighbourhood. Accordingly he proceeded forthwith to

Knaresborough : it was about the beginning of October, only a fortnight after the battle of Prestonpans, and, sending for Jack to his inn, Mr. Thornton told him of the state of affairs—that the French were coming to join the Scotch rebels—and that if the country were allowed to fall into their hands, no man's wife, daughter, nor sister would be safe.  Jack's loyalty was at once kindled. If no one else would join Mr. Thornton, he would! Thus enlisted—perhaps carried away by his love of daring adventure not less than by his feeling of patriotic duty—Metcalf immediately proceeded to enlist others, and in two days a hundred and forty men were obtained, from whom Mr. Thornton drafted sixty-four, the intended number of his company.  The men were immediately drilled and brought into a state of as much efficiency as was practicable in the time ; and when they marched off to join General Wade's army at Boroughbridge, the Captain said to them on setting out, " My lads! you are going to form part of a ring-fence to the finest estate in the world! "   Blind Jack played a march at the head of the company, dressed in blue and buff, and in a gold-laced hat.  The Captain said he would willingly give a hundred guineas for only one eye to put in Jack's head —he was such a useful, spirited, handy fellow.

On arriving at Newcastle, Captain Thornton's company was united to Pulteney's regiment, one of the weakest. The army lay for a week in tents on the Moor.  Winter had set in, and the snow lay thick on the ground ; but intelligence arriving that Prince Charles, with his High-landers, was proceeding southwards by way of Carlisle, General Wade gave orders for the immediate advance of the army on Hexham, in the hope of intercepting them by that route.  They set out on their march amidst hail and snow, and in addition to the obstruction caused by the weather, they had to overcome the difficulties occasioned by the badness of the roads.  The men were often three or four hours in marching a mile, the pioneers having to fill up ditches and clear away many obstructions in making a practicable passage for the artillery and baggage. The army was only able to reach Ovingham, a distance of little more than ten miles, after fifteen hours' marching. The night was bitter cold, the ground was frozen so hard that but few of the tent-pins could be driven, and the men lay down upon the earth amongst their straw. Metcalf, to keep up the spirits of his company —for sleep was next to impossible—took out his fiddle and played lively tunes whilst the men danced round the straw, which they set on fire.

Next day the army marched to Hexham; but the rebels having already passed southward, General Wade retraced his steps to Newcastle to gain the high road leading to Yorkshire, whither he marched in all haste; and for a time his army lay before Leeds on fields now covered with streets, some of which still bear the names of Wade-lane, Camp-road, and Camp-field, in consequence of the visit. On the retreat of Prince Charles from Derby, General Wade again proceeded to Newcastle, whilst the Duke of Cumberland hung upon the rear of the rebels along their line of retreat by Penrith and Carlisle. Wade's army proceeded by forced marches into Scotland, and at length came up with the Highlanders at Falkirk. Metcalf continued with Captain Thornton and his company throughout all these marchings and countermarchings, determined to be of service to his master if he could, and at all events to see the end of the campaign. At the battle of Falkirk he played his company to the field; but it was a grossly-mismanaged battle on the part of the Royalist General, and the result was a total defeat. Twenty of Thornton's men were made prisoners, with the lieutenant and ensign: Captain Thornton himself only escaped by taking refuge in a poor woman's house in the town of Falkirk, where he lay hid for many days; Metcalf returning to Edinburgh with the rest of the defeated army.

Some of the Dragoon officers, hearing of Jack's escape, sent for him to head-quarters at Holyrood, to question him about his Captain. One of them took occasion to speak ironically of Thornton's men, and asked Metcalf how *he* had contrived to escape. " Oh ! " said Jack, " I found it easy to follow the sound of the Dragoons' horses —they made such a clatter over the stones when flying from the Highlandmen." Another asked him how he, a blind man, durst venture upon such a service ; to which Metcalf replied, that had he possessed a pair of good eyes, perhaps he would not have come there to risk the loss of them by gunpowder. No more questions were asked, and Jack withdrew ; but he was not satisfied about the disappearance of Captain Thornton, and he determined on going back to Falkirk, within the enemy's lines, to get news of him, and perhaps to rescue him, if that were still possible.

The rest of the company were very much disheartened at the loss of their officers and so many of their comrades, and wished Metcalf to furnish them with the means of returning home. But he would not hear of such a thing, and strongly encouraged them to remain until, at all events, he had got news of the Captain. He then set out for Prince Charles's camp. On reaching the outposts of the English army, he was urged by the officer in command to lay aside his project, which would certainly cost him his life. But Metcalf was not now to be dissuaded, and he was permitted to proceed, which he did in company with one of the rebel spies, pretending that he wished to be engaged as a musician in the Prince's army. A woman whom they met returning to Edinburgh from the field of Falkirk, laden with plunder, gave Metcalf a token to her husband, who was Lord George Murray's cook, and this secured him an access to the Prince's quarters ; but, notwithstanding a most diligent search, he could hear nothing of his master. Unfortunately for him, a person who had seen him at Harrogate, pointed him out as a suspicious character, and he was seized and put in confinement for three days, after which he was tried by court martial; but as nothing could be alleged against him, he was acquitted, and shortly after made his escape from the rebel camp. On reaching Edinburgh, very much to his delight, he found Captain Thornton had safely arrived there before him.

On the 30th of January, 1746, the Duke of Cumberland reached Edinburgh, and put himself at the head of the Royal army, which proceeded northward in pursuit of the Highlanders. At Aberdeen, where the Duke gave a ball, Metcalf was found the only musician in camp who could play country dances, and he played to the company, standing on a chair, for eight hours—the Duke several times, as he passed him, shouting out "Thornton, play up!" Next morning the Duke sent him a present of two guineas; but as his Captain would allow him to receive no money while in his pay, Metcalf spent the gift, with the Captain's permission, in giving a treat to the Duke's two body servants. The battle of Culloden, so disastrous to the poor Highlanders, shortly after occurred; on which Captain Thornton, Metcalf, and the Yorkshire Volunteer Company, proceeded homewards. Metcalf's young wife had been in great fears for the safety of her blind, fearless, and almost reckless partner; but she received him with open arms, and his spirit of adventure being now considerably allayed, he determined to settle quietly down to the steady pursuits of business.

During his stay in Aberdeen, Metcalf had made himself familiar with the articles of clothing manufactured at that place, and he came to the conclusion that a profitable business might be carried on in buying them on the spot and selling them by retail to customers in Yorkshire. He accordingly proceeded to Aberdeen in the following spring, and bought a considerable stock of cotton and worsted stockings, which he found he could

readily dispose of on his return home.  His knowledge
of horseflesh—in which he was, of course, mainly guided
by his acute sense of feeling—also proved highly serviceable to him, and he bought considerable numbers of
horses in Yorkshire for sale in Scotland, bringing back
galloways in return.  It is supposed that at the same
time he carried on a profitable contraband trade in tea
and such like articles.  After this he began a new line
of business, that of common carrier between York and
Knaresborough, plying the first stage-waggon on that
road.  He made the journey twice a week in summer
and once a week in winter.  He also undertook the conveyance of army baggage, most other owners of carts
at that time being afraid of soldiers, regarding them as
a wild rough set, with whom it was dangerous to have
any dealings.  But Metcalf knew them better, and whilst
he drove a profitable trade in carrying their baggage
from town to town, they never did him any harm.  By
these means, he very shortly succeeded in realising a
considerable store of savings, besides being able to maintain his family in respectability and comfort.

Metcalf, however, had not yet entered upon the main
business of his life.  The reader will already have
observed how strong of heart and resolute of purpose he
was.  During his adventurous career he had acquired
a more than ordinary share of experience of the world.
Stone blind as he had been from his childhood, he had
not been able to study books, but he had carefully studied
men.  He could read characters with wonderful quickness, rapidly taking stock, as he called it, of those with
whom he came in contact.  In his youth, as we have
seen, he could follow the hounds on horse or on foot,
and managed to be in at the death with the most expert
riders.  His travels about the country as a guide to those
who could see, as a musician, soldier, chapman, fish-
dealer, horse-dealer, and waggoner, had given him a
perfectly familiar acquaintance with the northern roads.

He could measure timber or hay in the stack, and rapidly reduce their contents to feet and inches after a mental process of his own. Withal he was endowed with an extraordinary activity and spirit of enterprise, which, had his sight been spared him, would probably have rendered him one of the most extraordinary men of his age. As it was, Metcalf now became one of the greatest of its road-makers and bridge-builders.

About the year 1765 an Act was passed empowering a turnpike-road to be constructed between Harrogate and Boroughbridge. The business of contractor had not yet come into existence, nor was the art of road-making much understood; and in a remote country place such as Knaresborough the surveyor had some difficulty in finding persons capable of executing the necessary work. The shrewd Metcalf discerned in the proposed enterprise the first of a series of public roads of a similar kind throughout the northern counties, for none knew better than he did how great was the need of them. He determined, therefore, to enter upon this new line of business, and offered to Mr. Ostler, the master surveyor, to construct three miles of the proposed road between Minskip and Fearnsby. Ostler knew the man well, and having the greatest confidence in his abilities, he let him the contract. Metcalf sold his stage-waggons and his interest in the carrying business between York and Knaresborough, and at once proceeded with his new undertaking. The materials for metaling the road were to be obtained from one gravel-pit for the whole length, and he made his arrangements on a large scale accordingly, hauling out the ballast with unusual expedition and economy, at the same time proceeding with the formation of the road at all points; by which means he was enabled the first to complete his contract, to the entire satisfaction of the surveyor and trustees.

This was only the first of a vast number of similar projects on which Metcalf was afterwards engaged,

extending over a period of more than thirty years.  By
the time that he had finished the road, the building of
a bridge at Boroughbridge was advertised, and Metcalf
sent in his tender with many others.   At the same time
he frankly stated that, though he wished to undertake
the work, he had not before executed anything of the
kind.   His tender being on the whole the most favourable, the trustees sent for Metcalf, and on his appearing
before them, they asked him what he knew of a bridge.
He replied that he could readily describe his plan of the
one they proposed to build, if they would be good enough
to write down his figures.   " The span of the arch,
18 feet," said he, " being a semi-circle, makes 27 : the
arch-stones must be a foot deep, which, if multiplied
by 27, will be 486 ; and the basis will be 72 feet more.
This for the arch ; but it will require good backing, for
which purpose there are proper stones in the Old Roman
wall at Aldborough, which may be used for the purpose,
if you please to give directions to that effect."   It is
doubtful whether the trustees were able to follow his
rapid calculations ; but they were so much struck by his
readiness and apparently complete knowledge of the
work he proposed to execute, that they gave him the
contract to build the bridge ; and he completed it within
the stipulated time in a satisfactory and workmanlike
manner.

He next agreed to make the mile and a half of turnpike-road between his native town of Knaresborough and
Harrogate—ground with which he was more than ordinarily familiar.   Walking one day over a portion of
the ground over which the road was to be made, whilst
still covered with grass, he told the workmen that he
thought it differed from the ground adjoining it, and he
directed them to try for stone or gravel underneath ;
and, strange to say, not many feet down, the men came
upon the stones of an old Roman causeway, from which
he obtained much valuable material for the making of

his new road.   At another part of the contract there was
a bog to be crossed, and the surveyor thought it impossible to make a road over it.   Metcalf assured him that
he could readily accomplish it; on which the other
offered, if he succeeded, to pay him for the straight road
the price that he would have to pay if the road were
constructed round the bog.   Metcalf set to work accordingly, and had a large quantity of furze and ling laid
upon the bog, over which he spread layers of gravel.
The plan answered effectually, and when the materials
had become consolidated, it proved one of the best parts
of the road.

It would be tedious to describe in detail the construction of the various roads and bridges which Metcalf
subsequently executed, but a brief summary of the more
important will suffice.   In Yorkshire, he made the roads
between Harrogate and Harewood Bridge; between
Chapeltown and Leeds; between Broughton and Ad-
dingham; between Mill Bridge and Halifax; between
Wakefield and Dewsbury; between Wakefield and Doncaster; between Wakefield, Huddersfield, and Saddle-
worth (the Manchester road); between Standish and
Thurston Clough; between Huddersfield and Highmoor;
between Huddersfield and Halifax, and between Knaresborough and Wetherby.   In Lancashire also he made a
large extent of roads, which were of the greatest importance in opening up the communications of that county.
Previous to their construction, almost the only means
of communication between districts were by horse-tracks
and mill-roads, of sufficient width to enable a laden horse
to pass along them with a pack of goods or a sack of corn
slung across its back.   Metcalf's principal roads in Lancashire were those constructed by him between Bury
and Blackburn, with a branch to Accrington; between
Bury and Haslingden; and between Haslingden and
Accrington, with a branch to Blackburn.   He also made
some highly important main roads connecting Yorkshire

and Lancashire with each other at many parts: as, for instance, those between Skipton, Colne, and Burnley; and between Docklane Head and Ashton-under-Lyne. The roads from Ashton to Stockport and from Stockport to Mottram Langdale were also his work. He was, besides, extensively employed in the same way in the counties of Cheshire and Derby; constructing the roads between Macclesfield and Chapel-le-Frith; between Whaley and Buxton; between Congleton and the Red Bull (entering Staffordshire), and in various other directions. The total mileage of turnpike-roads thus constructed by him was about one hundred and eighty miles, for which he received in all about sixty-five thousand pounds. The making of these roads also involved the building of many bridges, retaining-walls, and culverts. We believe it was generally admitted of the works constructed by Metcalf that they well stood the test of time and use; and, with a degree of justifiable pride, he was afterwards accustomed to point to his bridges, when others were tumbling during floods, and boast that none of his had fallen.

This extraordinary man not only made the highways which were designed for him by other surveyors, but himself personally surveyed and laid out many of the most important roads which he constructed, in difficult and mountainous parts of Yorkshire and Lancashire. One who personally knew Metcalf thus wrote of him during his lifetime: "With the assistance only of a long staff, I have several times met this man traversing the roads, ascending steep and rugged heights, exploring valleys and investigating their several extents, forms, and situations, so as to answer his designs in the best manner. The plans which he makes, and the estimates he prepares, are done in a method peculiar to himself, and of which he cannot well convey the meaning to others. His abilities in this respect are, nevertheless, so great that he finds constant employment. Most of

the roads over the Peak in Derbyshire have been altered
by his directions, particularly those in the vicinity of
Buxton ; and he is at this time constructing a new one
betwixt Wilmslow and Congleton, to open a communi-

JOHN METCALF.   [By T. D. Scott.]

cation with the great London road, without being obliged
to pass over the mountains.   I have met this blind projector while engaged in making his survey.   He was
alone as usual, and, amongst other conversation, I made
some inquiries respecting this new road.   It was really
astonishing to hear with what accuracy he described its

course and the nature of the different soils through which it was conducted. Having mentioned to him a boggy piece of ground it passed through, he observed that ' that was the only place he had doubts concerning, and that he was apprehensive they had, contrary to his directions, been too sparing of their materials.' " [1]

Metcalf's skill in constructing his roads over boggy ground was very great; and the following may be cited as an instance. When the high-road from Huddersfield to Manchester was determined on, he agreed to make it at so much a rood, though at that time the line had not been marked out. When this was done, Metcalf, to his dismay, found that the surveyor had laid it out across some deep marshy ground on Pule and Standish Commons. On this he expostulated with the trustees, alleging the much greater expense that he must necessarily incur in carrying out the work after their surveyor's plan. They told him, however, that if he succeeded in making a complete road to their satisfaction, he should not be a loser; but they pointed out that, according to their surveyor's views, it would be requisite for him to dig out the bog until he came to a solid bottom. Metcalf, on making his calculations, found that in that case he would have to dig a trench some nine feet deep and fourteen yards broad on the average, making about two hundred and ninety-four solid yards of bog in every rood, to be excavated and carried away. This, he naturally conceived, would have proved both tedious as well as costly, and, after all, the road would in wet weather have been no better than a broad ditch, and in winter liable to be blocked up with snow. He strongly represented this view to the trustees as well as the surveyor, but they were immoveable. It

---

[1] " Observations on Blindness and on the Employment of the other Senses to supply the Loss of Sight." By Mr. Bew. — 'Memoirs of the Literary and Philosophical Society of Manchester,' vol. i., pp. 172-174. Paper read 17th April, 1782.

was, therefore, necessary for him to surmount the difficulty in some other way, though he remained firm in
his resolution not to adopt the plan proposed by the
surveyor.    After much cogitation he appeared again
before the trustees, and made this proposal to them :
that he should make the road across the marshes after
his own plan, and then, if it should be found not to
answer, he would be at the expense of making it over
again after the surveyor's proposed method.    This was
agreed to ; and as he had undertaken to make nine
miles of the road within ten months, he immediately set
to work with all despatch.

Nearly four hundred men were employed upon the
work at six different points, and their first operation was
to cut a deep ditch along either side of the intended
road, and throw the excavated stuff inwards so as to
raise it to a circular form.    His greatest difficulty was
in getting the stones laid to make the drains, there
being no firm footing for a horse in the more boggy
places.    The Yorkshire clothiers, who passed that way
to Huddersfield market—by no means a soft-spoken
race—ridiculed Metcalf's proceedings, and declared that
he and his men would some day have to be dragged out
of the bog by the hair of their heads !    Undeterred,
however, by sarcasm, he persistently pursued his plan
of making the road practicable for laden vehicles ; but
he strictly enjoined his men for the present to keep his
manner of proceeding a secret.

His plan was this.    He ordered heather and ling to
be pulled from the adjacent ground, and after binding it
together in little round bundles, which could be grasped
with the hand, these bundles were placed close together
in rows in the direction of the line of road, after which
other similar bundles were placed transversely over
them ; and when all had been pressed well down, stone
and gravel were led on in broad-wheeled waggons, and
spread over the bundles, so as to make a firm and level

way.  When the first load was brought and laid on, and
the horses reached the firm ground again in safety, loud
cheers were set up by the persons who had assembled
in the expectation of seeing both horses and waggons
disappear in the bog.  The whole length was finished
in like manner, and it proved one of the best, and even
the driest, parts of the road, standing in very little need
of repair for nearly twelve years after its construction.
The plan adopted by Metcalf, we need scarcely point
out, was precisely similar to that afterwards adopted by
George Stephenson, under like circumstances, when
constructing the railway across Chat Moss.  It consisted
simply in a large extension of the bearing surface,
by which, in fact, the road was made to float upon
the surface of the bog ; and the ingenuity of the expedient proved the practical shrewdness and mother-wit
of the blind Metcalf, as it afterwards illustrated the
promptitude as well as skill of the clear-sighted George
Stephenson.

Metcalf was upwards of seventy years old before he
left off road-making.  He was still hale and hearty,
wonderfully active for so old a man, and always full of
enterprise.  Occupation was absolutely necessary for
his comfort, and even to the last day of his life he could
not bear to be idle.  Whilst engaged in road-making in
Cheshire, he brought his wife to Stockport for a time, and
there she died, after thirty-nine years of happy married
life.  One of Metcalf's daughters became married to a
person engaged in the cotton business at Stockport, and,
as that trade was then very brisk, Metcalf himself commenced it in a small way.  He began with six spinning-
jennies and a carding-engine, to which he afterwards
added looms for weaving calicoes, jeans, and velveteens.
But trade was fickle, and finding that he could not sell
his yarns except at a loss, he made over his jennies
to his son-in-law, and again went on with his road-
making.  The last line which he constructed was one of

the most difficult he had ever undertaken,—that between Haslingden and Accrington, with a branch road to Bury. Numerous canals being under construction at the same time, employment was abundant and wages rose, so that though he honourably fulfilled his contract, and was paid for it the sum of 3500*l*., he found himself a loser of exactly 40*l*. after his two years' labour and anxiety. He completed the road in 1792, when he was seventy-five years of age, after which he retired to his farm at Spofforth, near Wetherby, where for some

METCALF'S HOUSE AT SPOFFORTH.
[By E. M. Wimperis, after a sketch by Thos. Sutcliffe, Leeds.]

years longer he continued to do a little business in his old line, buying and selling hay and standing wood, and superintending the operations of his little farm. During the later years of his career he occupied himself in dictating to an amanuensis an account of the incidents in his remarkable life ;[1] and finally, in the year 1810,

---

[1] 'The Life of John Metcalf.' Dedicated by him "to the Nobility and Gentry usually resorting to Harrogate Spa." Published by Edward Baines. Leeds, 1801. There are many persons still living in the neighbourhood of Knaresborough who well remember Metcalf. When our friend Mr. Sutcliffe visited the place he found many full of their reminiscences of Blind Jack. One person had been a roadmaker under him, and he says Jack would poke about with his long stick, and wherever there was a hollow

this strong-hearted and resolute man—his life's work over—laid down his staff and peacefully departed in the ninety-third year of his age; leaving behind him four children, twenty grand-children, and ninety great grand-children.

The roads constructed by Metcalf and others had the effect of greatly improving the communications of Yorkshire and Lancashire, and opening up those counties to the trade then flowing into them from all directions. But the administration of the highways and turnpikes being entirely local, their good or bad management depending upon the public spirit and enterprise of the gentlemen of the locality, it frequently happened that whilst the roads of one county were exceedingly good, those of the adjoining county were altogether execrable. Thus it was long before road-improvement penetrated the slow-going counties to the south of London. Almost within the memory of the existing generation there was no carriage-road between London and Horsham, but only a horse-track, impassable in winter. Even in the immediate vicinity of the metropolis the Surrey roads remained comparatively unimproved. Mr. Nash, one of the trustees, when examined before a Committee of the House of Commons in 1786, stated that the turnpike-road between Camberwell and the Elephant and Castle

---

place in the road he would find it out, and say " Here, let's ha' some in here!" One old lady told Mr. Sutcliffe that when her mother was first married she was standing at the door, when she saw Jack coming up the street; and, having heard that he could tell when he passed anyone, she remained perfectly still; but Jack had no sooner got near to her than he said, " Good morning, Mrs. Thornton," which very much astonished her, as she had been only married a few days. Shortly after, Jack met her husband, and accosted him with " Ye'v lit of a squary neece woman for a wife, Thornton." One old lady

of eighty-nine, the daughter of the landlord of the head and only inn in Spofforth in Jack's time, told Mr. Sutcliffe that when the "Arle o' Agrimony" [Egremont] first came to see his estate there, he provided Jack with a house rent-free, and ordered that he should have three dinners a-week at the inn on his Lordship's account. The lady in question generally waited on him on those occasions, and he would sometimes say to her, " Here, lass, come and have a game wi' me, an' thoo'l be able to say when thoo's an auld woman at thoo's played cards wi' Blind Jack."

had in several places been hardly passable, and the pitching in the holes and ruts had broken the perches of several carriages.

The roads through the interior of Kent were no better. When Mr. Rennie, the engineer, was engaged in surveying the Weald with a view to the cutting of a canal through it in 1802, he found the country almost destitute of practicable roads, although so near to the metropolis on the one hand and to the sea-coast on the other. The interior of the county was then comparatively untraversed, except by bands of smugglers, who kept the inhabitants in a state of constant terror. Sydney Smith, in reviewing those times, says that before the age of stone-breaking Macadam and of railways, it took him nine hours in travelling the forty miles between Taunton and Bath, during which he suffered between ten and twelve thousand severe contusions, whilst his clothes were rubbed to pieces by being jolted about in the stage-coach basket, which was without springs. "Whatever miseries I suffered," he adds, "there was no post to whisk my complaints for a single penny to the remotest corners of the empire; and yet, in spite of all these privations, I lived on quietly, and am now ashamed that I was not more discontented, and utterly surprised that all these changes and inventions did not occur two centuries ago."

In an agricultural report on the county of Northampton as late as the year 1813,[1] it is stated that the only way of getting along some of the main lines of road in rainy weather was by swimming! "For instance," says the reporter, "between Daventry and Banbury are several unpleasant, if not dangerous fords on the Charwell, which I crossed in July. I was in water for two hundred yards in one and for a considerable distance in others without knowing the bottom, or the road." In

---

[1] 'General View of the Agriculture of the County of Northampton. Drawn up for the consideration of the Board of Agriculture and Internal Improvement.' By William Pitt. London, 1813, p. 231.

the neighbourhood of the city of Lincoln the communications were little better, and there still stands upon what is called Lincoln Heath—though a heath no longer —a curious memorial of the past in the shape of Dunstan Pillar, a column seventy feet high, erected about the middle of last century in the midst of the then dreary, barren waste, for the purpose of serving as a mark to wayfarers by day and a beacon to them by night.[1] At that time the Heath was not only uncultivated, but it was also unprovided with a road across it. When the late Lady Robert Manners visited Lincoln from her residence at Bloxholm, she was accustomed to send forward a groom to examine some track, that on his return he might be able to report one that was practicable. Travellers frequently lost themselves upon this heath. Thus a family, returning from a ball at Lincoln, strayed from the track twice in one night, and they were obliged to remain there till morning. All this is now changed, and Lincoln Heath has become covered with excellent roads and thriving farmsteads. "This Dunstan Pillar," says Mr. Pusey, in his review of the agriculture of Lincolnshire, in 1843, "lighted up no longer time ago for so singular a purpose, did appear to me a striking witness of the spirit of industry which, in our own days, has reared the thriving homesteads around it, and spread a mantle of teeming vegetation to its very base. And it was certainly surprising to discover at once the finest farming I had ever seen and the only land lighthouse ever raised. Now that the pillar has ceased to cheer the wayfarer, it may serve as a beacon to encourage other landowners in converting their dreary moors into similar scenes of thriving industry."[2]

---

[1] The pillar was erected by Squire Dashwood in 1751; the lantern on its summit was regularly lighted till 1788, and occasionally till 1808, when it was thrown down and never replaced. The Earl of Buckingham afterwards mounted a statue of George III. on the top. When the King heard of it, he exclaimed, "What —what! Lincolnshire? All flats, fogs and fens—Eh, Eh!"

[2] 'Essay on the Agricultural Improvements of Lincolnshire.' By Philip Pusey, M.P.—'Journal of the Agricultural Society of England, 1843.'

In all this series of improvements, roads have played an important part. They have brought the population nearer to each other in all respects, and thoroughly opened up the national resources. The several stages of improvement by which the pack-horse has been superseded by the waggon, the waggon by the coach, and both by the railway train—in the course of which Bradshaw the highwayman has been succeeded by Bradshaw the railway guide printer—mark the steps of a silent revolution which has affected not only the physical but the moral condition of the entire population of the empire.

LAND LIGHTHOUSE ON LINCOLN HEATH.

[By Percival Skelton.]

ANCIENT BRITISH BRIDGE ON DARTMOOR.

[By Percival Skelton, after his original Drawing.]

# BRIDGES, HARBOURS,

### AND

# FERRIES.

---

## CHAPTER I.

### OLD BRIDGES.

IN a country such as Britain, full of running streams, bridges form an essential part of every system of roads connecting the various districts of the kingdom with each other. The west wind is constantly scattering showers over its undulating surface, the surplus waters finding their way to the sea along the valleys extending in all directions from the central high land districts; so that it is impossible to travel any distance in England without having to cross many rivers and rivulets, which, though easily fordable in summer, become impassable torrents in winter.

So long as the population was scanty and the intercourse between different parts of the country of a limited character, the necessity for bridges, by which the continuity of the tracks was preserved, was probably little felt. The shallow and broad parts of rivers, provided with a gravelly bottom, were naturally selected as the places for fords, which could be easily waded by men or horses when the water was low; and even in the worst case, when the waters were out, they could be crossed by swimming. Towns and villages sprang up at these fordable places, along the main lines of communication, the names of many of which survive to this day

and indicate their origin.    Thus, along the old line of
road between London and Dover, there was first Deep
Ford, now Deptford, at the crossing of the Ravens-
bourne—next Crayford on the river Cray—Dartford on
the Darent—and Aylesford on the Medway, upon the
line of the pilgrim's road between the west of England
and Becket's shrine at Canterbury.    In all other directions round London it was the same.    Thus, eastward,
there was Stratford[1] on the Lea, Romford on the Bourne,
and Chelmsford on the Chelmer.    Westward were Brentford and Twyford on the Brent, Watford on the Colne,
and Oxenford or Oxford on the Isis.[2]    And along the
line of the Great North Road, crossing as it did the large
streams descending from the high lands of the centre of
England towards the North Sea, the fords were very
numerous.    At Hertford the Maran was crossed, at
Bedford the Ouse, at Stamford the Welland, and so on
through the northern counties of England.

As population and travelling increased, the expedient
of the Bridge was adopted, to enable rivers of moderate
width to be crossed dryshod.    An uprooted tree thrown
across a narrow stream was probably the first bridge;
and he would be considered a potent man in his way
who laid down a couple of such trees, fixed upon them
a cross-planking, and so enabled foot-passengers and
pack-horses to cross from one bank to the other.
But these loose timber structures were very apt to be
swept away by the rains of autumn, and thus the continuous track would again become completely broken.
In a rough district, where rocky streams with rugged
banks had to be crossed, such interruptions must necessarily have led to considerable inconvenience, and
hence arose the idea of tying the rocky gorges together

---

[1] There are numerous bridges in
England at places called Stratford or
Stretford—literally the ford on the
street or road, the ford being afterwards superseded by the bridge.

[2] Oxenford was the spot at which
the Thames, then called the Isis, was
most easily fordable for cattle.—H.
Brandreth, Esq., in 'Archæologia,'
vol. xxvii., 97.

by means of stone bridges of a more solid and permanent character.

The first of such bridges in Britain were probably those erected across the streams of Dartmoor. The rivers of that district are rapid and turbulent in winter, and come sweeping down from the hills with great fury. The deep gorges worn by them in the rocks amidst which they run, prevented their being forded in the usual way; and the ordinary expedient of bridging the gaps in the track by means of felled trees thrown across, was found impracticable in a district where no trees grew. But there was an abundance of granite blocks, which not only afforded the means of forming solid piers, but were also of sufficient size to be laid in a tabular form from one pier to another, so as to constitute a solid enough road for horsemen and foot-passengers. Hence the Egyptian-looking Cyclopean bridges of Dartmoor—a series of structures most probably coeval with the building of Stonehenge, and of the greatest possible interest. One of the largest of these bridges is that crossing the East Dart, near Post Bridge, on the road between Moreton and Tavistock, of which we have above given a representation. Though the structure is rude, it is yet of a most durable character, otherwise it could not have withstood the fury of the Dart for full twenty centuries, as it most probably has done. The bridge is of three piers, each consisting of six layers of granite-slabs above the foundation. One of the side piers, by accident or design, has unfortunately been displaced, and the tabular slabs originally placed upon it now lie in the bottom of the river. Each of the table stones is about fifteen feet long and six feet wide, and the whole structure is held together merely by the weight of the blocks.

There are other more perfect specimens of these Cyclopean bridges in existence on Dartmoor, but none of a size equal to that above delineated. For instance, there is one of three openings, in a very complete state, in the neighbourhood of Sittaford Tor, spanning the North Teign : it is twenty-seven feet long, with a roadway seven feet wide, and, like the others, is entirely formed of granite blocks. There is another over the Cowsic, near Two Bridges, presenting five openings : this bridge is thirty-seven feet long and four feet broad, but it is only about three feet and a half above the surface ; nevertheless it has firmly withstood the moorland torrents of centuries. There is a fourth on the Blackabrook, consisting of a single stone or clam. We believe that no structures resembling these bridges have been found in any other part of Britain, or even in Brittany, so celebrated for its aboriginal remains. The only bridges at all approaching them in character are found in ancient Egypt, to which indeed they bear a striking resemblance.

Although the Romans were great bridge-builders, it is not certain that they erected any arched stone bridges[1] during their occupation of England, though it is probable that they built numerous timber bridges upon stone piers. The most important were those of Rochester, Newcastle, and London. Not many years since, when a railway-bridge was being built across the Medway at Rochester, the workmen came upon the foundations of the ancient structure in a place where no such foundations were looked for, and their solidity caused considerable interruption to the work. So at Newcastle, when the old bridge over the Tyne was taken down in 1771, the foundations of the piers, which were laid on piles of fine black oak, in a perfect state of preservation, were found to be of Roman masonry.

---

[1] Mr. Wright is, however, of opinion that some of the Roman bridges in England had arches; and he says Mr. Roach Smith has pointed out a very fine semi-circular arched bridge over the little river Cock, near its entrance into the Wharfe, about half-a-mile below Tadcaster, on the Roman road leading southward from that town (the ancient Calcaria), which he considered to be Roman. The masonry of this bridge is massive, and remarkably well preserved ; the stones are carefully squared and sharply cut, and in some of them the mason's mark, an R, is distinctly visible. The roadway was very narrow.— 'The Celt, the Roman, and the Saxon,' 2nd Ed., p. 187.

Similar bridges were erected at different points along the lines of the Roman military roads wherever a river had to be crossed; and it is probable that the town of Pontefract (Pons fractus) derived its name from a broken Roman bridge in that neighbourhood, the remains of which were visible in the time of Leland.

The first arched bridge of stone erected in England is said to have been the singular-looking structure still standing in the immediate neighbourhood of Croyland Abbey in the Fens. As the monks were in early times the

CROYLAND BRIDGE.

[From the Topographia Britannica]

principal agriculturists, gardeners, and land-reclaimers, so they were the principal church and bridge-builders. This triangular bridge at Croyland, however, could have been erected for no particularly useful purpose, but rather as a curiosity; and it has been conjectured that it was reared out of the offerings of pilgrims to the shrine of St. Guthlac, the saint of the Fens, as an emblem of the Trinity.[1] The bridge stands on three piers, from each of which springs the segment of a circular arch, all the segments meeting at a point in the centre. It is situated at the junction of the three principal streets of the little town, which was originally built on piles; and along those streets the waters of the Nene, the Welland, and the Catwater respectively, used to flow and meet under the bridge. Carrying out the Trinitarian illustration, each pier of the bridge was said to stand in a different county : one in Lincoln, the second in Cambridge, and the third in Northampton. The road over the bridge is so steep that horses can scarcely cross it, and they usually go under it; indeed the arches underneath are now quite dry. This curious structure is referred to in an ancient charter of the year 943, although the precise date of its erection is unknown. On the south-west wing, facing the London road, is a sitting figure, carved in stone, very much battered about the face by the mischievous boys of the place. The figure has a globe or orb in its hand. It is supposed to be a statue of King Ethelbald, though it is commonly spoken of in the village as Oliver Cromwell holding a penny loaf !

The first road-bridge of which we have any authentic account is that erected at Stratford over the river Lea, several miles to the east of London. The road into Essex by the Old Ford across the Lea is noticed as early as the seventh century, when it appears that the body of St. Erkenwald was stopped there by the flood while being conveyed from the abbey of Barking, where he

---

[1] The famous bridge at Croyland is the greatest curiosity in Britain, if not in Europe. It is of a triangular form, rising from three segments of a circle, and meeting at a point at top. It seems to have been built under the direction of the abbots, rather to excite admiration and furnish a pretence for granting indulgences and collecting money, than for any real use ; for though it stands in a bog, and must have cost a vast sum, yet it is so steep in its ascent and descent that neither carriages nor horsemen can get over it.—' History and Antiquities of Croyland Abbey.' Bibliotheca Topographica Britannica, No. 11.

died, for interment in London. It appears that many
lives were afterwards lost in crossing the Old Ford,
and amongst those who narrowly escaped drowning
was Matilda, Queen-Consort of Henry I. To prevent
this great danger to travellers, the good Queen directed
two bridges to be built over the two branches of the
Lea—one at Bow, the other at Channelsea, connected by
a gravel causeway; and she bequeathed certain manors
and a mill to the abbess of Barking for their maintenance and repair. The bridges were erected some time
between the years 1100, when Matilda became Queen,
and 1118, the year of her death; and they are supposed to have been named " de Arcubus," or the Bows,
because of their arched form. Stowe says, " the bridge
(of Stratford-le-Bow) was arched like a bow; a rare
piece of work, for before that the like had never been
seen in England."

Notwithstanding the ample endowment of the bridges,
and the additions made to it by successive benefactors,
their repairs seem to have been sadly neglected, and the
approaches were often found impassable. The crowns
of the arches became worn into deep ruts, and they
must shortly have fallen in, had not one Hugh Pratt,
who lived in the neighbourhood in the time of King
John, contrived, by begging aid from the passers by,
to keep the structures in repair. His son continued the
practice, and even obtained leave to levy tolls, amongst
which we find the following : " For every cart carrying
corn, wood, coal, &c., one penny; of one carrying tasel,
two pence; and of one carrying a dead Jew, eight
pence." [1] At a still later period, we find collections made
in all the churches throughout the City, for the purpose
of repairing Bow Bridge, as " a work of great necessity for
the passage of victual unto the inhabitants;" and in the

---

[1] Probably the last toll was imposed on the bodies of Jews in progress of removal for interment in a
Jewish burying-ground situated to
the eastward of the bridge.

reign of Elizabeth we find a letter, signed by Burleigh, Lincoln, Sussex, and other Lords of the Privy Council, to the Corporation, acknowledging that such collection had been made of the free-will of the citizens, and was not to be drawn into a precedent for compelling the citizens at any future time to be at the cost of repairing the said bridge.[1]

Bow Bridge was unquestionably a structure of great utility ; but though Stowe describes it as a rare piece of work, it possessed no great merit in an architectural point of view, as will be obvious from the following representation.

OLD BOW BRIDGE.   [From the Archæologia.]

This bridge, like most of the early structures, had large piers, occupying a great part of the waterway, and supporting small and low-arched openings, with high battlements for the enclosure of a roadway of the narrowest possible dimensions. The piers were provided

---

[1] 'Corporation of the City of London Journals,' 21 fol. 58 b. 20th July, 1580.

with large angular projections, not only to divide the
force of the current but to admit of spaces for foot-
passengers to retire into, and thus avoid danger from
carriages and horsemen when passing along the narrow
roadway.   Indeed, its extreme narrowness, notwithstanding the attempts made to widen it, eventually led
to the removal of the bridge, and the substitution of a
new one of a single arch on the same site some twenty
years ago.

The great convenience of bridges gradually led to
their erection along many of the principal routes through
the country.   In the first place they superseded fords;
and when the art of bridge-building had become more
advanced, they superseded ferries—always an inconvenient, and often a dangerous, method of crossing rapid
rivers.   The bridge brought the inhabitants of certain
districts into immediate connection with those on the
opposite bank of the river flowing between them, and
enabled them freely to hold intercourse and exchange
produce with each other; and the public advantages
of this improved means of communication were found
so great as to lead many benevolent and thoughtful
men, in those early days, to bequeath large sums of
money for the purpose of building and maintaining
bridges; in like manner as public benefactors, in after-
times, left money to build and endow churches and hos-
pitals.   Yet popular tradition in some places attributes
these structures to a very different origin.   Thus the
fine old bridge of three arches over the river Lune at
Kirkby Lonsdale, in Westmoreland, is said to have been
the work of the devil.[1]

The religious orders seem early to have taken in

---

[1] How this tradition could have
originated does not appear.   The
bridge is very lofty, and of excellent
workmanship.   It consists of three
semi-circular ribbed arches, the centre
one being much higher than the
others.   The roadway is, however,
inconveniently narrow, like all the
old bridges.   It is evidently of the
Norman period, and the erection of a
very clever architect.

hand the erection and maintenance of bridges, and we owe to the old Churchmen the finest structures of this kind still extant, many others having been superseded by modern works. An order called the Brothers of the Bridge was founded by St. Benezet, the builder of the noble bridge at Avignon early in the thirteenth century; and the brethren spread into England, and went from one work to another, building bridges and chapels thereon,—the provision of a bridge-chantry characterizing nearly all the early structures in this country. Indeed, the architecture of the early bridges in many respects resembled that of the early cathedrals. From the point at which the piers rose above the level of the stream, ribs of stone usually spanned the openings from one pier to the other, precisely similar to the Gothic arching of cathedrals and vaults of chapter-houses; and it is most probable that the bridges and cathedrals were built by the same class of workmen.

One of the finest of such bridges was that erected by Abbot Bernard over the Trent at Burton, until recently the longest in England. It was 1545 feet in length, and consisted of thirty-four arches, built of squared freestone,—a most useful and substantial structure. Another old bridge of the same period is that over the Wensum at Norwich, still called Bishop's Bridge, a sin-gular-looking old building of patched-up stone and flint, erected in 1295. It consists of three arches, inside of which are some grotesque heads and remains of old ornamental work. Fairs used formerly to be held on it at Easter and Whitsuntide, as was the practice on several other old bridges. At Leeds the weekly cloth-market was held on the old bridge at the foot of Briggate, some of the old arches of which are still in existence; the clothiers being summoned to assemble by the ringing of a bell in the old bridge-chapel, when they exposed their cloth for sale on the parapets. But the bridge was so narrow, and the market caused so great an obstruction, that at length a special cloth-hall was built, to which the clothiers removed about the end of last century.

The erection of Wade Bridge over the river Camel, in Cornwall, is an example of the origin of many of these structures in early times. The benevolent vicar of Egloshayle, lamenting the number of lives that were annually lost in crossing the ferry, determined to raise a fund sufficient to build a bridge, and success crowned his efforts. It was erected in 1485, and claimed the distinction, with Burton, of being the longest in England. It consisted of seventeen arches, and was a highly picturesque object, though it has since been replaced by a more convenient structure. The vicar must have been a man of great energy, for it is recorded of him that he designed the bridge and worked diligently upon it until it was finished.[1] At his death he left an endowment of 20*l.* a year towards its maintenance.

Rochester Bridge was an important part of the great highway between London and the Continent, and a Roman timber roadway on stone piers formed part of the ancient Watling-street. The bridge long continued to be of timber, and we find Simon de Montfort burning it down in 1264. Twenty years later, having been repaired in the interval, it was seriously damaged by the breaking up of the ice, the force of which, rushing down the Medway, carried away several of the piers. It was patched up from time to time until the reign of Edward III., when the gallant Sir Robert Knolles, who had raised himself by his valour from the rank of a private soldier to that of a commander in the royal army during the wars in France, returning to England, and determining to leave behind him some

---

[1] It is said he had great difficulty in securing foundations, owing to the sandy nature of the ground, until he had recourse to "packs of wool," which he placed under the piers.

The same tradition was handed down of London Bridge, but the expedient could not possibly have answered the purpose.

useful work by which his name should be held in kindly
remembrance by his countrymen, resolved upon the
erection of an arched stone bridge over the Medway,
and it was accordingly built at his charge and made
over by him to the public. It was completed in the
fifteenth year of the reign of Richard II., and was considered one of the finest bridges at that time in England. It had eleven arches, resting on substantial piers,
the foundations of which were blown up, not many
years since, by the mines sprung by the Royal Engineers,
at a considerable expenditure of gunpowder.[1] A chapel
was afterwards erected by Sir John Cobham at its east
end, where collections were made in the usual manner
for maintaining the structure. But it appears that the
monies thus collected had been insufficient, and the
bridge shortly fell into decay; for about a century after
its erection (in 1489) we find John Morton, Archbishop
of Canterbury, adopting the extraordinary expedient of
publishing a remission from purgatory for forty days, of
all manner of fines, to such persons as should give anything towards the repairs, as the bridge had by that
time become very much broken.

Bishop-Auckland Bridge over the Wear, and Newcastle
Bridge over the Tyne, were similar structures, maintained
by the voluntary offerings collected by the priests who
ministered in the chantries. The chapel was invariably
dedicated to some patron saint. That on old London
Bridge was dedicated to St. Thomas, on Bow Bridge
to St. Catherine, and others were dedicated to St.
Nicholas, the patron saint of sailors. Those chapels
were exceedingly picturesque objects, and were often
highly decorated. They were erected over one of the
piers, about the centre of the bridge, elongated for the

---

[1] The foundations seem to have
been obtained in the then usual manner, by throwing loose rubble and
chalk into the river, and surrounding
the several heaps with huge starlings,
which occupied a very large part of
the water-way, and consequently presented a serious obstruction to the
navigation of the Medway.

purpose ; and a brother stood at the door to receive the
offerings of the passers-by towards the repairs of the
bridge and the support of the services in the chantry.
Nearly all these old bridge-chapels have perished, but a
beautiful specimen has happily been preserved in the
chantry on Wakefield Bridge, of which the following is
a representation.

WAKEFIELD BRIDGE AND CHANTRY.

[By E. M. Wimperis, after an original Drawing by T. Sutcliffe, New Water Colour Society ]

This bridge is supposed to have been built by Edward
Duke of York, afterwards Edward IV., in memory of
his father and followers who fell at the battle of Wakefield, in the wars of the Roses ; and it is said to have
been richly endowed, that prayers might be offered up
there for the souls of the slain, and especially of poor little
Rutland. However this may be, the bridge chantry at
Wakefield, which has recently been renovated in excellent taste, is one of the most beautiful and interesting
of these ancient structures. The entrance to the chapel
is directly from the roadway, and it stands upon an elongated pier obviously erected for the purpose, and forming

part of the original structure. The bridge itself has undergone many changes, in order to adapt it to the improved modes of travelling. When chaises, stage-coaches, and waggons came into general use, the old erections were found altogether inadequate for the traffic. They were very narrow,[1] and often very steep; and though they had been well enough adapted for the foot-passenger, the horseman, and the pack-horse convoy, many of them did not admit of sufficient width for the convenient passage of wheeled vehicles. The picturesque gateways at the ends of old bridges—such as existed over the Monnou at Monmouth and over the Ouse at York, and a specimen of which still exists at Raglan Castle, as shown in the annexed cut—were also found to be a

RAGLAN CASTLE BRIDGE.

great obstacle to stage-coach travelling, as the arched gateways did not admit of the passage of a coach without danger to the outside passengers; and where it was not found practicable to turn the thoroughfare another way, they were at once demo-lished. The bridges them-selves were widened and enlarged; and though in many cases, as at Wake-field, the old piers were included in the new work, the original picturesque cha-racter of the bridge was in a great measure lost.

Notwithstanding the increased necessity for such

[1] De Quincey, in his 'Autobiographic Sketches,' says he has known of a case, even in the nineteenth century, where a post-chaise of the common narrow dimensions was obliged to retrace its route for four-teen miles on coming to a bridge in Cumberland built in some remote age when as yet post-chaises were neither known nor anticipated, and, unfor-tunately, too narrow by three or four inches to enable the vehicle to pass.

structures, the art of bridge-building seems to have fallen into decay until about the middle of last century; and whilst many of the erections of the Brothers of the Bridge continued to stand firm on their foundations, as they had done for centuries, the bridges of more modern construction were liable to be swept away by the first winter's flood. The only mode of securing foundations was the clumsy one of throwing loose stones promiscuously into the bed of the river, so as to find their own bearing, and then, on the top of these loose stones, to erect the stonework of the starlings. The piers were built up on the foundations thus rudely formed; but they were constantly liable, as may be readily imagined, to be unsettled, undermined, and carried away by any unusual pressure of water. No architect of eminence devoted himself to bridge-building; and although Inigo Jones furnished the design for the bridge of Llanrwst, over the Conway in Wales, in 1634, it was a work of a comparatively unimportant character, and the only one of the kind on which he seems to have been employed. In the plan of this bridge the pointed arch is no longer adopted, but three segmental arches, the middle of which is of the span of fifty-eight feet. The roadway approached a horizontal line, and was of a sufficient breadth to accommodate carriage traffic. On the whole, the design was of a very modern character, and was probably adopted, to a considerable extent, as a model by succeeding bridge-builders. The work, however, seems to have been so badly done, that it was shortly after found necessary to rebuild one of the arches; and to this day the bridge is known as "the shaking bridge," [1] not standing by any means firmly on its foundations. The people of the locality

---

[1] A tourist in North Wales says:— "While standing on the bridge, admiring the beautiful scenery, two or three men came and asked me in broken English 'whether I would like to have a shake.' On inquiry I found that the bridge will strongly vibrate by a person striking his back forcibly against the parapet of the centre arch." — Parry's 'Cambrian Mirror,' p. 134.

consider this a merit, as it certainly is a curiosity, and attribute the shaking of the bridge to the "very nice principles on which it is built." But that the bridge should shake or rock could have formed no part of Inigo Jones's design, and that it stands at all must be attributable mainly to the fact of its foundation being upon a rock, which cannot be undermined and washed away.

INIGO JONES'S BRIDGE AT LLANRWST.

[By E. M. Wimperis.]

# CHAPTER II.

## Old London Bridge.

THE erection of the old bridge across the Thames at London was the most formidable enterprise of the kind undertaken in England during the middle ages. It was a work of great difficulty and magnitude, in consequence of the rapid rise and fall of the tides in the river, but it was one of essential importance as connecting the fertile districts lying to the south of the Thames directly with the population of the metropolis.

As in all similar cases, the ferry (where the river could not be forded) preceded the bridge. The Romans first established a trajectus on the Thames, thus connecting their station in London with their military road to Dover. After the Romans the Saxons continued it, and the name of one of the masters of the ferry has descended to us in a tradition of a singular character.[1] This was

---

[1] The tradition is, that John Overy rented the ferry of the City, and what with hard work, great gains, and penurious living, he became exceedingly rich. His daughter Mary, beautiful and of a pious disposition, was sought in marriage by a young gallant, who was rather more ambitious of being the ferryman's heir than his son-in-law. It is related that the ferryman, in one of his fits of usury, formed a scheme of feigning himself dead for twenty-four hours, in the expectation that his servants would, out of propriety, fast until after his funeral. He was laid out as dead accordingly, his daughter consenting to the plan, against her better nature. The servants, instead of fasting, as the ferryman had anticipated, broke open the larder and fell to banqueting, until the dead man could bear it no longer, but rose up in his sheet to rate them. At this, one of the ferrymen, thinking it was the devil who stood before them, seized the butt-end of a broken oar and brained John Overy on the spot. Mary Overy's gallant, hearing of the news, rode up to town in all haste from the country; but, his horse stumbling, he was thrown, and "brake his neck." On which, Mary Overy is said to have founded the church which still bears her name, and made over her possessions to the college of priests which became there established. Whatever the truth of this tradition may be, the probability is that John Overy is merely another way of spell-

John Overy, the father of the foundress of St. Mary's church in Southwark.   The property in the ferry, with its revenues, having become the possession of the adjoining college of priests of St. Mary's, they determined on the bold enterprise of erecting a bridge of timber across the river.   The first mention of this structure is contained in the laws of Ethelred, where the tolls of vessels coming to Billingsgate *ad pontem* are fixed and defined:   William of Malmesbury states that, in 994, Sweyn, the Danish king, when sailing up the river to the attack of London, ran foul of the bridge with his ships, and in the contest which subsequently ensued between the Londoners on the north and the Danes on the south of the river, the bridge was destroyed.   It seems, however, to have been repaired by the time that Canute sailed up the Thames with his fleet several years later; for, finding the bridge to be an obstacle in his way, he adopted the bold expedient of cutting a wide ditch or canal from near Dockhead, at Redriff, through the marshes on the south side of the river, westward to the lower end of Chelsea Reach, through which he drew his ships and completed the blockade of the city.   Not long after, in 1091, the timber-bridge was entirely swept away by a flood; but the provision of so great a convenience was found indispensable, and William Rufus levied a heavy tax for its rebuilding.   Again, in 1097, a new timber-bridge rose upon the ruins of the old one; but fifty years later we find it destroyed by a fire which broke out in a tenement

---

ing John of the Ferry, and that the church of St. Mary Overy was originally St. Mary of the Ferry—ferries, like bridges, being invariably placed under the protection of some patron saint, that at London being dedicated to St. Mary.   The odd abbreviations to which old names have become subject are sufficiently illustrated by the adjoining St. Olave's Street, now become Tooley Street, and by another street in the City, originally called Saint Guthurum

Lane, now shrunk into Gutter Lane. But probably the largest abbreviation has been that effected in the name of Saint Bartholomew's Town, in Lincolnshire, which afterwards became known as Botolph's Town, and has finally shrunk into Bo'ston!   There are various towns called Overy at other old ferries in England, the name doubtless originating in similar circumstances; for instance, Burnham Overy, in Norfolk.

near London Stone, and burnt all down eastward as far as Aldgate, and from thence to the south bank of the river, including the bridge. It was again patched; but it was found so costly to maintain the structure whilst of wood, and it ran so much risk from fire and floods, that it was eventually determined to build a bridge of stone upon nearly the same site; and the work was accordingly begun by one Peter, the chaplain of St. Mary's, Cole-church, in the Poultry, in the year 1176.

One of the most important considerations in building a bridge across a deep and rapid river is the security of its foundations. Comparatively few of the older bridges failed from the unskilful construction of their arches, but many were undermined and carried away by floods where the piers were insecure. The period at which Old London Bridge was built is so remote, and the records left of the mode of conducting the work are so meagre, that it is impossible, even were it desirable, to give any detailed account of the building. Some writers have supposed that the whole course of the river was diverted in the line of Canute's canal above referred to, and that the bed of the Thames was thus laid dry to enable the foundations of the piers to be got in.[1] This ex-pedient has frequently been adopted in building bridges across streams of moderate size; but it is not probable that it was employed in this case. When the founda-tions of the old bridge were taken up, it was ascer-tained that strong elm piles had been driven deep into the bed of the river as closely as possible, over which long planks, strongly bolted, were placed, and on these (great stones having been thrown into the interior spaces) the bases of the piers were laid, the lowermost bedded in pitch, whilst outside of all was placed the pile-work,

---

[1] Stowe was of this opinion. See his 'Survey.' See also Dr. Wallis to Pepys, Oct. 24th, 1699, 'Pepys's Diary,' v. 375. For much antiqua-rian information on this and all other points relating to the structure, see Thompson's 'Chronicles of Old Lon-don Bridge,' a singularly curious book.

called starlings, for the purpose of breaking the rush of the water and protecting the foundation piles.

Another statement was long current—that London Bridge was built on wool-packs,—arising probably from the circumstance that a tax was levied by the King upon wool, skins, and leather, towards defraying the cost of its construction. The bridge was in a measure regarded as a national work, and for more than two centuries after its erection, tribute continued to be levied upon the inhabitants of the counties nearest the metropolis for its maintenance and repair. Liberal gifts and donations were also made with the same object, until at length the Bridge Estates yielded a large annual income.

OLD LONDON BRIDGE, 1559.

Not less than thirty-three years were occupied in the erection of this important structure. It was begun in the reign of Henry II., carried on during that of Richard I., and finished in the eleventh year of King John, 1209. Before then, however, the aged priest, its architect, died, and he was buried in the crypt of the chapel which had by that time been erected over the centre pier. At his death another priest, a Frenchman, called Isenbert, who had displayed much skill in constructing the bridges at Saintes and Rochelle, was recommended by the King as his successor. But his appointment was not confirmed by the Mayor and citizens of London, who deputed three of their own body to superintend the finishing of the work,—the chief difficulties connected with which had indeed already been surmounted.

The bridge, when finished, was a remarkable and curious work. That it possessed the elements of stability and strength was sufficiently proved by the fact that upon it the traffic of London was safely borne across the river for more than six hundred years. But it was an unsightly mass of masonry, so far as the bridge was concerned; although the overhanging buildings extending along both sides of the roadway, the chapel on the centre pier, and the adjoining drawbridge, served to give it an exceedingly picturesque appearance. One of the houses adjoining the drawbridge was dignified with the name of Nonsuch House : it was said to have been constructed in Holland and brought over in pieces, when it was set up without mortar or iron, being held together solely by wooden pegs.

The piers of the bridge were so close, and the arches so low, that at high water they resembled a long low series of culverts hardly deserving the name of arches. The piers were of various dimensions, in some cases almost as thick as the spans of the arches which they supported were wide. The structure might be compared to a very strong stone embankment built across the river, perforated by a number of small openings, through which the water rushed with tremendous force as the tide was rising or falling, the power thus produced being at a later period economised and employed in some of the arches to work water-engines. The bridge had not less than twenty arches, including the drawbridge, some of them being too narrow to admit of the passage of boats of any kind. This great obstruction of the stream, at a point where the river is about the narrowest, had the effect of producing a series of cataracts at the rise and fall of each tide, so that what was called " the roar of the bridge" was heard a long way off. The feat of "shooting the bridge" was in those days attended with considerable danger, and lives were frequently lost in the attempt. Hence prudent passengers, who took a boat for down river, usually landed above the bridge and walked to the nearest wharf below, where they again embarked. The more venturesome risked "shooting the bridge," and thus boats were often swamped and their passengers drowned. In 1428 John Mowbray, second Duke of Norfolk, when passing under one of the arches, ran his boat upon the pile-work, and had very nearly perished; but leaping on to one of the starlings, he was then hauled up to the bridge by ropes let down to him for the purpose. The risk attending this operation of shooting the bridge explains the old proverb, that " London Bridge was made for wise men to go over and fools to go under."

Perhaps the most singular features of the old bridge were its upper platform, consisting of two rows of houses with a narrow roadway between, the chapel and draw-bridge, and the turreted battlements at either end. The length of the roadway was 926 feet, and from end to end it was enclosed by the lofty timber-houses, which were held together by arches crossing overhead from one range to the other and thus keeping the whole in posi-tion. The street was narrow, dark, and dangerous. There were only three openings along it on either side, provided with balustrades, from which a view of the river and its shipping might be obtained, as well as of the rear of the houses themselves, which overhung the parapets and completely hid the arches from sight. On the centre pier was the chapel with its tower, and at the ends of the bridge were the gate-houses, on which the grim heads of traitors and unfortunate partisans were stuck upon poles until a comparatively recent period. Hentzner, a German traveller, counted above thirty heads displayed upon them as late as the year 1598.

The drawbridge was another curious feature. It occupied the fourteenth arch from the north end, and provided an opening of about thirty feet. It was used for purposes of defence as well as to provide for the passage of masted ships. When Jack Cade was told of the army marching against him, Shakespeare makes him say, " Let's go fight with them ; but first go and set London Bridge on fire." But Cade's project having failed, his head was taken off and placed upon a pole, amongst those of other traitors, over the southern gatehouse, with his face looking towards Kent. The bridge was also used as a place of public punishment. Persons found guilty of practising witchcraft were compelled to do penance there. No less a personage than Eleanor Cobham, Duchess of Gloucester, was exposed upon the bridge in 1440, for the alleged crime of witchcraft.

OLD LONDON BRIDGE, 1650.

[By E. M. Wimperis, after the Painting by Claude de Jongh.]

The bridge had a long history and many vicissitudes. It had scarcely been completed ere the timber-houses upon it were consumed by a great fire, and the bridge was thus at once stripped of its cumbrous load. But, as the revenues required for its maintenance and repair were in a great measure derived from the rental of the houses, which let for high sums, they were shortly after erected in even more cumbersome forms than before, and were for a long time principally inhabited by pin and needlemakers.

At a very early period the bridge showed signs of weakness and required constant patching. Before the end of its first century a patent was issued by Edward I., authorising its speedy repair, in order to prevent its sudden fall and "the destruction of innumerable people dwelling thereon." Tolls were authorised to be taken—for every man crossing, a farthing; for every horseman, a penny; and for every pack carried on a horse, one half-penny. There was not a word of vehicles, which did not as yet exist. The repairs then done to the structure do not seem to have been of much effect; for in 1281 five of the arches, with the buildings over them, were carried away by a flood following a thaw, and the repairs had to be begun again on a more extensive scale than before. At a subsequent period Stowe's gate, tower, and arches, at the Southwark side, also fell into the river. But after repeated patching, the bridge nevertheless continued to hang together for several centuries longer. It witnessed the processions of priests, the jousting of knights, the march of Kentish rebels, the triumphal march of Henry V. into the City after the battle of Agincourt, the funeral procession of the same monarch when borne to his royal tomb in Westminster Abbey, and the entrance to the metropolis of his successor after being crowned King of France at Notre Dame. Generation after generation of toiling men and women passed over the bridge, wearing its tracks deep with their feet, and sometimes moistening them with their tears. Still the old bridge stood on, almost down to our own day; for we shall find, in the lives of Smeaton and Rennie, that these eminent engineers, amongst others, were called upon from time to time to direct its repair; until at last the old structure, which had served its purpose so long, was condemned and taken down, and the magnificent New London Bridge erected in its stead.

It was long before any second bridge was built over the Thames near London.  The advantages derived from the current of traffic passing through the City from a district extending for fifty or sixty miles on either side the river, were felt to be of such importance that the citizens would not lightly part with them.  Bridges were regarded as the best feeders of towns and cities, and wherever one was erected, all the avenues by which it was approached became speedily converted into streets of valuable houses. At the two ends of the Thames Bridge were London and Southwark; at Tyne Bridge, Newcastle and Gateshead; and at the Medway Bridge, Rochester and Strood.  But London was extending westward with such rapid strides, and the population of Westminster as well as Lambeth' had so much increased, that the provision of an additional bridge for those districts, in course of time came to be regarded as a matter of absolute necessity.

A movement with this object seems to have been commenced in the reign of Charles II., but the project was vigorously resisted by the citizens of London.  They waited upon his Majesty in state, and implored him to oppose the measure ; and, on his compliance with their petition, their expression of gratitude towards him was as great as if he had delivered the City from a famine, or a plague, or a great fire, or some such overwhelming calamity.  It is not improbable that the citizens secured his Majesty's support by the offer of money, which he very much wanted at the time ; for we find from the records of the Common Council, of date the 25th October, 1664, that upon advancing, by way of loan, the sum of 100,000l. to Charles II., the citizens took occasion to thank his Majesty in the following terms for preventing the erection of the new bridge at Westminster :—

" And withal to represent unto his Majesty the City's great sense and apprehension of, and most humble thanks for, the great instance of his Majesty's good and favour towards them expressed in preventing of the new bridge proposed to be built over the river of Thames betwixt Lambeth and Westminster, which, as is conceived, would have been of dangerous consequence to the state of this city." [1]

A few years later, in 1671, a similar project was attempted, and a bill was brought into the House of Commons to enable a bridge to be erected over the Thames as far west as Putney. But the Corporation of London were again up in arms, protesting against the establishment of *any* bridge which should enable the traffic to pass from one side of the river to the other without going through the City. The debate on the subject, as recorded by Mr. Grey, is exceedingly curious, read by the light of the present day. Mr. Love declared the opinion of the Lord Mayor to be, " that if carts were to go over the proposed new bridge, London would be destroyed." Sir William Thompson opposed it because it " would make the skirts of London too big for the body," besides producing sands and shelves in the river, and affecting the below-bridge navigation, which would cause the ships to lie as low down as Woolwich ; whilst Mr. Boscawen opposed the bill, because, if conceded, there might be a claim set up for even a third bridge, at Lambeth, or some other point.[2] The bill was thrown out on these grounds by a majority of 67 to 54 ; and for nearly a hundred years more, London had no second bridge, notwithstanding that the old structure was so narrow that there was not room for two carts to pass each other ! Since that time, however, ten bridges have been thrown across the river between Putney and the City, and London is not yet destroyed,—indeed, the cry is still for more bridges.

The second bridge was built in 1738-50, nearly opposite the palace of Westminster. During the many cen-

---

[1] 'City of London Records,' jor. 45, 423.
[2] 'Debates of the House of Com-mons, from the year 1667 to 1694. Collected by the Hon. A. Grey. London, 1769.

turies that had elapsed since old London Bridge had been
erected, the science of bridge-building had made but
little progress in England.   The principal structures of
the sort were of wood.   Trees, merely squared, were
laid side by side, at right angles with the stream, supported on perpendicular piles, the roadway being planked
over and covered with gravel.   Old Battersea Bridge
was an example of the primitive structures by means
of which many of our wide rivers long continued to be
crossed.   Few were built of stone, and these, of a comparatively rude kind, were principally situated upon the
main lines of road ; but they were usually liable to be
swept away by the first heavy flood.   During the period
referred to, however, the science of construction had
made great progress in France, and from the practice of
French engineers our best models continued for some
time longer to be drawn.   Hence, when the sanction of
Parliament was at length obtained to a second bridge
being built across the Thames, Labelye, the French engineer, a native of Switzerland, was employed to design
and execute the work.

It will have been observed that the chief difficulty
with the early bridge-builders was in securing proper
foundations for their piers.   A common practice was to
sink baskets of small dimensions, full of stones, in the bed
of the river, and on these, when raised above water, the
foundations were laid.   But where the bottom was composed of loose, shifting material, such as sand, it will be
obvious that a firm basis could scarcely be secured by
such a method.   The plan adopted by Labelye, though
considered an improvement at the time, was even inferior
to the method employed by Peter of Colchurch in founding the piers of old London Bridge in the 13th century.   For, clumsy though the latter structure was, it
stood more than six hundred years, whilst Westminster
Bridge had not been erected a century before it exhibited
signs of giving way.

Labelye's method of founding his piers was as follows. He had a sufficient number of large caissons, or water-tight chests, prepared on shore, of such form as to fit close alongside of each other. They were then floated on rafts over the spots destined for the piers, where they were permanently sunk. The top of each caisson, when sunk, being above high-water mark, the masonry was commenced within it, and carried up to a level with the stream, when the timber sides were removed and the pier was left resting firmly on the bottom grating. The foundations were then protected by sheet-piling, that is, by a row of timbers driven firmly side by side into the earth all round the piers.

Westminster Bridge was originally intended for a wooden bridge, but the design was subsequently altered to one of stone, Labelye considering it necessary to have a great weight of masonry in order to keep his caissons at the proper level. To add to this weight the engineer added a lofty parapet, which Grosley, a French traveller, gravely asserted was placed there for the purpose of preventing the Londoners from committing suicide!

Not many years after Westminster Bridge had been opened, the London Common Council, in order to facilitate the passage of traffic across the Thames as near to the centre of the City as possible, applied to Parliament for powers to construct a bridge at Blackfriars; and the requisite Act having been passed, the works were commenced in 1760, and finished in 1769. The architect and engineer of Blackfriars Bridge was Robert Mylne, and a noble piece of masonry it was. The principal new feature in this structure was the elliptical arch,[1] which Mr. Mylne

---

[1] The disadvantage of the semicircular arch was that, though self-contained, it necessarily led to a great rise in the road over the bridge, which was steep at both sides. By means of the flat elliptical arch this disadvantage was obviated, and more water-way was afforded, with less rise in the bridge. But greater science was required to construct bridges of this sort, as the strength mainly depended upon the abutments, which bore the lateral pressure. When the span was extensive, and the arches of considerable flatness, the greatest care was also required in the selection of

was the first to introduce in England. The innovation
gave rise to a lively controversy at the time, in which
Dr. Johnson took part, in opposition to Mr. Mylne,
and in support of his friend Gwyn, who was the author
of a rival plan. Boswell, in his 'Life of Johnson,'
defends the design of Mylne, his countryman, and adds,
"it is well known 'that not only has Blackfriars Bridge
never sunk either in its foundation or in its arches, which
were so much the subject of contest, but any injuries
which it has suffered from the effects of severe frosts, have
been already, in some measure, repaired with sounder
stone, and every necessary renewal can be completed
at a moderate expense." This was written in 1791, only
twenty years after the bridge had been opened; and,
though it may have been true then, it is so no longer.
When the numerous heavy piers of old London Bridge
were removed, the velocity of the unimpeded tide, sweeping up and down the river twice in every twenty-
four hours, and the consequent increased scour of the
water along the bottom of the Thames above bridge,
soon began to tell upon the foundations both of Blackfriars and Westminster Bridges; and they exhibited the
unsightly appearance of numerous props and centerings
to prevent the further subsidence of their foundations.
Hence Labelye's bridge at Westminster has already been
removed, and the probability is that before long Mylne's
bridge at Blackfriars will share the same fate.

---

the stone, which must necessarily be
capable of resisting the severest com-
pression. Mylne overcame these various difficulties with great ability; and
had not the foundations of the structure
proved defective, Blackfriars Bridge
might have stood for a thousand years
and more.

## CHAPTER III.

### WILLIAM EDWARDS, BRIDGE BUILDER.

THE difficulties encountered by the early bridge builders cannot be better illustrated than by a brief account of the life of William Edwards, the architect of Pont-y-Prydd, a remarkable work erected at Newbridge, in South Wales, about the middle of last century.

Edwards was born in 1719, in a small farm-house in the parish of Eglwysilan, in Glamorganshire. His father died when William was only two years old; but his mother, who was an industrious, well-doing woman, kept on the farm, and piously and virtuously brought up her family. William's literary culture was confined to Welsh, which he could read and write from his early youth; but as he grew older he also learnt to read and write English, though more imperfectly. He had the character of being a very obstinate, stubborn, and self-willed boy—qualities which, under the guidance of rectitude and integrity, became developed into inflexible courage and resolution in his manhood. Until eighteen years of age he was regarded as a wild, headstrong fellow, with little promise of good in him; but he was gradually tamed and disciplined by hard work, and as he grew older he became thoughtful and sedate even beyond his years.

Edwards's ordinary employment was common farm-work; but at the same time he was a diligent self-educator, taking lessons in arithmetic from a neighbour in the evenings. It happened that, in the ordinary course of affairs, he had occasion to repair the dry stone walls about the farm. He took particular pleasure in this kind of work, and very soon became remarkably handy at it;

but he always longed to do better.  Some masons having come into the neighbourhood to build a smithy, Edwards would occasionally leave his farm-work and take his stand in the field over against which the masons were employed,

WILLIAM EDWARDS.  [By M. Morgan.]

eagerly watching them while they worked.  He admired the way in which they handled their tools and prepared the stones for the building.  One thing that he particularly noted was the way in which they dressed the rough blocks by means of the pointed end of the mason's hammer. He tried to do the same, but failed, his hammer-point not being steeled. He then inquired and ascertained the cause of his failure, and went to a smith and had a steeled point added to his hammer. With this he succeeded in dressing his stones much more neatly and quickly than he had been able to do before.

Practice and application, and the desire to excel, even in dry stone-wall building, inevitably carry a man onward; and Edwards soon became so expert in this sort of work, that he was extensively employed in repairing and building dry stone walls for the neighbouring farmers. His walls were observed to be so neat, so firm, and so serviceable, that he was everywhere in request, and his earnings were regularly added to the common stock of his mother and brothers, who carried on the business of the farm. He began to consider himself fitted for something better than continuing this rough sort of work; and he thought that, instead of being a mere builder of dry stone walls, he might even undertake to become a builder of houses.

An opportunity occurred of erecting a little workshop for a neighbour, and Edwards acquitted himself so well, that he gained much praise for his skill. Thus proving his ability in small things, he was shortly entrusted with the execution of works of greater importance. He had scarcely reached the age of twenty-one when he was employed to build an iron forge at Cardiff, and while carrying on the work he lodged with a blind man, named Rosser, by trade a baker. Rosser knew the English language, which as yet Edwards did not; and, what was more, the blind man could teach it to others. The young mason determined to take lessons of his landlord; and such was his assiduity and perseverance during his leisure hours, that he very shortly contrived to master the new language. When he had completed his contract, which he did to the entire satisfaction of his employers, he regularly entered upon the business of a house-builder on a considerable scale, and very shortly there was no building of any magnitude or importance in the neighbourhood—whether it were a mansion, a mill, or an iron forge—which he was not willing as well as competent to undertake.

During his leisure he took great pleasure in studying the ruins of Caerphilly Castle, near to where he lived.

RUINS OF CAERPHILLY CASTLE.

This castle was once the largest in the kingdom next to Windsor, and its ruins are still of great extent, covering an area of about thirty acres. Its walls are of prodigious thickness, and its leaning tower has stood for centuries, inclining as much as eleven feet out of the perpendicular, held together principally by the strength of its cement. This old castle was the college in which Edwards studied the principles of masonry; and he himself was accustomed to say that he had derived more advantage from wandering about the ruins, observing the methods adopted by the ancient builders, the manner in which they had hewed, dressed, and set their stones, than from all the other instruction he received. It was while employed in erecting a mill in his own parish that he first applied the knowledge he had gained by studying the ruins of Caerphilly, in the construction of an arch. The mill was finished to admiration, and professional builders pronounced Edwards's arch to be an excellent piece of masonry.

Employment now flowed in upon him, and when any work of more than ordinary difficulty was proposed, application was usually made to William Edwards. Hence, in 1746, when it was proposed to throw a bridge over the river Taff, he was employed to build it; and though he was only twenty-seven years old, and had not yet built any bridge, he had the courage at once to undertake the work. The bridge was built of three arches, in a style superior to anything of the kind that had been erected in the neighbourhood; the stones were excellently dressed and closely jointed; the arches were light and elegant, and supposed to be sufficiently substantial for the duty they had to perform; and as a whole the erection was much admired, and greatly added to the fame of its builder. It would appear, however, that Edwards had not sufficiently provided for the passage of the floods, which in certain seasons rush down from the Brecknock Beacon mountains with great impetuosity. Above Newbridge several rivers of considerable capacity, such as the Crue, the Bargold Taff, and the Cynon, besides numberless brooks descending rapidly from the high grounds, contribute to swell the torrent so as to render it almost irresistible. The piers of Edwards's new bridge unfortunately proved a serious obstruction in the way of a heavy flood which swept down the valley about two years and a half after the bridge had been completed. Trees were torn up by the roots and carried down the stream, lodging athwart the piers, where brushwood, haystacks, and field-gates, becoming firmly stuck amongst their branches, choked up the arches and fairly dammed the torrent. The waters rapidly accumulated above the bridge and rose to the parapets; the sides of the valley being steep, left no room for their escape, and the tremendous force finally swept away arches and piers together, carrying the materials far down the river.

This destruction of his first bridge was doubtless a terrible blow to the builder, who was bound in sureties to maintain it for a period of seven years.  But worse even than the loss of his time and labour was the failure of his work, the most distressing of all things to the man who takes a proper pride in his calling.  He resolved, however, to fulfil his contract, and began the building of a second bridge of only one arch, to avoid the defect which had proved the ruin of the first.  This second bridge, without piers, was a much more difficult work than the first, in consequence of the wide span of the arch, which was not less than 140 feet, the segment of a circle of 170 feet in diameter.  No such extensive span had yet been attempted in England; and even on the Continent, where the science of bridge-building was much better understood, the only bridges of larger span were of ancient construction, chiefly Roman.  Michael Angelo's beautiful bridge of the Rialto, at Venice, was the largest span attempted in modern times, and its width was only about 100 feet.  The result of Edwards's daring experiment proved its extreme difficulty.  He succeeded in finishing the arch, but had not added the parapets, when the tremendous pressure of the masonry over the haunches forced them down, the light crown of the bridge sprang up, the key stones were forced out, and a second time the labour of Edwards was lost, and his masonry lay a ruin at the bottom of the river.  Yet not altogether lost: for by failure he learnt experience, dearly bought though it had been.

The undaunted man determined to try again.  Twice he had failed, yet he was not utterly defeated in resources. He would try a new expedient, and he believed he should eventually succeed.  Fortunately his friends believed in him too, for they generously came forward and helped him with the means of building his third bridge, which proved a complete success, and the courage and skill of Edwards were crowned at last.  The plan which he adopted of more equally balancing the work and relieving the severe thrust upon the haunches, was to introduce three cylindrical holes or tunnels in the masonry at that part of the bridge. The same plan is found to have been adopted in some of the ancient bridges, and Perronet, the great French engineer, not only formed such tunnels over the haunches, but occasionally in the piers themselves. Where Edwards gained his information as to the expedient, or whether he had gathered it from his own bitter experience, is not known; but it answered

PONT-Y-PRYDD. [By Percival Skelton.]

his purpose. Three cylindrical holes were built over each haunch—the lowest and outermost nine feet in diameter, the next six feet, and the highest and innermost three feet. The arch, the same in width as that which fell four years before, was finished in 1755, and the beautiful "rainbow bridge" lightly spans the Taff at Newbridge to this day.

The singular inflexibility of purpose displayed by our engineer in grappling with and overcoming the difficulties encountered by him in the erection of his first bridge, became the subject of general interest throughout Wales.   When it was finished and opened for public traffic, and the news spread abroad that the extraordinary arch of Pont-y-Prydd at last stood firm as the rocks on which it rested, strangers flocked from all parts to view it, and the Welsh people, as was natural, became proud of their countryman.   Employment flowed in upon him, and he went on building bridge after bridge in all parts of South Wales.   Among the more important of the later works of Edwards were the large and handsome bridge over the river Usk, at the town of Usk, in Monmouthshire; one of three arches, over the river Tame, near Swansea; another, of one arch of 95 feet span, over the same river near Morriston; a third, with an arch of 80 feet, at Pont-cer-Tame, several miles higher up; and Bettws and Llandovery Bridges, in the county of Caermarthen, the latter of 84 feet span.   He also built Aberavon Bridge, in Glamorganshire, with an arch of 70 feet span; and Glasbury Bridge, of four arches, over the Wye, near Hay, in Brecknockshire, afterwards carried away by one of the floods so common in the district.

Edwards's strong judgment and quick observant faculties, ripened by experience, enabled him to introduce many improvements in his bridges as he grew older.   He flattened his arches, so as to render the passage of vehicles over them more easy than in the case of Pont-y-Prydd, the steepness on either side of which was found to be so great an obstacle that it was afterwards found necessary to supersede its use by a more level bridge erected on modern principles.   Hence his later works presented a considerable improvement in this respect upon his earlier ones; and while he continued to be equally careful in providing ample waterway under the arching, and to erect his bridges with a view to the greatest possible durability, he took increasing pains to provide a more capacious and level roadway over them, and render them in all ways more easy and convenient for public use.

Besides his numerous bridges, Edwards continued, during the remainder of his long life, to erect smelting-houses, forges, and buildings of various kinds for purposes of manufacture. Nor did his building business exclusively occupy his time, for, in addition to his trade or profession as a building engineer, he carried on the business of a farmer until the close of his life. Not even on Sundays did he cease from his labours; but, though the Sabbath was no day of rest for him, his labours then were all labours of love. In 1750 he became an ordained preacher amongst the Independents. Shortly after, he was chosen minister of the congregation to which he belonged, and he continued to hold the office for about forty years, until his death. He occasionally preached in the neighbouring meeting-houses: amongst others, in that of Mr. Rees, the father of Abraham Rees, editor of the well-known 'Encyclopedia.' This meeting-house was one of the numerous buildings erected by Edwards himself. He always preached in Welsh, and his discourses are said to have been simple, sensible, and full of loving-kindness. His fellow-countryman Malkin[1] says of him, that, though a Calvinist, he was one of a very liberal description; indeed, he carried his charity so far that many persons suspected he had changed his opinions, and for that reason spoke very unhandsomely of him. As he grew older he became increasingly charitable and tolerant of other men's views, avoiding points of doctrinal difference, but urging

---

[1] 'The Scenery, Antiquities, and Biography of South Wales.' By Benjamin Heath Malkin, Esq., M.A., F.A.S. 1807. Vol. i., p. 144.

and enforcing that the love of God and of our neighbour is the aim and end of all religion. Holding it to be the duty of every religious society to contribute liberally of their means to the support of their ministry, he regularly took the stipulated salary which his congregation allowed to their preachers, but distributed the whole of it amongst the poorer members of his church, often adding to it largely from his own means. This worthy Christian labourer died at the advanced age of seventy, respected and beloved by men of all parties, and he was buried in the churchyard of his native parish of Eglwysilan, amidst the graves of his children. Three of his sons were, like their father, eminent bridge-builders : David having constructed the fine five-arched bridge over the Usk at Newport, as well as the bridges at Llandilo, Edwinsford, Pontloyrig, Bedwas, and other places. Indeed, William Edwards may be said to have fairly inaugurated the revival of the art of bridge-building in England. After his time, it was taken up by Smeaton, Rennie, and Telford, and its progress will accordingly be found described in connection with the lives and works of those distinguished engineers.

# CHAPTER IV.

## Harbours and Lighthouses.

The maritime greatness of Britain is of as modern a character as its engineering, and has been mainly the creation of the last century. At a time when Spain, Holland, France, Genoa, and Venice were great maritime powers, England was almost without a fleet, the little trade which it carried on with other countries being conducted principally by foreigners. Our best ships were also built abroad by the Venetians or the Danes, but they were mostly of small tonnage, little bigger than modern herring-boats. In 1540 there were only four vessels belonging to the Thames of 120 tons burden.[1] Bristol, then next in importance to London, possessed several large foreign-built ships; but the principal craft belonging to that port were of only from 50 to 100 tons each. In Queen Elizabeth's time the whole shipping of Liverpool was only 223 tons; the largest vessel being of but 40 tons burden.[2] It is, however, astonishing to find what bold and daring things were done by the men who navigated these diminutive vessels. Sir Humphry Gilbert crossed the Atlantic and sailed along the coast of America in the *Squirrel* of only 10 tons. Martin Frobisher set out with two barques of 25 tons each to discover the North-West Passage. Sir Francis Drake's fleet, which left the English shores for the circumnavigation of the globe, consisted of five vessels, the largest

---

[1] So stated by one Wheeler, secretary to the English Company of Merchant Adventurers, as quoted in Macpherson's 'Annals of Commerce,' vol. ii., p. 85.

[2] Wedgwood and Bentley's pamphlet, entitled, 'A View of the Advantages of Inland Navigation.' London, 1765.

of which was not of 100 tons burden.[1]  In the year
1575 there were only one hundred and thirty-five ships
in all England above 100 tons.  The royal navy was
on a par with the mercantile; and at the time when the
Spanish Armada bore down upon the English coast, it
consisted of only twenty-three ships, eight of which were
under 120 tons.  There were only nine of 500 tons and
upwards, the ship of the greatest burden being of 1000
tons, carrying only forty guns.  The principal part of
the fleet which held at bay the Armada until the storms
had scattered it, were coasting-vessels of small burden,
belonging to Lyme, Weymouth, and other ports along
the southern coast.  Of the whole seventy-five vessels
which constituted the squadrons under the Lord Admiral and Sir Francis Drake, not fewer than sixty were
from 400 down to as low as 20 tons.  About the same
period, the small but flourishing republic of Venice possessed a fleet of more than three thousand vessels of
various kinds, carrying upwards of thirty-six thousand
seamen.

The English navy, however, made gradual progress.
In 1613 there were ten vessels of 200 tons belonging
to the port of London.  The suppression of the monopoly of the carrying trade, which had virtually been
enjoyed by the merchants of the Low Countries and
the Hanse Towns of Germany until the year 1552, had
the effect of giving a considerable impetus to English
shipping business; and by the year 1640 we find the
number of English ships and sailors more than trebled.
It would appear that not only had the greater part of the

---

[1] One of the last of Sir Francis
Drake's ships was used, until quite
recently, as a Thames barge.  It was
broken up only a few years ago.  Another interesting little vessel, the *Investigator*, of about 150 tons, used to
lie moored off Somerset House, where
it was used as one of the floating stations of the Thames River Police, but
has since been replaced by the *Royalist*.
The *Investigator* was the vessel in
which Captain Ross made his first
voyage to the Polar Seas in the search
for a North-West Passage.

foreign trade of England been previously conducted in
foreign ships, but even our coasts were fished by foreign
fishermen. The writer of a pamphlet published in 1614,
one Tobias Gentleman, a fisherman and mariner, pointed
out the great amount of wealth yearly taken out of his
Majesty's seas by the Hollanders, whereby they had
grown rich and powerful, possessed of a great fleet, and
were able to dictate terms to the Spaniards; whereas
the English coasting people were poor, idle, and negligent, and constrained even to beg bread of the "plump
Hollanders." Mr. Gentleman was indignant at seeing
the foreigners, whose industry and diligence he nevertheless greatly praised, using our seas as a rich treasury,
and drawing wealth from them as from a gold-mine. Six
hundred Dutch busses, of some six score tons each,
were employed in the herring fishery along the British
coast, from the mouth of the Thames as far north as
Shetland, besides numerous others in the cod-fishery,
protected by some twenty, thirty, and even forty ships
of war to prevent their being pillaged by the Dunkirkers,
who were the chief pirates of those times. That these
Dutchmen should come into our markets and sell us our
own fish, carrying away great quantities of our gold and
silver, whilst English ships lay up a-rotting, was a thing
that Mr. Gentleman thought was not coolly to be borne.
" It is much to be lamented," said he, "though we have
such a plentiful Country and Store of able and idle people,
that not one of His Majesty's Subjects is there to be seen,
all the whole Summer, to fish or to take one Herring : but
only the North-sea Boats of the Sea-coast Towns, that go
to take Cods; they do take so many as they need to bait
their Hooks, and no more. We are daily scorned by
these Hollanders for being so negligent of our Profit and
careless of our Fishing; and they do daily flout us that
be the poor Fishermen of England, to our Faces at sea,
calling to us and saying, *Ya* English, *ya zall or oud scone*

*dragien,* which in English is this : *You* English, *we will make you glad to wear our old shoes.*" [1]   From this curious tract it would appear that much even of our commonest English industry is of modern growth; and that the herring fishery, which it might be supposed was indigenous in England, is as modern as most other branches of employment.   Down to about the end of last century the only fishing was conducted close in shore, the fishermen shooting the nets from their small cobles; and it was not until the year 1787 that the Yarmouth men began the deep-sea herring fishery.[2]

Another remarkable feature of those early times was the piracy which prevailed around the English coasts. The seas were quite as unsafe as the roads, and a system of plundering passing ships was as common as that of robbing mail-coaches.   Sea-roving doubtless ran in the blood of the coast population, themselves the descendants of the pirate Northmen.   There were many daring spirits amongst them, and when a bold leader started up and fitted out a ship to make a dash at Spanish galleons, or a descent on the French coast, he had never a lack of desperadoes to follow him,—thorough-going seamen, equally ready to brave the storm and the battle—to face the hurricanes of the Atlantic in an open boat, or to fight against any odds.   Hence Scaliger, when describing the English of that day, said of them, " They make excellent sailors and pirates,"—" Nulli melius piraticam exercent quam Angli."

We have seen the London merchants [3] and guilds making a common purse to fit out a fleet under Raleigh, sending it to sea to capture Spanish galleons, and afterwards dividing the proceeds of the prizes taken.   Similar ventures were often made, both before and after Raleigh's

---

[1] 'Harleian Miscellany,' vol. iii., 378-90.

[2] ' An Historical Account of the Herring Fishery on the North-East Coast of England ' (small pamphlet). By Dr. Cortis.   Feb., 1858.

[3] See ' Life of Sir Hugh Myddelton,' ante, p. 103.

time. In Richard II.'s reign, one Philpot hired a thousand men and sent them to sea, where they captured fifteen rich Spanish vessels.[1] Harry Page, of Poole, ravaged the coasts of Spain, France, and Flanders, bringing home the plunder of many churches, numerous prisoners, and prizes laden with rich cargoes. But the piratical propensity was not only displayed against our continental neighbours, but by the seagoing population of one town against those of another. In 1342 Yarmouth and Hull sent out a piratical fleet against London and Bristol; and ports as near each other as Lyme and Dartmouth, in the adjoining counties of Dorset and Devon, waged deadly feud and strove to capture each other's vessels.[2] The sailors of the Cinque Ports were at war with those of Yarmouth, and in Edward I.'s reign regular safe conducts were granted to certain Cinque Ports vessels requiring to visit that port, as if it were an enemy's. The Yarmouth men were even at war with those of Lowestoft, Camden relating of them that "they often engaged their neighbours, the Lestoffenses, or men of Lowestoff, in sea-fights, with great slaughter on both sides." Robert de Battayle, of Winchelsea, plundered a passing ship belonging to some merchants of Sherborne; but the feat must have been regarded as creditable, as a few years later his townsmen chose him for their mayor. At the end of the sixteenth century three noted pirates—Hamilton, Twittie, and Purser—ravaged the coast of the south-western counties. In 1582 Purser attacked the ships, both English and French, riding in Weymouth harbour, and carried off a Rochelle ship of sixty tons. But Weymouth itself sent out piratical vessels, which picked up many rich prizes.

Down even to the middle of the seventeenth century piracy was quite common along the Devonshire coast, as

---

[1] Roberts's 'Social History of the Southern Counties of England.'
[2] Ibid., p. 74.

Mr. Roberts shows from the entries in church and corporation records of the time.    The weakness of the Royal
navy is sufficiently obvious from the fact that Turks and
Algerines sailed along the Channel, up the Severn, and
into the Irish Sea, capturing ships; whilst the Dunkirk
pirates assailed with impunity the east coast towns from
Dover to Berwick-upon-Tweed.    The Emperor of Morocco
was even bribed to cease from his piratical expeditions,
and protect British trade; and the bribe continued to be
paid until the year 1690.    When piracy was at length
put a stop to along the English coasts (and Mr. Roberts
avers that sea-robbers were masters of the Channel at
times as late as the reigns of James I. and Charles I.[1]),
the more desperate pirates took service under the Turks,
while many sailed away to the West India Islands and
turned buccaneers.    Hugh Miller, in his autobiography,
speaks of his great grandfather, John Feddes, as "one
of the last of the buccaneers," and states that the house
in which he himself was born "had been built, he had
every reason to believe, with Spanish gold." [2]

Such being the early state of British shipping, there
was very little need of harbours.    The natural inlets
all round the coast, and more particularly the navigable tidal rivers, were found amply sufficient for the
accommodation of the ships of comparatively small burden
by means of which our trade was then carried on.    London
possessed a great advantage in her fine river, the Thames,
up which the natural power of the tide lifted vessels of
the largest burden into the heart of the land, and lowered
others down again to the sea, twice in every twenty-four
hours.    The river served as harbour, dock, and depôt in
one, and provided ample waterway, with abundant quay
accommodation, which served all the purposes of trade
down almost to our own day.

---

[1] Roberts's ' Social History,' p. 68.
[2] Hugh Miller's ' Schools and School-masters.'

Among the early ports, Bristol ranked next in importance to London; it also was provided with a convenient river, the Avon, up which ships were floated by the tide to port. At the siege of Calais, in Edward III.'s time, Bristol furnished almost as many ships and mariners as London; and it went on increasing in importance down to the end of the seventeenth century, at which time Liverpool had scarcely sprung into existence, and was as yet little better than a fishing village. Before the art of engineering had advanced so far as to enable harbour walls to be built in deep water, these tidal rivers sufficiently answered the purpose of harbours. Hence London on the Thames, Bristol on the Avon, Hull on the river Hull, Chester (the principal shipping-port for Ireland) on the Dee, Gloucester on the Severn, Boston on the Witham, and Newcastle on the Tyne. At Bristol the ships lay upon the mud at low water, the course of the river Froom having been turned, in early times, in order to make "a softe and whosy (oozy) harboure for grete shippes;" and the habit of lying on the mud made the Bristol ships so bulge and swell out, that until quite recently "a Bristol hog" could be recognised by the practised sailor's eye far off at sea. Bristol was only provided with floating docks at the beginning of the present century, long after Liverpool had overcome the difficulties of the Mersey and provided for itself a system of docks now considered superior to everything else of the kind in the kingdom.

The ample line of the British coast, broken by innumerable deep water bays and inlets, also afforded considerable convenience for the shipping of early times. The small size of the craft enabled them to be beached with ease, and the utmost that was done in the way of harbour works was to empty large stones roughly into the sea so as to form a breakwater or a pier at the harbour head. But the sea was found a fickle and dangerous neighbour, and those early works were often washed away.   Mr. Roberts gives the rough representation shown in the annexed cut, of the mode of constructing the ancient pier at Lyme Regis, and most probably the same method was pursued elsewhere.

ANCIENT MANNER OF CONSTRUCTING PIERS

The rocks which lay upon the shore were floated over the line of the proposed sea-work by means of casks, and dropped into their places, after which—or, in certain cases, before the stones were sunk—strong oak piles were driven into the ground along either side to hold them together.   Great reliance was placed on timber, and especially upon oak. The Cobb or harbour at Lyme Regis was so successfully put together in this way, that Queen Mary ordered the workmen to be impressed and forwarded to Dover, to execute a similar work for the protection of the harbour at that place.   They were next employed at Hastings, where they reared a pier of huge rocks edgeways without timber.   But the seas of the ensuing winter completely overthrew the structure; and again, in 1597, the workmen erected another pier, using much timber in cross-dogs, bars, and braces.   The work was thirty feet high, " bewtyfull to behold, huge, invariable, and unremoveable in the judgment of all beholders;" but on the next All Saints day a storm upon a spring tide scattered the whole,[1] and to this day Hastings is without a pier.

Among the numerous fine natural harbours on the south coast were those of Portsmouth, Plymouth, Weymouth, Falmouth, and Dartmouth, all situated at the mouths of rivers or bays, as their names indicate.   None of them had piers until a comparatively recent date, the only landing-places at Portsmouth and Southampton being on " the Hard."   The Cinque Ports, on the coast of

[1] Roberts's ' Social History of the Southern Counties,' 305.

Kent, were mostly beach harbours, and were constantly
liable to be choked by the movement of the shingle up
channel, so that Winchelsea, Romsey, and Hythe thus
became completely lost.    For the same reason Dover
was always a port most difficult to be preserved.    The
shingle, rolled up along the coast by the prevailing
south-westerly winds, from time to time blocked up the
port by a bank which extended from east to west, until
the pent-up inland waters collecting behind it forced
their way to sea, and thus maintained an opening always
more or less partial.    Various attempts were made to
preserve the harbour in early times, the most important
improvements being those conducted by Sir John
Thompson, master of the Maison Dieu in the reign of
Henry VIII.    He enclosed a small basin with a quay
by driving two rows of piles into the sea bottom as far
out as the Mole Rock, and filling in the interstices with
blocks of stone and chalk.    The stones were floated along
shore from Folkestone by means of empty casks, as at
Lyme.    It is said that not less than 50,000*l.* were expended
on these works; but the imperfect manner in which
they were constructed may be inferred from the fact that
the sea very soon made several breaches in the wall and
the pier, and the beach accumulated as before all round
the bay, so that a boat drawing only four feet of water
could scarcely enter the harbour.    Foreign engineers
were then called in—amongst others Ferdinand Poins, a
Fleming, and Thomas Diggs, who had studied harbour-
construction in the Netherlands; and various additions
were made by them to the works in the reigns of Elizabeth and James.    The harbour was always, however, in
danger of becoming silted up down to our own times; and
the best means of improving it has formed the subject of
repeated reports of Perry, Smeaton, Rennie, and Telford.
Indeed it is doubtful, notwithstanding the enormous
expenditure which has been incurred in the construction
of the modern works at Dover Harbour, whether the

problem of the preservation of the haven be even yet satisfactorily solved.

Along the eastern coast of England the early harbours were few and bad. Thoresby relates that in his time (1682) Whitby, in Yorkshire, possessed a harbour formed by a rough quay projecting at the mouth of the river; but he adds that there was no other haven for ships between that place and Yarmouth, in Norfolk. The last-mentioned port has, like Dover, been the subject of much unavailing engineering in early times, arising from the peculiar difficulties of its situation. It stands on the banks of the rivers Yare and Burr, from the former of which it received its name. It was always liable to be silted up by the sands which abound along shore. Nevertheless it continued to maintain a trade, and down to Henry VIII.'s reign, and even later, it was regarded as the most important maritime town on the east coast. But the channels leading to it were so liable to become choked up, that its prosperity was very irregular, and sometimes its navigation was all but lost. The Yarmouth people were reduced to even greater straits than ordinary in the reign of Elizabeth, on which they adopted the then usual expedient of sending abroad for an engineer of reputation to recover their navigation, and Joyse Johnson, a celebrated man in his day, came over from Holland to direct the works. He caused a strong pier of piles to be formed, which had the effect of directing the current in such a manner, in a north-easterly direction, as to give relief for a time; though the difficulty was by no means surmounted, for we still find the inhabitants fighting against the sea-banks which hemmed them in, all through the reigns of James and Charles, and through the time of the Commonwealth; until eventually a south pier was formed, the continuation of which, in a fine curve, was carried up the river, and formed an extensive wharf, affording considerable accommodation and security for shipping. The original

north pier was subsequently abandoned, and a new
north pier was erected, on a plan chiefly intended to
assist in warping ships into the harbour. The following cut gives an indication of the nature of this
curious and interesting old haven.

YARMOUTH OLD PIER.  [By R. P. Leitch.]

The lighting up of the coast by means of beacons and
lighthouses, for the purpose of insuring greater safety
to shipping approaching our coasts by night, received
very little attention in early times; our lighthouses
being amongst the triumphs of modern engineering.
So long as our mercantile navy was comparatively
insignificant and the amount of our foreign trade but
small, the lighting up of our shores after dark was of
much less importance than it is now.

The idea of the lighthouse is very old, and the ancient
commercial nations were familiar with its use, erecting
a Pharos on any dangerous part of a much-frequented
coast.   The Romans were the first to introduce the
practice in England, and on the summit of Dover Mount

still stands the Roman Pharos, which is supposed to have been used to light vessels from the coasts of France to their station at Portus Rutupiæ (now Richborough) near Sandwich, or to Regulbium (now known as the Reculvers) on the Thames.

PHAROS, DOVER CASTLE.

The old English adopted a similar practice ; but their beacons were of a more rough and homely character. Lambarde says, " before the time of King Edward III. they were made of great stacks of wood ; but about the eleventh yeere of his raigne it was ordained that in our shyre they should be high standards, with their pitch-pots." [1] These beacons were, however, oftener used to alarm the country on the approach of danger than for the purpose of lighting the coasts, though there is good reason to believe that the same sort of beacons were employed for the latter purpose at a more recent period. Professor Faraday says, the first idea of a lighthouse was the candle in the cottage window, guiding the husband across the water or the pathless moor. In the dark the main point was a steady light, and it mattered not whether it was given forth by pitch-pots, coals, or oil. But wood, being the article readiest at hand, was most generally used. The Tour de Cordouan, situated off the coast of France,

---

[1] William Lambarde's ' Perambulation of Kent.' Speaking of Dungeness Point, Lambarde says : " Before this neshe lieth a flat into the sea, threatening great danger to sailors. In the reign of Edward III. it was first ordered that beacons in this country should have their pitch-pots, and that they should no longer be made of wood-stacks or piles, as they be yet in Wiltshire and elsewhere." On this Holloway (' History of Romney Marsh ') observes : " This must imply that either a beacon was now first erected on the Ness Point, or that there had previously been one composed of wood, and for which a pitch-pot was now introduced, as being considered preferable."

was lit up by oak billets brought from the Gascon forests, and until a comparatively recent period the lighthouses at Spurn Point and on the Isle of May were lit up by coal-chauffers.

OLD ENGLISH BEACON.

[By R. P. Leitch. The Design of the Standard from Roberts's ' Social History.']

The importance of insuring greater safety to ships at sea led to steps being taken with that object by the early monarchs; and in the year 1515 Henry VIII. incorporated the Trinity House, for the purpose of protecting commerce and navigation by licensing and regulating pilots, and erecting beacons, lighthouses, and buoys around the coast. The only step taken, however, to carry out these important objects, was merely the granting of leases by the Crown, for a definite number of years, to private persons willing to find the means of building and maintaining lights, in consideration of which, authority was given them to levy tolls on passing shipping. Very little was actually done to insure the greater safety of the coast by means of lights. The first erected was on Dungeness Point, in the reign of James I.; but it would appear that it was the practice, about the same time, to light up some parts of the coast of Cornwall, for we are informed, in the Travels of the Grand Duke Cosmo in England, about two centuries ago, that the Plymouth shipping "paid fourpence per ton for the lights which were in the light-houses at night."[1]   We also find from the records of the Corporation of Rye, that a light was hung out from the south-east angle of the castellated building in that town, called the Ypres Tower, as a guide for vessels entering the harbour in the night-time, and that not being found sufficient, another light was ordered by the Corporation "to be hung out o' nights on the south-west corner of the church, for a guide to vessels entering the port."   A light-pot used also to be hung out from the spire of old Arundel Church for the purpose of guiding vessels entering the harbour of Littlehampton after dark, and we are informed that the iron support of the rude apparatus is still to be seen.[2]   That lights were used for the guidance of ships may also be learnt from the practice which then prevailed among the wreckers along the Cornish coast of displaying *false* lights, and thus luring passing vessels to their destruction; the shipwreck season being long regarded as the harvest season in Cornwall.   With the increase of navigation, the erection of lighthouses at the more dangerous parts of the coast became a matter of urgent necessity; and it was such necessity, as we shall afterwards find, which brought to light the genius of Smeaton.

---

[1] 'Travels of Cosmo the Third, Grand Duke of Tuscany, through England' (1668-9). London, 1821.

[2] The tower of Hadley Church, near Chipping Barnet in Middlesex, was similarly used in ancient times, but as a *land* beacon. The iron cage in which the pitch-pot was placed is still there. It is said that a lamp used formerly to be hung from the old steeple of All Saints, York, for the purpose of guiding travellers at night over the forest of Galtres, and the hook of the pulley by which the lamp was raised is still in its place. Lantern lights were also hung from the steeple of Bow Church, London, Stowe says, "whereby travellers to the city might have the better sight thereof, and not miss their way."

Until the Eddystone Lighthouse of that engineer, the only stone lighthouse in Europe erected out at sea was the fine Tour de Cordouan, on a flat rock off the mouth

TOUR DE CORDOUAN.

of the Garonne in the Bay of Biscay. It was finished and lit up more than two hundred and fifty years ago; and though one of the earliest, it continues one of the most splendid structures of the kind in existence. It replaced a lighthouse founded by the English on the rock in 1362-71, whilst the Black Prince was Governor of Guienne. The stone building was begun by Louis de Foix, one of the architects of the Escurial, in 1584, in the time of Henry III., and was continued all through the reign of Henry IV., being finally completed in 1611, in the reign of Louis XIII. Its height originally was 169 feet French; but in 1727 it was raised to the height of 175 feet French, or 186½ feet English. The building exhibits that taste for magnificence in construction which attained its meridian in France under Louis XIV. The tower does not receive the shock of the waves, but is protected at the base by a wall of circumvallation, which encloses the apartments for the attendants. It is not conical like the Eddystone, but is constructed in three successive stages, angular in the interior, and consequently more susceptible of decoration than the simple and solid structures of Smeaton, Rennie, and Stevenson. The Tour de Cordouan is further memorable as the first lighthouse in which a revolving light was ever exhibited.

# CHAPTER V.

## Ferries and Navigable Rivers.

Notwithstanding the great want of roads and bridges which we have found to exist in the earlier periods of English history, comparatively little use was made of the abundant facilities for inland navigation which the rivers of the country presented.  The trade of the kingdom being comparatively small, the strings of packhorses, and afterwards the heavy waggons drawn by horses or oxen, proved sufficient for its accommodation. The goods carried were mostly of a light character—the cutlery and ironware of Birmingham and Sheffield, the cloths of the villages of Wilts and Somerset, and the cottons (or coatings) and baizes of Manchester and the neighbourhood.  The light articles brought from abroad to the ports of London and Bristol were in like manner distributed through the country by packhorse or waggon.  The chief difficulty was in transporting food and fuel.  But as corn was mostly sent to London by sea, and the city lay fronting the ports of Holland, almost at the door of Europe, there were usually abundant facilities for supplying its large and rapidly increasing population.  The tide lifted daily into the heart of the country fleets of ships laden with stores from all parts of the world, and London tended more than ever to become the metropolis of Europe.

The difficulties of sending coal from Newcastle to London in early times seem, however, to have been considerable.  For a long period a strong prejudice existed against the use of " sea coal."  Edward I. issued a proclamation against it, and a man was actually hanged during his reign for committing the crime of burning it within the limits of the City. But as the forests became consumed for the production of "charre coal" for domestic purposes and for iron-smelting,[1] there was no alternative but to fall back upon the rich stores of coal found in the northern parts of England. Then it was that the Newcastle coal shipping-trade sprang into importance, and ever since has proved the principal nursery of our seamen. The fleets of colliers entering the Thames, added to the other shipping, caused a great throng of vessels in the river; and what with the coal-lighters and merchandise-barges, which formed the communication between the vessels lying in the Pool or down the river, and the warehouses and coalyards on shore, it became a very crowded and often a very confused scene. The merchandise, thus borne from the vessels to the warehouses, became liable to serious depredations; and the losses from this cause, as well as the crowding of the river, at length led to the provision of floating docks at various points, and to a further vast development of the port of London.

The Thames was not only the harbour but the great highway of the metropolis. The city lay mostly along the line of the river, and the streets and roads for a long time continued so bad, that passengers desiring to proceed eastward or westward almost invariably went by boat.

---

[1] The destruction of the woods was a topic of lamentation with the poets of the time. George Withers, in 1634, tells us with what feelings he beheld

The havoc and the spoyle,
Which, ev'n within the compass of my dayes,
Is made through every quarter of this Ile—
In woods and groves, which were this kingdom's praise.

Stowe, also, in his 'Annals,' says: "At this present, through the great consuming of wood as aforesaid, and the neglect of planting of woods, there is so great scarcity of wood throughout the whole kingdom that not only the city of London, all haven towns, and in very many parts within the land, the inhabitants in general are constrained to make them fires of sea-coal or pit-coal, even in the chambers of honourable personages; and through necessity, which is the mother of all arts, they have of very late years devised the making of iron, the making of all sorts of glass, and burning of brick, with sea-coal or pit-coal. Within thirty years last, the nice dames of London would not come into any house or room where sea-coals were burned, nor willingly eat of the meat that was either sod or roasted with sea-coal fire."—Stowe's 'Annals,' by Horner. London, 1632, p. 1025.

There were also ferry-boats constantly plying from side to side of the river, and so long as London Bridge presented the only means of crossing by coach or on foot, the number of persons daily using the ferries was necessarily very considerable. A horse-ferry plied between Lambeth Palace and Millbank, the tolls of which belonged to the Archbishop of Canterbury, and there was another across the river at Hungerford, both being rendered comparatively unnecessary when the second bridge was erected at Westminster. The extent of the river traffic may be inferred from the circumstance stated by Stowe, that in his time the Watermen's Company could at any time furnish twenty thousand men for the fleet. But as the streets of the metropolis were improved, as more bridges were built, and when the use of coaches had extended—against which the watermen strongly protested—their numbers rapidly diminished, until at length they have almost become extinct.

What was called the Long Ferry, however, continued to be used until a comparatively recent period. In early times, the Continental Route was by river to Gravesend, and thence by road to Dover. Gravesend Manor belonged to the Abbot of Tower-hill, who, " finding that by the continual recourse to and from Calais, the passage by water between London and Gravesend was much frequented, both for the great ease, good, cheap, and speedy transportation (requiring not one whole tide), made offer to the young King Richard the Second, that if he would be pleased to grant unto the inhabitants of Gravesend and Milton the privilege that none should transport any passengers by water from Gravesend to London but they only, in their own boats, then should they, of these two parishes, undertake to carry all such passengers, either for twopence each one with his farthell (a truss of straw) or otherwise, making the whole fare or passage worth four shillings." [1]   To this proposal the King consented, and

---

[1] ' A Perambulation of Kent.' By William Lambarde, of Lincoln's Inn, Gent.  London, 1656, p. 534.

hence the route to and from the Continent long continued to be by Gravesend—Ambassadors to the King's Court usually taking boat there for London; and probably a more noble entrance into the great Capital of the kingdom could not well have been selected. The comfort of the long ferry for the commoner sorts of people could not have been great, the passengers being required to bring with them their respective trusses of straw to lie on, whilst they were sometimes under the necessity of landing in the mud a mile or two short of their destination when the tide was low, and either wade their way to shore or pay for being carried on the backs of the mud-larks. The boatmen rendered themselves liable to a penalty only if they landed the passengers more than two miles short of their destination!

Fielding has left an account, in his 'Voyage to Lisbon,' of the tediousness and discomfort of voyaging about the middle of last century. His ship was fixed to sail from opposite the Tower Wharf at a certain time, and Fielding, ghastly and ill, was rowed off to it in a wherry, running the gauntlope through rows of sailors and watermen, who jeered and insulted him as he passed. The ship, however, did not set out for several days, and Fielding was compelled to spend the intervening time in the confines of Wapping and Redriff. The vessel at length sailed, and reaching Gravesend anchored there until the evening of the following day. Next day they sailed for the Nore, and the day after that they anchored off Deal, and lay there for a week. It took four days more to beat down Channel to Ryde, where Fielding was landed in the mud fifteen days after his embarcation at the Tower; and a long, long time elapsed before the termination of his voyage at Lisbon.

When coaches began to run upon the improved road between London and Dover, passing by Blackheath and Dartford to Rochester and Canterbury, the principal part of the continental traffic was diverted from Gravesend, though the comfort of the journey does not seem to have

been much improved.  Smollett gives a rather dismal
account of his progress from London to Boulogne in
1763, which presents a curious contrast to the facilities
of travelling by the modern Boulogne steam route.  After
tediously grumbling his way through Rochester and
Canterbury, fleeced by every innkeeper on the road, he
at last reached Dover in very bad temper.  He pronounced the place to be a den of thieves, where the people
live by piracy in time of war, and by smuggling and
fleecing strangers in time of peace.  He did them the
justice, however, to admit, that " they make no distinction between foreigners and natives.  Without all doubt
a man cannot be much worse lodged and worse treated in
any part of Europe ; nor will he in any other place meet
with more flagrant instances of fraud, imposition, and
brutality.  One would imagine they had formed a general
conspiracy against all those who go to or return from the
Continent."  But Smollett's troubles had scarcely yet
begun, as he found to his cost before he reached Boulogne.
He sent for the master of the packet-boat—a comfortless
tub, called a Folkestone cutter—and hired it to carry him
across the Strait for six guineas, the master demanding
eight.  " We embarked," he says, " between six and
seven in the evening, and found ourselves in a most
wretched hovel.  The cabin was so small that a dog
could hardly turn in it, and the beds put me in mind of
the holes described in some catacombs, in which the
bodies of the dead were deposited, being thrust in with
the feet foremost : there was no getting into them but
endways, and indeed they seemed so dirty, that nothing
but extreme necessity could have obliged me to use them.
We sat up all night in a most uncomfortable situation,
tossed about by the sea, cold and cramped, and weary
and languishing for want of sleep.  At three in the
morning the master came down and told us we were just
off the harbour of Boulogne ; but the wind blowing off
shore he could not possibly enter, and therefore advised

us to go ashore in the boat." Smollett went on deck, when the master pointed out through the spray raised by the scud of the sea where the harbour's mouth lay. The passengers were so impatient to get on shore, that after paying the captain and "gratifying the crew" (which was no easy matter in those days), they committed themselves to the ship's boat to be rowed on shore. They had scarcely, however, got half way to land, before they perceived a boat coming off to meet them, which the captain pronounced to be the French boat, and that it would be necessary to shift from the one small boat to the other in the open sea, " it being a privilege of the boatmen of Boulogne to carry all passengers ashore." Smollett then proceeds,—" This was no time nor place to remonstrate. The French boat came alongside, half filled with water, and we were handed from the one to the other. We were then obliged to lie upon our oars till the captain's boat went on board, and returned from the ship with a packet of letters : we were then rowed a long league in a rough sea, against wind and tide, before we reached the harbour, where we landed, benumbed with cold, and the women excessively sick. From our landing-place we were obliged to walk very near a mile to the inn where we purposed to lodge, attended by six or seven men and women, bare-legged, carrying our baggage. This boat cost me a guinea, besides paying exorbitantly the people who carried our things; so that the inhabitants of Dover and Boulogne seem to be of the same kidney, and indeed they understand one another perfectly well." [1] The passage of the ferry between England and France continued much the same until a comparatively recent period; Fowell Buxton relating that as late as the year 1817 the packet in which he sailed from Dover to Boulogne drifted about in the Channel for two days and two nights, and only reached the port of Calais when every morsel of food on

[1] Smollett's ' Travels through France and Italy.' Letter I., June 23rd, 1763.

board had been consumed. Steam has entirely altered
this state of things, as every traveller knows ;. and the
same passage is now easily and regularly made four times
a day, both ways, in about two hours.

The passage of ferries in the northern parts of England
was equally tedious, uncomfortable, and often dangerous.
In ' A Tour through England in 1765,' it is stated that
at Liverpool passengers were carried to and from the
ferry-boats which plied three times a day to the opposite
shore, " on the backs of men, who waid knee-deep in the
mud to take them out of the boats." Between Hull and
Barton a packet plied once a day across the Humber, the
travellers wading to the boats through a long reach of
mud ; but whether the voyage would occupy two hours
or a day, no one could predict when embarking. If the
weather looked threatening, the travellers would take
up their abode at the miserable inn on the Barton side
until the wind abated. Now the voyage is regularly and
frequently performed every day, to and from New Holland, in less than half an hour. The ferry of the Frith
of Forth was also a formidable affair, and a voyage to
Fife was often full of peril. The passage to Kinghorn
or Burntisland was made in an open boat or a pinnace,
and the boatmen usually waited, it might be for hours,
until sufficient passengers had assembled to go across.
The difficulty of passing the Forth ferries was experienced by Mr. Rennie as late as 1808, when returning
across the Frith from Pettycur, where he had been
examining the harbour with a view to its improvement
for the packet-boats which plied between there and
Leith. " The wind blew fresh," he says, " from about
three points westward of south, and after beating about
in the Frith for nearly three hours, we were obliged to
return to Pettycur ; and, to save time, I went round by
Queen's Ferry," a place nine miles to the westward, from
whence it was three miles across the Forth, and then
other nine miles to Edinburgh ; the distance directly

across from Pettycur being only seven miles. This state of things, we need scarcely add, has been entirely altered by the facilities afforded by modern steam-navigation.

The passage of the Bristol Channel was equally uncertain and dangerous. Gilpin gives a graphic account of the perils of his voyage across from Cardiff in 1770, in his ' Observations on the River Wye.' On descending towards the beach he heard the ferryman winding his horn, as a signal to bring down the horses. The old ferry-boat was usually furnished with falling ends for the admission of cattle and heavy articles; and when the ferry was across a river, there was usually a chain passing along the side of the boat on pulleys, and fixed to each bank, by which it was hauled across. But from Cardiff to the other side of the Bristol Channel was several miles, and it was accordingly rather of the nature of a voyage. The same morning on which Gilpin crossed, the ferry-boat had made one ineffectual attempt to make the further side at high water; but after toiling three hours against the wind, it had been obliged to put back. When the horses were all on board, the horn again sounded for the passengers. " A very multifarious company assembled," says Gilpin, " and a miserable walk we had to the boat, through sludge, and over shelving, slippery rocks. When we got to it we found eleven horses on board and above thirty people; and our chaise (which we had intended to convert into a cabin during the voyage) slung into the shrouds. The boat, after some struggling with the shelves, at length gained the channel. After beating about near two hours against the wind our voyage concluded, as it began, with an uncomfortable walk through the sludge to high-water mark." The passage of this ferry was often attended with loss of life, when the tide ran strong and the wind blew up channel. Moreover, the ferrymen were by no means skilful in the management of the boat. A British admiral, who arrived at one of these ferries, and in-

tended to cross, observing the boat as she worked her
way from the other side, declared that he durst not trust
himself to the seamanship of such fellows as managed
her; and, turning his horse, he rode some fifty miles
round by Gloucester!

Whilst the internal communications of the country
were so imperfect as we have shown them to be, commercial intercourse between different districts was in a
great measure prohibited. The roads of the kingdom
were for a long time but the reflex of its trade. So long
as corn, fuel, wool, iron, and manufactured articles had
to be transported mainly on horses' backs, it is clear that
the progress of commerce must necessarily have been but
slow. The cost of transport of the raw materials required
for food, manufactures, and domestic purposes, formed so
large an item as greatly to check their use; and before
they could be multiplied and made to enter largely into
the general consumption, it was absolutely necessary that
greater facilities for transporting them should be provided. The improvement of the roads towards the
latter half of the last century certainly afforded increased
facilities for internal trade; but they were still in a very
imperfect state, and further provisions with the same
object were still urgently needed. It mattered not that
England was provided with convenient natural havens
situated on the margin of the world's great highway,
the ocean, and with fine tidal rivers capable of accom-
modating ships of the largest burden. Unless the
country inland could be effectually connected with these
ports and tidal rivers, the general extension of commerce
and its civilizing influences upon the community must
necessarily be in a great measure prevented. Hence,
from an early period, attention became directed to the
improvement of those natural means of communication
provided by the numerous large rivers flowing through
the country.

The great commercial republic of Holland may be said to have led the way in Europe in the construction of a system of artificial water-roads, which placed the whole country in direct communication with the seaports. Centuries before any attempt of the kind had been made in England,[1] Holland had provided itself with a magnificent system of canals; and the industrial energy of its people had enabled the nation to attain a remarkable degree of vigour, prosperity, and power. France also had constructed a system of canals, connecting the Loire and the Seine, the Loire and the Saone, and, by means of the great canal of Languedoc, the Atlantic Ocean with the Mediterranean Sea. This latter magnificent work was begun almost a century before England had attempted to make a single canal. Even Sweden and Russia were long before England in undertaking such works, and Peter the Great constructed his grand canal between the Don and the Volga about half a century before England had entered upon her career of canal-making.

In the reign of James I. several Acts of Parliament were passed enabling rivers to be improved, so as to facilitate the passage of boats and barges carrying merchandise. Thus, in 1623, we have found Sir Hugh Myddelton engaged upon a Committee on a bill then under consideration " for the making of the river of Thames navigable to Oxford." In the same year Taylor, the water poet, pointed out to the inhabitants of Salisbury that their city might be effectually relieved of its poor by having their river made navigable from thence to Christchurch.[2] The progress of improvement, however, must have been slow; for urgent appeals continued to be addressed to Parliament and the public, for a century later, on the same subject.

---

[1] It is right, however, to state that the large drains cut by the early churchmen in the Cambridge Fens seem to have been employed for purposes of inland navigation—Bishop Morton's Leam having been thus used between Peterborough and the sea as early as the fifteenth century.

[2] ' A Discovery by Sea from London to Salisbury.' London, 1623.

In 1656 we find one Francis Mathew addressing Cromwell and his Parliament [1] on the immense advantage of opening up a water-communication between London and Bristol. But he only proposed to make the rivers Isis and Avon navigable to their sources, and then either to connect their heads by means of a short sasse or canal of about three miles across the intervening ridge of country, or to form a fair stone causeway between the heads of the two rivers, across which horses or carts might carry produce between the one and the other. His object, it will be observed, was mainly the opening up of the existing rivers; "and not," he says, "to have the old channel of any river to be forsaken for a shorter passage." Mathew fully recognised the formidable character of his project, and considered it quite out of the question for private enterprise, whether of individuals or any corporation, to undertake to execute it; but he ventured to think that it might not be too much for the power of the State to construct the three miles of canal and carry out the other improvements suggested by him, with a reasonable prospect of success. The scheme was, however, too daring for Mathew's time, and another century elapsed before a canal was made in England.

A few years later, in 1677, a curious work was published by one Andrew Yarranton, gentleman,[2] in which he strongly pointed out what the Dutch had done by means of inland navigation, and what England ought to do as the best means of excelling the Dutch without fighting them. The main feature of his scheme was the improvement of our rivers so as to render them navigable, and the inland country thus more readily accessible to commerce. For in England, said he, there are large

---

[1] 'Of the Opening of Rivers for Navigation; the benefit exemplified by the two Avons of Salisbury and Bristol, with a Mediterranean Passage by Water for Billanders of Thirty Ton between Bristol and London, with the results.' London, 1656.

[2] 'England's Improvement by Sea and Land : To outdo the Dutch without fighting, to pay Debts without Moneys, to set at work all the Poor of England with the growth of our own Land,' &c. London, 1677.

rivers well situated for trade, great woods, good wool and large beasts, with plenty of iron stone, and pit coals, with lands fit to bear flax, and with mines of tin and lead; and besides all these things in it, England has a good air. But to make these advantages available, the country, he held, must be opened up by navigation. First of all, he proposed that the Thames should be improved to Oxford, and connected with the Severn by the Avon to Bristol—these two rivers, he insisted, being the master rivers of England. When this has been done, says Mr. Yarranton, all the great and heavy carriage from Cheshire, all Wales, Shropshire, Staffordshire, and Bristol, will be carried to London and recarried back to the great towns, especially in the winter time, at half the rates they now pay, which will much promote and advance manufactures in the counties and places above named. "If I were a doctor," he says, "and could read a Lecture of the Circulation of the Blood, I should by that awaken all the City: For London is as the Heart is in the Body, and the great Rivers are as its Veins; let them be stopt, there will then be great danger either of death, or else such Veins will apply themselves to feed some other part of the Body, which it was not properly intended for: For I tell you, Trade will creep and steal away from any place, provided she may be better treated elsewhere." But he goes on—"I hear some say, You projected the making Navigable the River Stoure in Worcestershire: what is the reason it was not finished? I say it was my projection, and I will tell you the reason it was not finished. The River Stoure and some other Rivers were granted by an Act of Parliament to certain Persons of Honour, and some progress was made in the work; but within a small while after the Act passed it was let fall again. But it being a brat of my own, I was not willing it should be Abortive; therefore I made offers to perfect it, leaving a third part of the Inheritance to me and my heirs for ever, and we came to an agreement. Upon which I fell on, and made it compleatly Navigable from Sturbridge to Kederminster; and carried down many hundred Tuns of Coales, and laid out near one thousand pounds, and then it was obstructed for Want of Money, which by Contract was to be paid." [1]

There is no question that this "want of money" was the secret of the little progress made in the improvement of the internal communications of the country, as well as the cause of the backward state of industry generally. England was then possessed of little capital and less spirit, and hence the miserable poverty, starvation, and beggary which prevailed to a great extent amongst the lower classes of society at the time when Mr. Yarranton wrote, and which he so often refers to in the course of his book. For the same reason most of the early Acts of Parliament for the improvement of navigable rivers remained a dead letter: there was not money enough to carry them out, modest though the projects usually were. Among the few schemes which were actually carried out about the beginning of the eighteenth century, was the opening up of the navigation of the rivers Aire and Calder, in Yorkshire. Though a work of no great difficulty, Thoresby speaks of it in his diary as one of vast magnitude. It was certainly, however, one of great utility, and gave no little impetus to the trade of that important district.

It was, indeed, natural that the demand for improvements in inland navigation should arise in those quarters where the communications were the most imperfect and where good communications were most needed, namely, in the manufacturing districts of the north of England. On the western side of the island Liverpool was then rising in importance, and the necessity became urgent for opening up its water-communications with the interior. By the assistance of the tide, vessels were enabled to reach as high up the Mersey as Warrington;

[1] Yarranton's ' England's Improvement by Sea and Land,' p. 66.

but there they were stopped by the shallows, which it was necessary to remove to enable them to reach Manchester and the adjacent districts. Accordingly, in 1720, an Act was obtained empowering certain persons to take steps to make navigable the rivers Mersey and Irwell from Liverpool to Manchester. This was effected by the usual contrivance of wears, locks, and flushes, and a considerable improvement in the navigation was thereby effected. Acts were also obtained for the improvement of the Weaver navigation, the Douglas navigation, and the Sankey navigation, all in the same neighbourhood; and the works carried out proved of much service to the district.

But these improvements, it will be observed, were principally confined to clearing out the channels of existing rivers, and did not contemplate the making of new and direct navigable cuts between important towns or districts. It was not until about the middle of last century that English enterprise was fairly awakened to the necessity of carrying out a system of artificial canals throughout the kingdom; and from the time when canals began to be made, it will be found that the industry of the nation made a sudden start forward. Abroad, monarchs had stimulated like undertakings, and drawn largely on the public resources for the purpose of carrying them into effect; but in England such projects are usually left to private enterprise, which follows rather than anticipates the public wants. In the upshot, however, the English system, as it may be termed—which is the outgrowth in a great measure of individual energy—does not prove the least efficient; for we shall find that the English canals, like the English railways, were eventually executed with a skill, despatch, and completeness, which imperial enterprise, backed by the resources of great states, was unable to surpass or even to equal. How the first English canals were made, how they prospered, and how the system extended, will appear from the following biography of James Brindley, the father of canal engineering in England.

*James Brindley.*

Engraved by W. Holl, after the portrait by F. Parsons.

Published by John Murray, Albemarle Street, 1861.

WORSLEY BASIN.

[ By Percival Skelton, after his original Drawing. ]

# LIFE OF JAMES BRINDLEY.

## CHAPTER I.

### THE WHEELWRIGHT'S APPRENTICE.

IN the third year of the reign of George I., whilst the British Government were occupied in extinguishing the embers of the Jacobite rebellion which had occurred in the preceding year, the first English canal engineer was born in a remote hamlet in the High Peak of Derby, in the midst of a rough country, then inhabited by quite as rough a people.

The nearest town of any importance was Macclesfield, where a considerable number of persons were employed, about the middle of last century, in making wrought buttons in silk, mohair, and twist—such being then the staple trade of the place. Those articles were sold throughout the country by pedestrian hawkers, most of whom lived in the wild country called "The.Flash," from a hamlet of that name situated between Buxton, Leek, and Macclesfield. They squatted on the waste lands and commons in the district, and were notorious for their wild, half-barbarous manners, and brutal pastimes. Travelling about from fair to fair, and using a cant or slang dialect, they became generally known as "Flash men," and the name still survives. Their numbers so grew, and their encroachments on the land became so great, that it was at length found necessary to root them out; but for some time no bailiff was found sufficiently bold to attempt to serve a writ in the district. At length an officer was found who undertook to arrest several of

them, and other landowners, taking courage, followed
the example ; when those who refused to become tenants
left, to squat elsewhere ; and the others then consented
to settle down to the cultivation of their farms.    Another set of travelling rogues belonging to the same

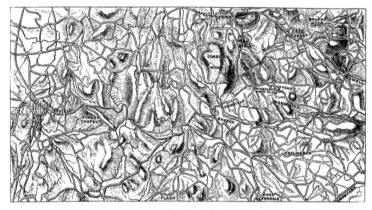

BRINDLEY'S NATIVE DISTRICT    [Ordnance Survey.]

neighbourhood was called the "Broken Cross Gang,"
from a place called Broken Cross, situated to the south-
east of Macclesfield.    Those fellows consorted a good deal
with the Flash men, frequenting markets and travelling
from fair to fair, practising the pea-and-thimble trick,
and enticing honest country people into the temptation
of gambling.    They proceeded to more open thieving
and pocket-picking, until at length the magistrates of
the district took active measures to root them out of
Broken Cross, and the gang became broken up.    Such
was the district and such the population in the neighbourhood of which our hero was born.

James Brindley first saw the light in a humble cottage
standing about midway between the hamlet of Great
Rocks and that of Tunstead, in the liberty of Thornsett,
some three miles to the north-east of Buxton.    The
house in which he was born, in the year 1716, has long
since fallen to ruins—the Brindley family having been
its last occupants.    The walls stood long after the roof

had fallen in, and at length the materials were removed
to build cowhouses; but in the middle of the ruin there
grew up a young ash tree, forcing up one of the flags
of the cottage-floor. It looked so healthy and thriving
a plant, that the labourer employed to remove the stones
for the purpose of forming the pathway to the neighbouring farm-house, spared the seedling, and it grew
up into the large and flourishing tree, six feet nine
inches in girth, standing in the middle of the Croft,
and now known as "Brindley's Tree." This ash tree is
nature's own memorial of the birth-place of the engineer,
and it is the only one as yet erected to the genius of
Brindley.

BRINDLEY'S CROFT.[1]

[By Percival Skelton, after a Sketch by Mrs. Fleming.]

Although the enclosure is called Brindley's Croft, this
name was only given to it of late years by its tenant, in
memory of the engineer who was born there. The state-

---

[1] The site of the Croft is very elevated, and commands an extensive
view as far as Topley Pike, between
Bakewell and Buxton, at the top of
what is called the Long Hill. Topley
Pike is behind the spectator in looking at the Croft in the above aspect.
The rising ground behind the ash
tree is called Wormhill Common,
though now enclosed. The old road
from Buxton to Tideswell skirts the
front of the rising ground.

ment made in Mr. Henshall's memoir of Brindley,[1] to the
effect that Brindley's father was the freehold owner of
his croft, does not appear to have any foundation; as the
present owner of the property, Dr. Fleming, informs us
that it was purchased, about the beginning of the present
century, from the heirs of the last of the Heywards,
who became its owners in 1688. No such name as
Brindley occurs in any of the title-deeds belonging to
the property; and it is probable that the engineer's
father was an under-tenant, and merely rented the
old cottage in which our hero was born. There is no
record of his birth, nor does the name of Brindley occur
in the register of the parish of Wormhill, in which the
cottage was situated; but registers in those days were
very imperfectly kept, and part of that of Wormhill
has been lost.

It is probable that Brindley's father maintained his
family by the proceeds of his little croft, and that he was
not much, if at all, above the rank of a cottier. It is
indeed recorded of him that he was by no means a steady
man, and was fonder of sport than of work. He went
shooting and hunting, when he should have been labouring; and if there was a bull-running within twenty
miles, he was sure to be there. The Bull Ring of the
district lay less than three miles off, at the north end of
Long Ridge Lane, which almost passed his door; and of
that place of popular resort Brindley's father was a
regular frequenter. These associations led him into bad
company, and very soon reduced him to poverty. He
neglected his children, not only setting before them a
bad example, but permitting them to grow up without
education. Fortunately, Brindley's mother in a great
measure supplied the father's shortcomings: she did what
she could to teach them what she knew, though that was
but small; but, perhaps more important still, she encou-

---

[1] Kippis's 'Biographia Britannica,' Art. Brindley.

raged them in the formation of good habits by her own steady industry.[1]

The different members of the family, of whom James was the eldest, were thus under the necessity of going out to work at a very early age to provide for the family wants.   James worked at any ordinary labourer's employment which offered until he was about seventeen years old.   His mechanical bias had, however, early displayed itself, and he was especially clever with his knife, making models of mills, which he set to work in little mill-streams of his contrivance.   It is said that one of the things in which he took most delight when a boy, was to visit a neighbouring grist-mill and examine the water-wheels, cog-wheels, drum-wheels, and other attached machinery, until he could carry away the details in his head; afterwards imitating the arrangements by means of his knife and such little bits of wood as he could obtain for the purpose.   We can thus readily understand how he should have turned his thoughts in the direction in which we afterwards find him employed, and that, encouraged by his mother, he should have determined to bind himself, on the first opportunity that offered, to the business of a millwright.

The demands of trade were so small at the time, that Brindley had no great choice of masters; but at the village of Sutton, near Macclesfield, there lived one Abraham Bennett, a wheelwright and millwright, to whom young Brindley offered himself as apprentice; and in the year 1733, after a few weeks' trial, he became bound to that master for the term of seven years. Although the employment of millwrights was then of a very limited character, a great deal of valuable practical

---

[1] Brindley's father seems afterwards to have somewhat recovered himself; for we find him, in 1729, purchasing an undivided share of a small estate at Lowe Hill, within a mile of Leek, in Staffordshire, where he had before gone to settle; and he contrived to realise the remaining portion before his death, and to leave it to his son James.   None of the Brindley family remained at Wormhill, and the name has disappeared in the district.

information was obtained whilst carrying on their business.   The millwrights were as yet the only engineers.
In the course of their trade they worked at the foot-lathe,
the carpenter's bench, and the anvil, by turns; thus
cultivating the faculties of observation and comparison,
acquiring practical knowledge of the strength and
qualities of materials, and dexterity in the handling of
tools of many different kinds.   In country places, where
division of labour could not be carried so far as in the
larger towns, the millwright was compelled to draw
largely upon his own resources, and to devise expedients
to meet pressing emergencies as they arose.   Necessity
thus made them dexterous, expert, and skilful in mechanical arrangements, more particularly those connected
with mill-work, steam-engines, pumps, cranes, and such
like.   Hence millwrights in those early days were looked
upon as a very important class of workmen.   The nature
of their business tended to render them self-reliant, and
they prided themselves on the importance of their
calling.   On occasions of difficulty the millwright was
invariably resorted to for help; and as the demand for
mechanical skill arose, in the course of the progress of
manufacturing and agricultural industry, the men trained
in millwrights' shops, such as Brindley, Meikle, Rennie,
and Fairbairn, were borne up by the force of their practical skill and constructive genius into the highest rank
of skilled and scientific engineering.

Brindley, however, only acquired his skill by slow
degrees.   Indeed, his master thought him slower than
most lads, and even stupid.   Bennett, like many well-
paid master mechanics at that time, was of intemperate
habits, and gave very little attention to his apprentice,
leaving him to the tender mercies of his journeymen,
who were for the most part a rough and drunken set.
Much of the lad's time was occupied in running for beer,
and when he sought for information he was often met
with a rebuff.   Skilled workmen were then very jealous

of new hands, and those who were in any lucrative employment usually put their shoulders together to exclude those who were out. Brindley had thus to find out nearly everything for himself, and he only worked his way to dexterity through a succession of blunders. He was frequently left in sole charge of the wheelwrights' shop—the men being absent at jobs in the country, and the master at the public-house, from which he could not easily be drawn. Hence, when customers called at the shop to get any urgent repairs done, the apprentice was under the necessity of doing them in the best way he could, and that often very badly. When the men came home and found tools blunted and timber spoiled, they abused Brindley and complained to the master of his bungling apprentice's handiwork, declaring him to be a mere " spoiler of wood." On one occasion, when Bennett and the journeymen were absent, he had to fit in the spokes of a cart-wheel, and was so intent on completing his job that he did not find out that he had fitted them all in the wrong way until he had applied the gauge-stick. Not long after this occurrence, Brindley was left by himself in the shop for an entire week, working at a piece of common enough wheelwright's work, without any directions; and he made such a " mess " of it, that on the master's return he was so enraged that he threatened, there and then, to cancel the indentures and send the young man back to farm-labourer's work, which Bennett declared was the only thing for which he was fit.

Brindley had now been two years at the business, and in his master's opinion had learnt next to nothing; though it shortly turned out that, notwithstanding the apprentice's many blunders, he had really groped his way to much valuable practical information on matters relating to his trade. Bennett's shop would have been a bad school for an ordinary youth, but it proved a prolific one for Brindley, who was anxious to learn, and determined to make a way for himself if he could not find one. He must have had a brave spirit to withstand the many difficulties he had to contend against, to learn dexterity through blunders, and success through defeats. But this is necessarily the case with all self-taught work-men; and Brindley was mainly self-taught, as we have seen, even in the details of the business to which he had bound himself apprentice.

In the autumn of 1735 a small silk-mill at Maccles-field, the property of Mr. Michael Daintry, sustained considerable injury from a fire at one of the gudgeons inside the mill, and Bennett was called upon to execute the necessary repairs. Whilst the men were employed at the shop in executing the new work, Brindley was sent to the mill to remove the damaged machinery, under the directions of Mr. James Milner, the superin-tendent of the factory. Milner had thus frequent occa-sion to enter into conversation with the young man, and was struck with the pertinence of his remarks as to the causes of the recent fire and the best means of avoiding similar accidents in future. He even applied to Bennett, his master, to permit the apprentice to assist in executing the repairs of certain parts of the work, which was reluctantly assented to. Bennett closely watched his "bungling apprentice," as he called him; but Brindley, encouraged by the superintendent of the mill, succeeded in satisfactorily executing his allotted portion of the repairs, not less to the surprise of his master than to the mortification of his men. Many years after, Brindley, in describing this first successful piece of mill-work which he had executed, observed, "I can yet remember the delight which I felt when my work was fixed and fitted complete; and I could not understand why my master and the other workmen, instead of being pleased, seemed to be dissatisfied with the insertion of every fresh part in its proper place."

The completion of the job was followed by the usual

supper and drink at the only tavern in the town, then
on Parsonage Green. Brindley's share in the work was
a good deal ridiculed by the men when the drink began
to operate; on which Mr. Milner, to whose intercession
his participation in the work had been entirely attributable, interposed and said, "I will wager a gallon of the
best ale in the house, that before the lad's apprenticeship is out he will be a cleverer workman than any
here, whether master or man." We have not been informed whether the wager was accepted; but it was
long remembered, and Brindley was so often taunted
with it by the workmen, that he was not himself allowed
to forget that it had been offered. Indeed, from that
time forward, he zealously endeavoured so to apply himself as to justify the prediction, for it was nothing less,
of his kind friend Mr. Milner; and before the end of his
third year's apprenticeship his master was himself constrained to admit that Brindley was not the "fool" and
the "blundering blockhead" which he and his men had
so often called him. Very much to the chagrin of the
latter, and to the surprise of Bennett himself, the neighbouring millers, when sending for a workman to execute
repairs in their machinery, would specially request that
"the young man Brindley" should be sent them in preference to any other of the workmen. Some of them
would even have the apprentice in preference to the
master himself. At this Bennett was greatly surprised,
and, quite unable to understand the mystery, he even
went so far as to inquire of Brindley where he had
obtained his knowledge of mill-work! Brindley could
not tell; it "came natural-like;" but the whole secret
consisted in Brindley working with his head as well as
with his hands. The apprentice had already been found
peculiarly expert in executing mill repairs, in the course
of which he would frequently suggest alterations and
improvements, more especially in the application of the
water-power, which no one had before thought of, but

which proved to be founded on correct principles, and worked to the millers' entire satisfaction. Bennett, on afterwards inspecting the gearing of one of the mills repaired by Brindley, found it so securely and substantially fitted, that he even complained to him of his style of work. " Jem," said he, " if thou persist in this foolish way of working, there will be very little trade left to be done when thou comes out of thy time : thou knaws firmness of wark's th' ruin o' trade." Brindley, however, gave no heed whatever to the unprincipled suggestion, and considered it the duty and the pride of the mechanic always to execute the best possible work.

Among the other jobs which Brindley's master was employed to execute about this time, was the machinery of a new paper-mill proposed to be erected on the river Dane. The arrangements were to be the same as those adopted in the Smedley paper-mill on the Irk, and at Throstle-Nest on the Irwell, near Manchester; and Bennett went over to inspect the machinery at those places. But Brindley was afterwards of opinion that he must have inspected the taverns in Manchester much more closely than the paper-mills in the neighbourhood ; for when he returned, the practical information he brought with him proved almost a blank. Nevertheless, Bennett could not let slip the opportunity of undertaking so lucrative a piece of employment in his special line, and, ill-informed though he was, he set his men to work upon the machinery of the proposed paper-mill.

It very soon appeared that Bennett was altogether unfitted for the performance of the contract which he had undertaken. The machinery, when made, would not fit; it would not work; and, what with drink and what with perplexity, Bennett soon got completely bewildered. Yet to give up the job altogether would be to admit his own incompetency as a mechanic, and must necessarily affect his future employment as a millwright. He and his men, therefore, continued distractedly to persevere in their operations, but without the slightest appearance of satisfactory progress. About this time an old hand, who happened to be passing the place at which the men were at work, looked in upon them and examined what they were about, as a mere matter of curiosity. When he had done so, he went on to the nearest public-house and uttered his sentiments on the subject very freely. He declared that the job was a farce, and that Abraham Bennett was only throwing his employer's money away. The statement of what the "experienced hand" had said, was repeated until it came to the ears of young Brindley. Concerned for the honour of his shop as well as for the credit of his master —though he probably owed him no great obligation on the score either of treatment or instruction—Brindley formed the immediate resolution of attempting to master the difficulty so as to enable the work to be brought to a satisfactory completion.

At the end of the week's work Brindley left the mill without saying a word of his intention to any one, and instead of returning to his master's house, where he lodged, he took the road for Manchester. Bennett was in a state of great alarm lest he should have run away; for Brindley, now in the fourth year of his apprenticeship, had reached the age of twenty-one, and the master feared that, taking advantage of his legal majority, he had left his service never to return. A messenger was despatched in the course of the evening to his mother's house; but he was not there. Sunday came and passed —still no word of young Brindley: he must have run away! On Monday morning Bennett went to the paper-mill to proceed with his fruitless work; and lo! the first person he saw was Brindley, with his coat off, working away with greater energy than ever. His disappearance was soon explained. He had been to Smedley Mill to inspect the machinery there with his own eyes, and clear up his master's difficulty. He had walked the twenty-five miles thither on the Saturday night, and on the following Sunday morning he had waited on Mr. Appleton, the proprietor of the mill, and requested permission to inspect the machinery. With an unusual degree of liberality Mr. Appleton gave the required consent, and Brindley spent the whole of that Sunday in the most minute inspection of the entire arrangements of the mill. He could not make notes, but he stored up the particulars carefully in his head; and believing that he had now thoroughly mastered the difficulty, he set out upon his return journey, and walked the twenty-five miles back to Macclesfield again.

Having given this proof of his determination, as he had already given of his skill in mechanics, Bennett was only too glad to give up the whole conduct of the contract thenceforth to his apprentice; Brindley assuring him that he should now have no difficulty in completing it to his satisfaction. No time was lost in revising the whole design; many parts of the work already fixed were rejected by Brindley, and removed; others, after his own design, were substituted; several entirely new improvements were added; and in the course of a few weeks the work was brought to a conclusion, within the stipulated time, to the satisfaction of the proprietors of the mill.

There was now no longer any question as to the extraordinary mechanical skill of Bennett's apprentice. The old man felt that he had been in a measure saved by young Brindley, and thenceforth, during the remainder of his apprenticeship, he left him in principal charge of the shop. Thus for several years Brindley maintained his old master and his family in respectability and comfort; and when Bennett died, Brindley carried on the concern until the work in hand had been completed and the accounts wound up; after which he removed from Macclesfield to begin business on his own account at the town of Leek, in Staffordshire.

## CHAPTER II.

### BRINDLEY AS MASTER WHEELWRIGHT AND MILLWRIGHT.

BRINDLEY had now been nine years at his trade, seven as apprentice and two as journeyman; and he began business as a wheelwright at Leek at the age of twenty-six. He had no capital except his skill, and no influence except that which his character as a steady workman gave him. Leek was not a manufacturing place at the time when Brindley began business there in 1742. It was but a small market town, the only mills in the neighbourhood being a few grist-mills driven by the streamlets flowing into the waters of the Dane, the Churnet, and the Trent. These mills usually contained no more than a single pair of stones, and they were comparatively rude and primitive in their arrangement and construction.

Brindley at first obtained but a moderate share of employment. His work was more strongly done, and his charges were consequently higher, than was customary in the district; and the agricultural classes were as yet too poor to enable them to pay the prices of the best work. He gradually, however, acquired a position, and became known for his skill in improving old machinery or inventing any new mechanical arrangement that might be required for any special purpose. He was very careful to execute anything committed to him within the stipulated time, and he began to be spoken of as a thoroughly reliable workman. Thus his business gradually extended to other places at a distance from Leek, and more especially into the Staffordshire Pottery districts, shortly about to rise into importance under the fostering energy of Josiah Wedgwood.

At first he kept neither apprentices nor journeymen, but felled his own timber and cut it up himself, with such assistance as he could procure on the spot. As his business increased he took in an apprentice, and then a journeyman, to carry on the work in the shop while he was absent; and he was often called to a considerable distance from home, more particularly for the purpose of being consulted about any new machinery that was proposed to be put up. Nor did he confine himself to mill-work. He was ready to undertake all sorts of machinery connected with the pumping of water, the draining of mines, the smelting of iron and copper, and the various mechanical arrangements connected with the manufactures rising into importance in the adjoining counties of Cheshire and Lancashire. Whenever he was called upon in this way, he endeavoured to introduce improvements; and to such an extent did he carry this tendency, that he became generally known in the neighbourhood by the name of " The Schemer."

A number of Brindley's memoranda books [1] are still in existence, which show the various character of his employment during this early part of his career. It appears from the entries made in them, that he was not only employed in repairing and fitting up silk-throwing mills at Macclesfield, all of which were then driven by water, but also in repairing corn-mills at Congleton, Newcastle-under-Lyne, and various other places, besides those in the immediate neighbourhood of Leek, where he lived. We believe the pocket memoranda books, to which we refer, were the only records which Brindley kept of his early business transactions; the rest he carried in his memory, which by practice became remarkably retentive. Whilst working as an apprentice at Macclesfield, he had taught himself the art of writing; but he never mastered

---

[1] In the possession of Joseph Mayer, Esq., of Liverpool, who has kindly permitted the author to inspect the whole of his valuable Brindley manuscripts, so curiously illustrative of the early part of his career.

it thoroughly, and to the end of his life he wrote with difficulty, and almost illegibly. His spelling was also very bad; and what with the bad spelling and what with the hieroglyphics in which he wrote, it is sometimes very difficult to decypher the entries made by him from time to time in his memoranda books.

We find him frequently at Trentham, entering, on one occasion, a " Loog of Daal 20 foot long," and at another time fitting a pump for " Arle Gower," the Earl being one of Brindley's first patrons. The log of deal, it afterwards appears, was required for a flint-mill of a Mr. Tibots—" a mow [new ?] invontion," as Brindley enters it in his book — of which more hereafter. On May 18, 1755, he enters " Big Tree to cut 1 day," and he seems to have felled the tree, and, some months after, to have cut it up himself, entering so many days at two shillings a day for the labour. When he had to travel some distance, he set down sixpence a day extra for expenses. Thus on one occasion he makes this entry : " For Mr. Kent corn mill of Codan looking out a shaft neer Broun Edge 1 day 0 : 2 : 6."

THE POTTERIES DISTRICT
[Ordnance Survey.]

Between Leek and Trentham lay the then small pottery village of Burslem, which Brindley had frequent occasion to pass through in going to and from his jobs for

the Earl.  The earthenware then manufactured at Burslem was of a very inferior sort, consisting almost entirely
of brown ware ; and the quantity turned out was so small
that it was hawked about on the backs of the potters
themselves, or sold by higglers, who carried it from
village to village in the panniers of their donkeys.  The
brothers Elers, the Dutchmen, erected a potwork of an
improved kind near Burslem, at the beginning of the
century, in which they first practised the art of salt-
glazing, brought by them from Holland.  The next
improvement introduced was the use of powder of flints,
used at first as a wash or dip, and afterwards mixed with
tobacco-pipe clay, from which an improved ware was
made, called " Flint potters."  The merit of introducing
this article is usually attributed to William Astbury, of
Shelton, who, when on a journey to London, stopping
at an inn at Dunstable, noticed the very soft and delicate
nature of some burnt flint-stones when mixed with water
(the hostler having used the powdered flint as a remedy
for a disorder in his horses' eyes), and from thence he is
said to have conceived the idea of applying it to the
purposes of his trade.  In first using the calcined flints,
Mr. Astbury's practice was to have them pounded in an
iron mortar until perfectly levigated ; and being but
sparingly used, this answered the demand for some time.
But when the use of flint became more common, this
tedious process would no longer suffice.  The brothers
John and Thomas Wedgwood carried on the pottery
business in a very small way, but were nevertheless
hampered by the short supply of flint powder, and
it was found necessary to adopt some means of increasing the quantity.  In their emergency the potters
called " The Schemer " to their aid ; and hence we find
him frequently occupied in erecting flint-mills, in Burslem
and the neighbourhood, from that time forward.  The
success which attended his efforts brought Brindley not
only fame, but business.  It happened that, while thus

occupied, Mr. John Edensor Heathcote, owner of the
Clifton estate near Manchester, was married to one of
the daughters of Sir Nigel Gresley, of Knypersley, in
the neighbourhood of Burslem, and that the marriage
festivities were in progress, when the remarkable ingenuity of the young millwright of Leek was accidentally
mentioned in the hearing of Mr. Heathcote one day at
dinner.  The Manchester man, in the midst of pleasure,
did not forget business; and it occurred to him that
this ingenious mechanic might be of use to him in contriving some method for clearing his Clifton coal-mines
of the water by which they had so long been drowned.
The old methods of the gin-wheel and tub, and the chain-
pump, had been tried, but entirely failed to keep the
water under : if this Brindley could but do anything to
help him in his difficulty, he would employ him at once ;
at all events, he would like to see the man.

Brindley was accordingly sent for, and the whole
case was laid before him.  Mr. Heathcote described
as minutely as possible the nature of the locality, the
direction in which the strata lay, and exhibited a plan
of the working of the mines.  Brindley was perfectly
silent for a long time, seemingly absorbed in a consideration of the difficulties to be overcome ; but at length his
countenance brightened, his eyes sparkled, and he briefly
pointed out a method by which he thought he should be
enabled, at no great expense, effectually to remedy the
evil.  His explanations were considered so satisfactory,
that he was at once directed to proceed to Clifton, with
full powers to carry out his proposed plan of operations.
This was, to call to his aid the fall of the river Irwell,
which formed one boundary of the estate, and pump out
the water from the pits by means of the greater power
of the water in the river.  With this object Brindley
contrived and executed his first tunnel, which he drove
through the solid rock for a distance of six hundred
yards, and in this tunnel he led the river on to the breast

of an immense water-wheel fixed in a chamber some
thirty feet below the surface of the ground, from the
lower end of which the water, after exercising its power,
flowed away into the lower level of the Irwell. The
expedient, though bold, was simple, and it proved effective. The machinery was found fully equal to the emergency; and in a very short time Brindley's wheel and
pumps, working night and day, so cleared the mine of
water as to enable the men to get the coal in places from
which they had long been completely "drowned out."

We are not informed of the remuneration which the
engineer received for carrying out this important work;
but from the entries in his memorandum book it is probable that all he obtained was only his workman's wage
of two shillings a-day. Notwithstanding his ingenuity
and hard-working energy, Brindley never seems, during
the early part of his career, to have earned more than
about one-third the wage of skilled mechanics in our
own time; and from the insignificant sums charged by
him for expenses, it is clear that he was satisfied to live in
the fashion of an ordinary labourer. What modern engineers will receive ten guineas a-day for doing, he,
with his strong original mind, was quite content to do
for two shillings. But eminent constructive skill seems
to have been lightly appreciated in those days, if we
may judge by the money value attached to it.[1] To this,
however, it must be added, that at the time of which

---

[1] Long before Brindley's time, Inigo
Jones was paid only eight shillings and
fourpence a-day as architect and surveyor of the Whitehall Banqueting
House, and forty-six pounds a-year
for house-rent, clerks, and incidental
expenses; whilst Nicholas Stowe,
the master mason, was allowed but
four and tenpence a-day. When the
Duchess of Marlborough was afterwards engaged in resisting the claims
of one of her Blenheim surveyors, she
told him indignantly "that Sir Christopher Wren, while employed upon
Saint Paul's, was content to be dragged
up to the top of the building three
times a-week in a basket, at the great
hazard of his life, for only 200l. a-
year"—the actual amount of his salary
as architect of that magnificent Cathedral. Brindley, however, fared worse
still, and for a long time does not seem
to have risen above mere mechanic's
pay, even whilst engaged in constructing the celebrated canal for the Duke
of Bridgewater, which laid the foundation of so many gigantic fortunes.

we speak, the people of the country were comparatively poor—manufacturers as well as landowners. In Macclesfield and the neighbourhood, where the inventions of men such as Brindley have issued in so extraordinary a development of wealth, the operations of trade were as yet in their infancy, and had numerous obstructions and difficulties to contend against. Perhaps the greatest difficulty of all was the absence of those facilities for communicating between one district and another, without which the existence of trade is simply impossible; but we shall shortly find Brindley also entering upon this great work of opening up the internal communications of the country, with an extraordinary degree of ability and success.

By the middle of last century, Macclesfield and the neighbouring towns were gradually rising out of the small button-trade, and aiming at greater things in the way of manufacture. In 1755 Mr. N. Pattison of London, Mr. John Clayton, and a few other gentlemen, entered into a partnership to build a new silk-mill at Congleton, in Cheshire, on a larger scale than had yet been attempted in that neighbourhood. Brindley was employed to execute the water-wheel and the commoner sort of mill-work about the building; but the smaller wheels and the more complex parts of the machinery, with which it was not supposed Brindley could be acquainted, were entrusted to a master joiner and millwright, named Johnson, who also superintended the progress of the whole work. The superintendent required Brindley to work after his mere verbal directions, without the aid of any plan; and Brindley was not even allowed to inspect the models of the machinery required for the proposed mill. He thus worked at a great disadvantage, and the operations connected with the construction of the intended machinery were very shortly found in a state of great confusion. The proprietors had reason to suspect that their superintendent was not equal to the

enterprise which he had undertaken. At first he endeavoured to assure them that all was going right; but at
last, after various efforts, he was obliged to confess his
incompetency and his inability to complete the work.

The proprietors, becoming alarmed, then sent for
Brindley and told him of their dilemma. " Would *he*
undertake to complete the works?" He asked to see the
model and plans which the superintendent engineer had
proposed to follow out. But on being applied to, the
latter positively refused to submit his designs to a common millwright, as he alleged Brindley to be. The
proprietors were almost in despair, and their only reliance
now was on Brindley's genius. " Tell me," he said,
" what is the precise operation that you wish to perform,
and I will endeavour to provide you with the requisite
machinery for doing it; but you must let me carry out
the work in my own way." To this they were only too
glad to assent; and having been furnished with the
requisite powers, he forthwith set to work. His intelligent observation of the process of manufacture in the
various mills he had inspected, his intimate practical
knowledge of machinery of all kinds then in use, and his
fertility of resources in matters of mechanical arrangement, enabled him to perform even more than he had
promised; and he not only finished the mill to the complete satisfaction of its owners, but added a number of
new and skilful improvements in detail, which afterwards
proved of the greatest value. For instance, he adapted
lifts to each set of rollers and swifts, by means of which
the silk was enabled to be wound upon the bobbins
equably, instead of in wreaths as in other mills; and he so
arranged the shafting as to throw out of gear and stop
either the whole or any part of the machinery at will—
an arrangement subsequently adopted in the throstle of
the cotton-spinning machine, and though common enough
now, then thought perfectly marvellous. And, in order
that the tooth-and-pinion wheels should fit with perfect

precision, he expressly invented machinery for their manufacture—a thing that had not before been attempted—
all such wheels having, until then, been cut by hand at
great labour and cost. By means of this new machinery,
as much work, and of a far better description, could be
cut in a day as had before occupied at least a fortnight.
The result was, that the new silk-mill, when finished,
was found to be one of the most complete and economical
arrangements of manufacturing machinery that had up
to that time been erected in that neighbourhood.

After the Congleton silk-mill had been completed, we
find Brindley engaged in erecting flint-mills in the Potteries, of a more powerful and complete kind than any
that had before been tried, but which were rendered
necessary by the growing demands of the earthenware-
manufacture. One of the largest was that erected for
Mr. Thomas Baddely, at a place called Machins' of the
Mill, near Tunstall. We find these entries in Brindley's
pocket-book :—" March 15, 1757. With Mr. Badley to
Matherso about a now flint mill upon a windey day 1 day
3s. 6d. March 19 draing a plann 1 day 2s. 6d. March 23
draing a plann and to sat out the wheel race 1 day 4s."
This new mill was driven by water-power, and the wheel
both worked the pumping apparatus by which the adjoining coal-mine was drained, and the stamping machinery
for pounding and grinding the flints. The wheel, which
was of considerable diameter, was fixed in a chamber
below the surface of the ground, and the water was conveyed to it from the mill-pool through a small trough
opening upon it at its breast, which kept the paddle-
boxes of the descending part constantly filled, without
any waste whatever, and thus, by the rotation of the
wheel, the pumps and stampers were effectually worked.
The main shaft was more than two hundred yards from
the mill ; and to work the pumps Brindley then invented
the slide rods, which were moved horizontally by a crank
at the mill, and gave power to the upright arm of a crank-

lever, whose axis was at the angle, and the lift at the other extremity. In course of time, as improvements were introduced in the grinding of flints, the stamping apparatus was detached from the machinery; but this water-wheel continued its constant and silent operation of pumping out the mines for full forty years after the death of its inventor; and when it was at length broken up, about the year 1812, the pump-trees, which consisted of wooden staves firmly bound together with ashen hoops, were found to be lined with cow-hides, the working buckets being also covered with leather—a contrivance of which the like, it is believed, has not before been recorded.[1]

About the same time Brindley was requested by Mr. John Wedgwood to erect a windmill for a similar purpose on an elevated site adjoining the town of Burslem, called the Jenkins; this being one of the first, if not the very first, experiments made of the plan of grinding the calcined flints in water, which in this case was pumped by the action of the machinery from a well situated within the mill itself.   This invention, which was of considerable importance, has by some been attributed to Brindley, whose ingenious mind was ever ready to suggest improvements in whatever process of manufacture came under his notice.   It was natural that he should closely watch the operation of flint-grinding, having to construct and repair the greater part of the machinery used in the process; and he could not fail to notice the distressing consequences resulting from inhaling the fine particles with which the air of the flint-mills was laden. Hence the probability of his suggesting that the flints should be ground in water, as calculated not only to prevent waste and preserve the purity of the air, but also to facilitate the operation of grinding,—a simple enough suggestion, but, as the result proved, a most valuable one.   With this object he invented an improved

---

[1] 'History of the Borough of Stoke-upon-Trent.' By John Ward. 1853. P. 164.

mill, which consisted of a large circular vat, about thirty
inches deep, having a central step fixed in the bottom, to
carry the axis of a vertical shaft. The moving power
was applied to this shaft by a crown cog-wheel placed
on the top. At the lower part of the shaft, at right
angles to it, were four arms, upon which the grinding-
stones were fixed, large blocks of stone of the same kind
being likewise placed in the vat. These stones were a
very hard silicious mineral, called "Chert," found in
abundance in the neighbourhood of Bakewell, in Derbyshire. The broken flints being introduced to the vat
and completely covered with water, the axis was made
to revolve with great velocity, when the calcined flints
were easily reduced to an impalpable powder. This
contrivance of Brindley's proved of great value to Wedgwood, and it was shortly after adopted throughout the
Potteries, and continues in use to this day.

Being thus extensively occupied in the invention and
erection of machinery driven by one power or another,
it was natural that Brindley's attention should have been
attracted to the use of steam power in manufacturing
operations. Wind and water had heretofore been almost
the exclusive agents employed for the purpose; but far-
seeing philosophers and ingenious mechanics had for centuries been feeling their way towards the far greater
power derived from the pent-up force of vaporised water;
and engines had actually been contrived which rendered it
probable that the problem would ere long be solved, and
a motive agent invented, which should be easily con-
trollable, and independent alike of wind, tides, and water-
falls. Reserving for another place the history of the
successive stages of this great invention, it will be sufficient for our present purpose merely to indicate, briefly,
the direction of Brindley's labours in this important field.
It appears that Newcomen had as early as the year 1711
erected an atmospheric engine for the purpose of drawing
water from a coal-mine in the neighbourhood of Wolver-

hampton; and after considerable difficulties had been
experienced in its construction and working, the engine
was at length pronounced the most effective and economical that had yet been tried.    Other engines of a similar
kind were shortly after erected in the coal districts of the
north of England, in the tin and copper mines of Cornwall, and in the lead mines of Cumberland, for the
purpose of pumping water from the pits.    Brindley, like
other contrivers of power, felt curious about this new
invention, and proceeded to Wolverhampton to study
one of Newcomen's engines erected there.    He was
greatly struck by its appearance, and, with the irrepres-
sible instinct of the inventor, immediately set about
contriving how it might be improved.    He found the
consumption of coal so great as to preclude its use excepting where coal was unusually abundant and cheap,
as, for instance, at the mouth of a coal-pit, where the
fuel it consumed was the produce and often the refuse of
the mine itself; and he formed the opinion that unless the
consumption of coal could be reduced, the extended use
of the steam-engine was not practicable, by reason of its
dearness, as compared with the power of horses, wind, or
water.

With this idea in his head, he proceeded to contrive
an improved engine, the main object of which was to
ensure greater economy in fuel.    In 1756 we find him
erecting a steam-engine for one Mr. Broade, at Fenton
Vivian, in Staffordshire, in which he adopted the expedient, afterwards tried by James Watt, of wooden cylinders made in the manner of coopers' ware, instead of
cylinders of iron.    He also substituted wood for iron
in the chains which worked at the end of the beam.
Like Watt, however, he was under the necessity of
abandoning the wooden cylinders; but he surrounded
his metal cylinders with a wooden case, filling the intermediate space with wood-ashes; and by this means,
and using no more injection of cold water than was

necessary for the purpose of condensation, he succeeded
in reducing the waste of steam by almost one-half.
Whilst busy with Mr. Broade's engine, we find from
the entries in his pocket-book that Brindley occasionally
spent several days together at Coalbrookdale, to superintend the making of the boiler-plates, the pipes, and
other iron-work. Returned to Fenton Vivian, he proceeded with the erection of his engine-house and the
fitting of the machinery, whilst, during five days more,
he appears to have been occupied in making the hoops
for the cylinders. It takes him five days to get the
" great leavor fixed," thirty-nine days to put the boiler
together, and thirteen days to get the pit prepared ; and
as he charges only workman's wages for those days, we
infer that the greater part of the work was done by his
own hands. He even seems to have himself felled the
requisite timber for the work, as we infer from the
entry in his pocket-book of "falling big tree $3\frac{1}{2}$ days."

The engine was at length ready after about a year's
work, and was set a-going in November, 1757, after
which we find these significant entries : " Bad louk
[luck] five days ;" then, again, " Bad louk " for three
days more ; and, after that, " Midlin louk ;" and so
on with " Midlin louk " until the entries under that
head come to an end. In the spring of the following
year we find him again striving to get his " engon at
woork," and it seems at length to have been fairly
started on the 19th of March, when we have the entry
" Engon at woork 3 days." There is then a stoppage
of four days, and again the engine works for seven
days more, with a sort of " loud cheer " in the words
added to the entry, of " driv a-Heyd !" Other intervals
occur, until, on the 16th of April, we have the words
" at woor good ordor 3 days," when the entries come
to a sudden close. The engine must certainly have
given Brindley a great deal of trouble, and almost
driven him to despair, as we now know how very im-

perfect an engine with wooden hooped cylinders must
have been ; and we are not therefore surprised at the
entry which he honestly makes in his pocket-book on
the 21st of April, immediately after the one last mentioned, when the engine had, doubtless, a second time
broken down, " to Run about a Drinking, 0 : 1 : 6." Perhaps he intended the entry to stand there as a warning
against giving way to future despair ; for he underlined
the words, as if to mark them with unusual emphasis.

Brindley did not long give way to this mood, but set
to work upon the contrivance and erection of another
engine upon a new and improved plan.    What his plan
was, may be learnt from the specification lodged in the
Patent Office, on the 26th December, 1758, by " James
Brindley, of Leek, in the county of Stafford, Millwright." [2]   In the arrangement of this new steam-engine
he provided that the boiler should be made of brick or
stone arched over, and the stove over the fireplace of
cast-iron, fixed within the boiler.    The feeding-pipe for
the boiler was to be made with a clack, opening and
shutting by a float upon the surface of the water in the
boiler, which would thus be self-feeding.    The great
chains for the segments at the extremity of the beams
were of wood ; and the pumps were also of wooden staves
strongly hooped together.    Brindley seems, indeed, to
have long retained his early predilection as a millwright
for wood, and to have preferred it to iron wherever its
use was practicable.    His plans were, however, sub-

---

[1] We find the following memorandum in Brindley's pocket-book, relating
to the expense of working the engine
in the year 1760 :—

Miss Clare Maria Broad⁵ fire engine at fentan
vivian.

First yeer's work and repare
    night and day  ..  ..  ..  ..  £164
Do. turn back ..  ..  ..  ..  ..  025

Due for tᵉ first yeer  ..  ..  ..  139
Due for the second yeer ..  ..  ..  102

[2] He describes it as " A Fire-Engine for Drawing Water out of Mines,
or for Draining of Lands, or for Supplying of Cityes, Townes, or Gardens with Water, or which may be
applicable to many other great and
usefull Purposes, in a better and more
effectual Manner than any Engine or
Machine that hath hitherto been made
or used for the like Purpose."—' Speci-
fications of Patents,' No. 730.

jected to modification and improvement from time to
time, as experience pointed out; and in the course of a
few years, brick, stone, and wood were alike discarded
in favour of iron; until, in 1763, we find Brindley
erecting a steam-engine for the Walker Colliery, at
Newcastle, wholly of iron, manufactured at Coalbrookdale, which was pronounced the most "complete and
noble piece of ironwork" that had up to that time been
produced.[1] But by this time Brindley's genius had
been turned in another direction; the invention of
the steam-engine being now safe in the hands of Watt,
who was perseveringly occupied in bringing it to completion.

---

[1] Stuart's ' Anecdotes of Steam-Engines,' p. 626.

CHAPTER III.

THE BEGINNINGS OF CANALS—THE DUKE OF BRIDGEWATER.

VERY little had as yet been done to open up the inland navigation of England, beyond dredging and clearing out in a very imperfect manner the channels of several of the larger rivers, so as to admit of the passage of small barges. Several attempts had been made in Lancashire and Cheshire, as we have already shown, to open up the navigation of the Mersey and the Irwell from Liverpool to Manchester. There were similar projects for improving the Weaver from Frodsham, where it joins the Mersey, to Winford Bridge above Northwich; and the Douglas, from the Ribble to Wigan. About the same time like schemes were started in Yorkshire, with the object of opening up the navigation of the Aire and Calder to Leeds and Wakefield, and of the Don from Doncaster to near Sheffield. One of the Acts passed by Parliament in 1737 is worthy of notice, as probably the beginning of the Bridgewater Canal enterprise : we allude to the Act for making navigable the Worsley Brook to its junction with the river Irwell, near Manchester. A similar Act was obtained in 1755, for making navigable the Sankey Brook from the Mersey, about two miles below Warrington, to St. Helens, Gerrard Bridge, and Penny Bridge. In this case the canal was constructed separate from the brook, but alongside of it; and at several points locks were provided to adapt the canal to the level of the lands passed through.

The same year in which application was made to Parliament for powers to construct the Sankey Canal,

the Corporation of Liverpool had under their consideration a much larger scheme—no less than a canal to unite
the Trent and the Mersey, and thus open a water-communication between the ports of Liverpool and Hull.
It was proposed that the line should proceed by Chester,
Stafford, Derby, and Nottingham.   A survey was made,
principally at the instance of Mr. Hardman, a public
spirited merchant of Liverpool, and for many years one
of its representatives in Parliament.   Another survey
was made at the instance of Earl Gower, afterwards
Marquis of Stafford, and it was in making this survey
that Brindley's attention was first directed to the business
of canal engineering.   We find his first entry relating to
the subject was on the 5th of February, 1758—" novo-
cion [navigation] 5 days ;" the second, a little better
spelt, on the 19th of the same month—" a bout the novogation 3 days ;" and afterwards—" surveing the novogation from Long brigg to Kinges Milles 12 days ½." It
does not, however, appear that the scheme made much
progress, or that steps were taken at that time to bring
the measure before Parliament; and Brindley continued to pursue his other employments, more especially
the erection of " fire-engines " after his new patent.   This
continued until the following year, when we find him in
close consultation with the Duke of Bridgewater relative
to the construction of his proposed canal from Worsley
to Manchester.

The early career of this distinguished nobleman was of
a somewhat remarkable character.   He was born in 1736,
the fifth and youngest son of Scroop, third Earl and first
Duke of Bridgewater, by Lady Rachel Russell.   He lost
his father when only five years old, and all his. brothers
died by the time that he had reached his twelfth year, at
which early age he succeeded to the title of Duke of
Bridgewater.   He was a weak and sickly child, and his
mental capacity was thought so defective, that steps were
even in contemplation to set him aside in favour of the

next heir to the title and estates.  His mother seems almost entirely to have neglected him.  In the first year of her widowhood she married Sir Richard Lyttleton, and from that time forward took the least possible notice of her boy.  He did not give much promise of surviving his consumptive brothers, and his mind was considered so incapable of improvement, that he was left in a great measure without either domestic guidance or intellectual discipline and culture.  Horace Walpole writes to Mann in 1761 : " You will be happy in Sir Richard Lyttleton and his Duchess ; they are the best-humoured people in the world."  But the good humour of this handsome couple was mostly displayed in the world of gay life, very little of it being reserved for home use.  Possibly, however, it may have been even fortunate for the young Duke that he was left so much to himself, and to profit by the wholesome neglect of special nurses and tutors, who are not always the most judicious in their bringing up of delicate children.

At seventeen, the young Duke's guardians, the Duke of Bedford and Lord Trentham, finding him still alive and likely to live, determined to send him abroad on his travels—the wisest thing they could have done.  They selected for his tutor the celebrated traveller, Robert Wood, author of the well-known work on Troy, Baalbec, and Palmyra ; afterwards made Under-Secretary of State by the Earl of Chatham.  Wood was an accomplished scholar, a persevering traveller, and withal a man of good business qualities.  His habits of intelligent observation could not fail to be of service to his pupil, and it is not unnatural to suppose that the great artificial watercourses and canals which they saw in the course of their travels had some effect in afterwards determining the latter to undertake the important works of a similar character by which his name became so famous.  During their residence in Italy the Duke and his tutor visited all the galleries, and Mr. Wood sat to Mengs for his portrait, which still forms part of the Bridgewater collection. The Duke also purchased works of sculpture at Rome; but that he himself entertained no great enthusiasm for art is evident from the fact related by the late Earl of Ellesmere, that these works remained in their original packing-cases until after his death.[1]

Returned to England, he seems to have led the usual life of a gay young nobleman of the time, with plenty of money at his command. In 1756, when he was only twenty years of age, he appears from the 'Racing Calendar' to have kept race-horses; and he occasionally rode them himself. Though in after life a very bulky man, he was so light as a youth, that on one occasion, Lord Ellesmere says a bet was jokingly offered that he would be blown off his horse. Dressed in a livery of blue silk and silver, with a jockey cap, he once rode a race against His Royal Highness the Duke of Cumberland, on the long terrace at the back of the wood in Trentham Park, the seat of his relative, Earl Gower. During His Royal Highness's visit, the large old green-house, since taken down, was hastily run up for the playing of skittles; and prison-bars and other village games were instituted for the recreation of the guests. Those occupations of the Duke were varied by an occasional visit to his racing-stud at Newmarket, where he had a house for some time, and by the usual round of London gaieties during the season.

A young nobleman of tender age, moving freely in circles where were to be seen some of the finest specimens of female beauty in the world, could scarcely be expected to pass heart-whole; and hence the occurrence of the event in his London life which, singularly enough, is said to have driven him in a great measure from society, and induced him to devote himself to the con-

---

[1] 'Essays in History, Biography, Geography, Engineering,' &c.   By   the late Earl of Ellesmere.  London, 1858.  P. 226.

struction of canals! We find various allusions in the
letters of the time to the rumoured marriage of the
young Duke of Bridgewater. One rumour pointed to
the only daughter and heiress of Mr. Thomas Revell,
formerly M.P. for Dover, as the object of his choice.[1]
But it appears that the lady to whom he became the
most strongly attached was one of the Gunnings—the
comparatively portionless daughters of an Irish gentleman, who were then the reigning beauties at court.
The object of the Duke's affection was Elizabeth, the
youngest daughter, and perhaps the most beautiful of
the three. She had been married to the fourth Duke of
Hamilton, in Keith's Chapel, Mayfair, in 1752, " with a
ring of the bed-curtain, half-an-hour after twelve at
night," [2] but the Duke dying shortly after, she was now
a gay and beautiful widow, with many lovers in her
train. In the same year in which she had been clandes-
tinely married to the Duke of Hamilton, her eldest
sister was married to the sixth Earl of Coventry.

The Duke of Bridgewater paid his court to the young
widow, proposed, and was accepted. The arrangements
for the marriage were in progress, when certain rumours
reached his ear reflecting seriously upon the character of
Lady Coventry, his intended bride's elder sister, who
was certainly more fair than she was wise. Believing
the reports, he required the Duchess to desist from
further intimacy with her sister, a condition which her
high spirit would not brook, and, the Duke remaining

---

[1] Thomas and Maria Revell were
both servants in the family of Mr.
Nightingale, of Knibsworth. They
afterwards married, and took a farm
at Shingay, under my Lord Orford,
who, taking a liking to their two
eldest sons, Thomas and Russell, gave
them an English education, and got
them both places in the Victualling
Office. The eldest, Thomas, was
M.P. for Dover, and, dying in 1752
at Bath, was buried, as I think, at or
near Leatherhead, Surrey, leaving an
only daughter behind him, to whom
he left about 120,000l. or 130,000l. It
is thought she is to be married to the
present Duke of Bridgewater, her
cousin.—'The Cole MSS.' (British
Museum), vol. ix., 113.

[2] 'Walpole to Mann,' Feb. 27th,
1752.

PORTRAIT OF THE YOUNG DUKE.

[By T D. Scott.]

firm, the match was broken off. From that time forward he is said never to have addressed another woman in the language of gallantry.[1] The Duchess of Hamilton, however, did not remain long a widow. In the course of a few months she was engaged to, and afterwards married, John Campbell, subsequently Duke of Argyll. Horace Walpole, writing of the affair to Marshal Conway, January 28th, 1759, says : "You and M. de Bareil do not exchange prisoners with half as much alacrity as Jack Campbell and the Duchess of Hamilton have exchanged hearts. .. . . It is the prettiest match in the world since yours, and everybody likes it but the Duke of Bridgewater and Lord Conway. What an extraordinary fate is attached to these two women! Who could have believed that a Gunning would unite the two great houses of Campbell and Hamilton ? For my part, I expect to see my Lady Coventry Queen of Prussia. I would not venture to marry either of them these thirty years, for fear of being shuffled out of the world prematurely to make room for the rest of their adventures."

The Duke of Bridgewater, like a wise man, seems to have taken refuge from his disappointment in active and useful occupation. Instead of retiring to his beautiful seat at Ashridge, we find him straightway proceeding to his estate at Worsley, on the borders of Chat Moss, in Lancashire, and conferring with John Gilbert, his land-steward, as to the practicability of cutting a canal by which the coals found upon his Worsley estate might be readily conveyed to market at Manchester.

Manchester and Liverpool at that time were improving towns, gradually rising in importance and increasing in population. The former place had long been noted for

---

[1] Chalmers, in his 'Biographical Dictionary,' vol. xiii., 94, gives another account of the rumoured cause of the Duke's subsequent antipathy to women; but the above statement of the late Earl of Ellesmere, confirmed as it is by certain passages in Walpole's Letters, is more likely to be the correct one.

its manufacture of coarse cottons or coatings made of wool, in imitation of the goods known on the Continent by that name. The Manchester people also made fustians, mixed stuffs, and small wares, amongst which leather-laces for women's bodice, shoe-ties, and points were the more important. But the operations of manufacture were still carried on in a clumsy way, entirely by hand. The wool was spun into yarn by means of the common spinning-wheel, for the spinning-jenny had not yet been discovered, and the yarn was woven into cloth by the common hand-loom. There was no whirr of engine-wheels then to be heard ; for Watt's steam-engine had not yet been invented. The air was free from smoke, except what arose from household fires, and there was not a single factory-chimney in Manchester. In 1724 Dr. Stukeley says Manchester contained no fewer than 2400 families, and that their trade

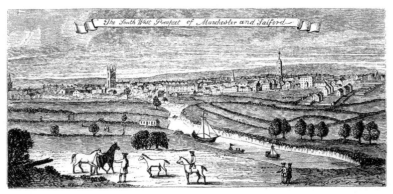

VIEW OF MANCHESTER IN 1740.
[Fac-simile of an Engraving of the period by J. Harris, published by Robert Whitworth.]

was "incredibly large " in tapes, ticking, girth-webb, and fustians. In 1757 the united population of Manchester and Salford was only 20,000 ;[1] it is now, after the lapse of a century, 460,000 ! The Manchester

---

[1] Aikin's 'Description of the Country from Thirty to Forty Miles round Manchester.' London, 1795.

manufacturer was then a very humble personage compared with his modern representative. He was part chapman, part weaver, and part merchant—working hard, living frugally, principally on oatmeal,[1] and contriving to save a little money. As trade increased, its operations became more subdivided, and special classes and ranks began to spring into importance. The manufacturers sent out riders to take orders, and chapmen with gangs of pack-horses to distribute the goods and bring back wool in exchange. The chapmen used packhorses because the roads were as yet mostly impracticable for waggons, and it was more difficult then to reach a village twenty miles out of Manchester than it is to make the journey from thence to London now. Indeed, the only coach to London plied but every second day, and it was four days and a-half in making the journey, there being a post only three times a week.[2] The roads in most districts of Lancashire were what were called "mill roads," along which a horse with a load of oats upon its back might proceed towards the mill where they were to be ground. There was no private carriage kept by any person in business in Manchester until the year 1758, when the first was set up by some specially luxurious individual. But wealth led to increase of expenditure, and Aikin mentions that there was " an evening club of the most opulent manufacturers, at which the expenses of each person were fixed

---

[1] Dr. Aikin, in 1795, gave the following description of the Manchester manufacturer in the first half of the eighteenth century : " An eminent manufacturer in that age," said he, " used to be in his warehouse before six in the morning, accompanied by his children and apprentices. At seven they all came in to breakfast, which consisted of one large dish of water-pottage, made of oatmeal, water, and a little salt, boiled thick, and poured into a dish. At the side was a pan or basin of milk, and the master and apprentices, each with a wooden spoon in his hand, without loss of time, dipped into the same dish, and thence into the milk-pan, and as soon as it was finished they all returned to their work." What a contrast with the " eminent manufacturer" of our own day !

[2] March 3rd, 1760, the Flying Machine was started, and advertised to perform the journey, " if God permit," in three days, by John Hanforth, Matthew Howe, Samuel Granville, and William Richardson. Fare inside, 2l. 5s.; outside, half-price.

at fourpence-halfpenny—fourpence for ale, and a half-penny for tobacco." The progress of luxury was further aided by the holding of a dancing assembly once a-week in a room situated at the middle of the now fashionable street called King Street, the charge for which was half-a-crown the quarter; the ladies having their maids to come with lanterns and pattens to conduct them home; "nor," adds Aikin, "was it unusual for their partners also to attend them." [1]

The imperfect state of the communications leading to and from Manchester rendered it a matter of some difficulty at certain seasons to provide food for so large a population. In winter, when the roads were closed, the place was in the condition of a beleaguered town; and even in summer, the land about Manchester itself being comparatively sterile, the place was badly supplied with fruit, vegetables, and potatoes, which, being brought from considerable distances slung across horses' backs, were so dear as to be beyond the reach of the mass of the population. The distress caused by this frequent dearth of provisions was not effectually remedied until the canal navigation became completely opened up. Thus a great scarcity of food occurred in Manchester and the neighbourhood in 1757, which the common people attributed to the millers and corndealers; and unfortunately the notion was not confined to the poor who were starving, but was equally entertained by the well-to-do classes who had enough to eat. An epigram by Dr. Byrom, the town clergyman, written in 1737, on two millers (tenants of the School corn-mills), who, from their spare habits, had been nicknamed "Skin" and "Bone," was now revived, and tended to fan the popular fury. It ran thus :—

> " Bone and Skin, two millers thin,
>     Would starve the town, or near it;
> But be it known to Skin and Bone,
>     That Flesh and Blood can't bear it."

---

[1] Aikin, p. 187.

The result of the popular hunger was, that a great commotion occurred, which at length broke out in open outrage, and a riot took place in 1758, long after remembered in Manchester as the "Shude Hill fight," in which several lives were unhappily lost.[1]

For the same reason the supply of coals was scanty in winter; and though abundance of the article lay underground, within a few miles of Manchester, in nearly every direction, those few miles of transport, in the then state of the roads, were an almost insurmountable difficulty. The coals were sold at the pit mouth at so much the horse-load, weighing 280 lbs., and measuring two baskets, each thirty inches by twenty, and ten inches deep; that is, as much as an average horse could carry on its back.[2]    The price of the coals at the pit mouth was 10$d$. the horse-load; but by the time the article reached the door of the consumer in Manchester, the price was usually more than doubled, in consequence of the difficulty and cost of conveyance.    The carriage alone amounted to about nine or ten shillings the ton.    There was as yet no connection of the navigation of the Mersey and Irwell with any of the collieries situated to the eastward of Manchester, by which a supply could reach the town in boats; and although the Duke's collieries were only a comparatively short distance from the Irwell, the coals had to be carried on horses' backs or in carts from the pits to the river to be loaded, and after reaching Manchester they had again

---

[1] In 1715 the first London baker settled in Manchester, Mr. Thomas Hatfield, known by his styptic. His apprentices took the mills in the vicinity, and in time reduced the inhabitants to the necessity of buying flour of them. Monopolies at length took place in consequence of these changes, which, at different times, produced riots; one of which, occasioned by a large party of country people coming to Manchester in order to destroy the mills, ended in the loss of several lives, at a fray known by

the name of Shude Hill fight, in the year 1758. Since that time until the present [1795] the demand for corn and flour has been increasing to a vast amount, and new sources of supply have been opened from distant parts by the navigations, so that monopoly or scarcity cannot be apprehended.—Aikin's 'Manchester.'

[2] This "load" is still used as a measure of weight, though the practice of carrying all sorts of commodities on horses' backs, in which it originated, has long since ceased.

to be carried to the doors of the consumers,—so that there was little if any saving to be effected by that route. Besides, the minimum charge insisted on by the Mersey Navigation Company of 3s. 4d. a ton for even the shortest distance, proved an effectual barrier against any coal reaching Manchester by the river.

The same difficulty stood in the way of the transit of goods between Manchester and Liverpool. By road the charge was 40s. a ton, and by river 12s. a ton; that between Warrington and Manchester being 10s. a ton : besides, there was great risk of delay, loss, and damage by the way. Some idea of the tediousness of the river-navigation may be formed from the fact, that the boats were dragged up and down stream exclusively by the labour of men, and that horses and mules were not employed for this purpose until after the Duke's canal had been made. It was, indeed, obvious that unless some means could be devised for facilitating and cheapening the cost of transport between the seaport and the manufacturing towns, there was little prospect of any considerable further development being effected in the industry of the district.

Such was the state of things when the Duke of Bridgewater turned his attention to the making of a water-road for the passage of his coal from Worsley to Manchester. The Old Mersey Company would give him no facilities for sending his coals by their navigation, but levied the full charge of 3s. 4d. for every ton he might send to Manchester by river even in his own boats. He therefore perceived that to obtain a vend for his article, it was necessary he should make a way for himself; and it became obvious to him that if he could but form a water-road or canal between the two points, he would at once be enabled to secure a ready sale for all the coals that he could raise from his Worsley pits.

## CHAPTER IV.

### THE BRIDGEWATER CANAL FROM WORSLEY TO MANCHESTER.

WE have already stated that, as early as 1737, an Act had been obtained by the Duke's father, to enable the Worsley Brook to be made navigable to the point at which it entered the Irwell. But the enterprise seemed to be too difficult, and its cost too great; so the powers of the Act were allowed to expire without anything being done to carry them out. The young Duke now determined to revive the Act in another form, and in the early part of 1759 he applied to Parliament for the requisite powers to enable him to cut a navigable canal from Worsley Mill eastward to Salford, and to carry the same westward to a point on the river Mersey, called Hollin Ferry. He introduced into the bill several important concessions to the inhabitants of Manchester. He bound himself not to exceed the freight of 2s. 6d. per ton on all coals brought from Worsley to Manchester, and not to sell the coal so brought from the mines to that town at more than 4d. per hundred, which was less than half the then average price. It was clear that, could such a canal be made and the navigation opened up as proposed, it would prove a great public boon to the inhabitants of Manchester, and it was hailed by them as such accordingly. The bill was well supported, and it passed the legislature without opposition, receiving the Royal assent in March, 1759.

The Duke gave further indications of his promptitude and energy, in the steps which he adopted to have the works carried out without loss of time. He had no intention of allowing the powers of this Act to remain a dead letter, as the former had done. Accordingly, no sooner had it passed than he set out for his seat at Worsley to take the requisite measures for constructing the canal. The Duke was fortunate in having for his land-agent a very shrewd, practical, and enterprising person, in John Gilbert, whom he consulted on all occasions of difficulty. Mr. Gilbert was the brother of Thomas Gilbert, the originator of the Gilbert Unions, then agent to the Duke's brother-in-law, Lord Gower. That nobleman had for some time been promoting the survey of a canal to unite the Mersey and the Trent, on which Brindley had been employed, who was thus well known to Gilbert as well as to his brother. We find from an entry in his pocket-book, that the millwright had sundry interviews with Thomas Gilbert on matters of business previous to the passing of the first Bridgewater Canal Bill, though there is no evidence that Brindley was employed in making the survey. Indeed, it is questionable whether any survey was made of the first scheme,—engineering projects being then submitted to Parliamentary Committees in a very rough state; levels being guessed at rather than surveyed and calculated; and merely general powers taken enabling such property to be purchased as might by possibility be required for the execution of the works —the prices of land and compensation for damage being assessed by a local committee appointed by the Act for the purpose.

When the Duke proceeded to consider with Gilbert the best mode of carrying out the proposed canal, it very shortly appeared that the plan originally contemplated was faulty in many respects, and that an application must be made to Parliament for further powers. By the original Act it was intended to descend from the level of the coal-mines at Worsley by a series of locks into the river Irwell. This, it was found, would necessarily involve both a heavy cost in the construction and working of the canal, as well as considerable delay

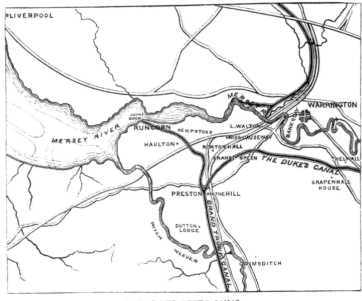

MAP OF THE DUKE'S CANAL.

[Western Part.]

in the conduct of the traffic, which it was most desirable
to avoid. Neither the Duke nor Gilbert had any practical knowledge of engineering ; nor, indeed, were there
many men in the country at that time who knew
much of the subject. For it must be remembered that
this canal of the Duke's was the very first project in
England for cutting a navigable trench through the dry
land, and carrying merchandise across the country in it
independent of the course of the existing streams.

It was in this emergency that Gilbert advised the
Duke to call to his aid James Brindley, whose fertility
of resources and skill in overcoming mechanical difficulties had long been the theme of general admiration
in the district. Doubtless the Duke was as much impressed by the native vigour and originality of the unlettered genius thus introduced to him, as were all with
whom he was brought in contact. Certain it was that
the Duke showed his confidence in Brindley by en-

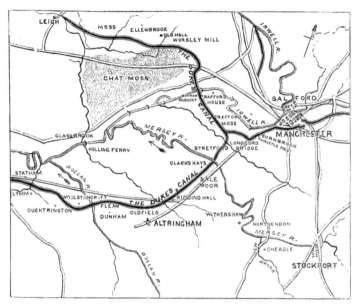

MAP OF THE DUKE'S CANAL.

[Eastern Part.]

trusting him with the conduct of the proposed work;
and, as the first step, he was desired to go over the
ground at once, and give his opinion as to the best
plan to be adopted for carrying it out with dispatch.
Brindley, accordingly, after making what he termed an
" ochilor [ocular] servey or a ricconitoring," speedily
formed his conclusion, and came back to the Duke with
his advice.  It was that, instead of carrying the canal
down into the Irwell by a flight of locks, and so up
again on the other side to the proposed level, it should be
carried right over the river, and constructed on one
entire level throughout.  But this, it was clear, would
involve a series of formidable works, the like of which
had never before been attempted in England.  In the first
place, the low ground on the north side of the Irwell
would have to be filled up by a formidable embankment,
and united with the land on the other bank by means of
a large aqueduct of stone.  Would it be practicable or

possible to execute works of such magnitude ? Brindley expressed so strong and decided an opinion of their practicability, that the Duke became won over to his views, and determined again to go to Parliament for the requisite powers to enable the design to be carried out.

Many were the deliberations which took place about this time between the Duke, Gilbert, and Brindley, in the Old Hall at Worsley, where the Duke had now taken up his abode. We find from Brindley's pocket-book memoranda, that in the month of July, 1759, he had taken up his temporary quarters at the Old Hall; and from time to time, in the course of the same year, while the details of the plan were being prepared with a view to the intended application to Parliament, he occasionally stayed with the Duke for several weeks together. He made a detailed survey of the new line, and at the same time, in order to facilitate the completion of the undertaking when the new powers had been obtained, he proceeded with the construction of the sough or level at Worsley Mill, and such other portions of the work as could be executed under the original powers.

During the same period Brindley travelled backwards and forwards a great deal, on matters connected with his various business in the Pottery district. We find, from his record, that he was occupied at intervals in carrying forward his survey of the proposed canal through Staffordshire, visiting with this object the neighbourhood of Newcastle-under-Lyme, Lichfield, and Tamworth. He also continued to give his attention to mills, water-wheels, cranes, and fire-engines, which he had erected, or required repairs, in various parts of the same district. In short, he seems at this time to have been fully employed as a millwright; and although, as we have seen, the remuneration which he received for his skill was comparatively small, being a man of frugal habits he had saved a little money; for about this time we find him able to raise a sum of 543*l.* 6*s.* 8*d.*, being his fourth share of the purchase-money of the Turnhurst estate, situated near Golden Hill, in the county of Stafford.   It appears, however, from his own record that he borrowed the principal part of this amount from his friend Mr. Launcelot, of Leek; showing that, amongst his townsmen and neighbours, who knew him best, he stood in good credit and repute.   His other partners in the purchase were Mr. Thomas Gilbert (Earl Gower's agent), Mr. Henshall (afterwards his brother-in-law), and his brother John Brindley.   The estate was understood to be full of minerals, the knowledge of which had most probably been obtained by Brindley in the course of his surveying of the proposed Staffordshire canal; and we shall afterwards find that the purchase proved a good investment.

At length the new plans of the canal from Worsley to Manchester were completed and ready for deposit; and on the 23rd of January, after a visit to the Duke and Gilbert at the Hall, we find the entry in Brindley's pocket-book of "Sot out for London."   On the occasion of his visits to London, Brindley adopted the then most convenient method of travelling on horseback, the journey usually occupying five days.   We find him varying his route according to the state of the weather and of the roads.   In summer he was accustomed to go by Coventry, but in winter he made for the Great North Road by Northampton, which was usually in better condition for winter travelling.

The second Act passed without opposition, like the first, early in the session of 1760.   It enabled the Duke to carry his proposed canal *over* the river Irwell, near Barton Bridge, some five miles westward of Manchester, by means of a series of arches, and to vary its course accordingly; whilst it further authorised him to extend a short branch to Longford Bridge, near Stretford,—that to Hollin Ferry, authorised by the original Act, being abandoned.   In the mean time the works near

Worsley had been actively pushed forward, and considerable progress had been made by the time the additional powers had been obtained. That part of the canal which lay between Worsley Mill and the public highway leading from Manchester to Warrington had been made; the sough or level between Worsley Mill and Middlewood, for the purpose of supplying water to the canal, was considerably advanced; and operations had also been begun in the neighbourhood of Salford and on the south of the river Irwell.

The most difficult part of the undertaking, however, was that authorised by the new Act; and the Duke looked forward to its execution with the greatest possible anxiety. Although aqueducts of a far more formidable description had been executed abroad, nothing of the kind had until then been projected in this country; and many regarded the plan of Brindley as altogether wild and impracticable. The proposal to confine and carry a body of water within a water-tight trunk of earth upon the top of an embankment across the low grounds on either side of the Irwell, was considered foolish and impossible enough; but to propose to carry ships upon a lofty bridge, over the head of other ships navigating the Irwell which flowed underneath, was laughed at as the dream of a madman. Brindley, by leaving the beaten path, thus found himself exposed to the usual penalties which befall originality and genius.

The Duke was expostulated with by his friends, and strongly advised not to throw away his money upon so desperate an undertaking. Who ever heard of so large a body of water being carried over another in the manner proposed? Brindley was himself appealed to; but he could only repeat his conviction as to the entire practicability of his design. At length, by his own desire and to allay the Duke's apprehensions, another engineer was called in and consulted as to the scheme. To Brindley's surprise and dismay, the person consulted concurred in the view so strongly expressed by the public. He characterised the plan of the Barton aqueduct and embankment as instinct with recklessness and folly; and after expressing his unqualified opinion as to the impracticability of executing the design, he concluded his report to the Duke thus: "I have often heard of castles in the air; but never before saw where any of them were to be erected." [1]

It is to the credit of the Duke that, notwithstanding these strongly adverse opinions, he continued to give his confidence to the engineer whom he had selected to carry out the work. Brindley's common-sense explanations, though they might not remove all his doubts, nevertheless determined him to give him the full opportunity of carrying out his design; and he was accordingly authorised to proceed with the erection of his "castle in the air." Its progress was watched with great interest, and people flocked from all parts to see it.

The Barton aqueduct is about two hundred yards in length and twelve yards wide, the centre part being sustained by a bridge of three semicircular arches, the middle one being of sixty-three feet span. It carries the canal over the Irwell at a height of thirty-nine feet above the river—this head-room being sufficient to enable the largest barges to pass underneath without lowering their masts. The bridge is entirely of stone blocks, those on the faces being dressed on the front, beds, and joints, and cramped with iron. The canal, in passing over the arches, is confined within a puddled [2] channel to prevent leakage, and is in as good

---

[1] We have heard the name of Smeaton mentioned as that of the engineer consulted on the occasion, but we are unable to speak with certainty on the point. Excepting Smeaton, however, there was then no other engineer in the country of recognised eminence in the profession.

[2] The process of puddling is of considerable importance in canal engineering. Puddle is formed by a mixture of well-tempered clay and sand reduced to a semi-fluid state, and rendered impervious to water by manual labour, as by working and chopping it about with spades. It is usually applied in three or more strata to a depth or thickness of about three feet; and care is taken at each operation so to work the new layer of puddling stuff as to unite it with the stratum immediately beneath. Over the top course a layer of common soil is usually laid. It is only by the

a state now as on the day on which it was completed. Although the Barton aqueduct has since been thrown into the shade by the vastly greater works of modern

BARTON AQUEDUCT.

[By Percival Skelton, after his original Drawing.]

engineers, it was unquestionably a very bold and ingenious enterprise, if we take into account the time at which it was erected. Humble though it now appears, it was the parent of the magnificent aqueducts of Rennie and Telford, and of the viaducts of Stephenson and Brunel, which rival the greatest works of any age or country.

The embankments formed across the low grounds on

careful employment of puddling that the filtration of the water of canals into the neighbouring lower lands through which they pass can be effectually prevented.

either side of the Barton viaduct were also considered very formidable works at that day.  A contemporary writer speaks of the embankment across Stretford Meadows as an amazing bank of earth 900 yards long, 112 feet in breadth across the base, 24 feet at top, and 17 feet high. The greatest difficulty anticipated, was the holding of so large a body of water within a hollow channel formed of soft materials.  It was supposed at first that the water would soak through the bank, which its weight would soon burst, and wash away all before it.  But Brindley, in the course of his experience, had learnt something of the powers of clay-puddle to resist the passage of water. He had already succeeded in stopping the breaches of rivers flowing through low grounds by this means; and the thorough manner in which he finished the bed of this canal, and made it impervious to water, may be cited as not the least remarkable illustration of the engineer's practical skill, taking into account the early period at which this work was executed.  Not the least difficult part of the undertaking was the formation of the canal across Trafford Moss, where the weight of the embankment pressed down and "blew up" the soft oozy stuff on either side; but the difficulty was again overcome by the engineer's specific of clay-puddle, which proved completely successful.  Indeed, the execution of these embankments by Brindley was regarded at the time as something quite as extraordinary in their way as the erection of the Barton aqueduct itself.

The rest of the canal between Longford and Manchester, being mostly on sidelong ground, was cut down on the upper side and embanked up on the other by means of the excavated earth.  This was comparatively easy work; but a matter of greater difficulty was to accommodate the streams which flowed across the course of the canal, and which were provided for in a highly ingenious manner.  For instance, a stream called Cornbrook was found too high to pass under the canal at its natural level. Accordingly, Brindley contrived a weir, over which the stream fell into a large basin, from whence it flowed into a smaller one open at the bottom. From this point a culvert, constructed under the bed of the canal, carried the waters across to a well situated on its further side, where the waters, rising up to their natural level, again flowed away in their proper channel. A similar expedient was adopted at the Manchester terminus of the canal, at the point at which it joined the waters of the Medlock. It was a principle of Brindley's never to permit the waters of any river or brook to intermix with those of the canal except for the purpose of supply ; as it was clear that in a time of flood such intermingling would be a source of great danger to the navigation. In order, therefore, to provide for the free passage of the Medlock without causing a rush into the canal, a weir was contrived, 366 yards in circumference, over which its waters flowed into a lower level, and from thence into a well several yards in depth, down which the whole river fell. It was received at the bottom in a subterranean passage, by which it passed into the river Irwell, near at hand. The weir was very ingeniously contrived, though it was afterwards found necessary to make considerable alterations and improvements in it, as experience suggested, in order effectually to accommodate the flood-waters of the Medlock. Arthur Young, when visiting the canal, shortly after it was opened up to Manchester, says, "The whole plan of these works shows a capacity and extent of mind which foresees difficulties, and invents remedies in anticipation of possible evils. The connection and dependence of the parts upon each other are happily imagined ; and all are exerted in concert, to command by every means the wished-for success."[1]

Brindley's labours, however, were not confined to the

---

[1] 'Six Months' Tour through the North of England,' vol. iii., p. 258. Ed. 1770.

construction of the canal, but his attention seems to have been equally directed to the contrivance of the whole arrangements and machinery by which it was worked. The open navigation between Worsley Mill and Manchester was 10¼ miles in length. At Worsley, where a large basin was excavated of sufficient capacity to contain a great many boats, and to serve as a head for the navigation, the canal did not stop, but entered the bottom of the hill by a subterraneous channel which extended for a great distance,—connecting the different workings of the mine, and enabling the coals readily to be transported in boats to their place of sale. In Brindley's time, this subterraneous canal, hewn out of the rock, was only about a mile in length, but it now extends to nearly forty miles in all directions underground.[1] Where the tunnel passed through earth or coal, the arching was of brick-work; but where it passed through rock, it was simply hewn out. This tunnel acts not only as a drain and water-feeder for the canal itself, but as a means of carrying the facilities of the navigation through the very heart of the collieries; and it will readily be seen of how great a value it must have proved in the economical working of the navigation, as well as of the mines, so far as the traffic in coals was concerned.

---

[1] Worsley-basin lies at the base of a cliff of sandstone, some hundred feet in height. [See the cut at p. 306.] Luxuriant foliage overhangs its precipitous side, and beyond is seen the graceful spire of Worsley church. In contrast to this bright nature above, lies the almost stagnant pool beneath. The barges are deeply laden with their black freight, which they have brought from the mine through the two low, semi-circular arches opening at the base of the rock, such being the entrances to the underground canals. The smaller aperture is the mouth of a canal of only half a mile in length, serving to prevent the obstruction which would be caused by the entrance and egress of so many barges through a single passage. The other archway is the entrance of a wider channel, extending nearly six miles in the direction of Bolton, and from which various other canals diverge in different directions. The barges are narrow and long, each conveying about ten tons of coal. They are drawn along the tunnels by means of staples fixed along the sides. When they are empty, and consequently higher in the water, they are so near the roof that the bargemen, lying on their backs, can propel them with their feet.

At every point Brindley's originality and skill were at work. He invented the cranes for the purpose of more readily loading the boats with the boxes filled with the Duke's "black diamonds." He also contrived and laid down within the mines a system of underground railways, all leading from the face of the coal (where the miners were at work) to the wells which he had made at different points in the tunnels, through which the coals were shot into the boats waiting below to receive them. At Manchester, where they were unloaded for sale, the contrivances which he employed were equally ingenious. It was at first intended that the canal should terminate at the foot of Castle Hill, up which the coals were dragged by their purchasers from the boats in wheelbarrows or carts. But the toil of dragging the loads up the hill was found very great; and, to remedy the inconvenience, Brindley contrived to extend the canal for some way into the hill, opening a shaft from the surface of the ground down to the level of the water. The barges having made their way to the foot of this shaft, the boxes of coal were hoisted to the surface by a crane, worked by a box waterwheel of 30 feet diameter and 4 feet 4 inches wide, driven by the waterfall of the river Medlock. In this contrivance Brindley was only adopting a modification of the losing and gaining bucket, moved on a vertical pillar, which he had before successfully employed in drawing water out of coal-mines. By these means the coals were rapidly raised to the higher ground, where they were sold and distributed, greatly to the convenience of those who came to purchase them.

Brindley's practical ability was equally displayed in planning and building a viaduct and in fitting up a crane —in carrying out an embankment or in contriving a coal-barge. The range and fertility of his constructive genius were extraordinary. For the Duke, he invented water-weights at Rough Close, riddles to wash coal for

the forges, rising dams, and numerous other contrivances
of well-adapted mechanism. At Worsley he erected a
steam-engine for draining those parts of the mine which
were beneath the level of the canal, and consequently
could not be drained into it; and he is said to have erected,
at a cost of only 150*l.*, an engine which until that time
no one had known how to construct for less than 500*l.*
At the mouth of one of the mines he erected a water-
bellows for the purpose of forcing fresh air into the
interior, and thus ventilating the workings.[1] At the
entrance of the underground canal he designed and built
an over-shot mill of a new construction, driven by a wheel
twenty-four feet in diameter, which worked three pair of
stones for grinding corn, besides a dressing or boulting
mill, and a machine for sifting sand and mixing mortar.[2]
Brindley's quickness of observation and readiness in
turning circumstances to account, were equally displayed
in the mode by which he contrived to obtain an ample
supply of lime for building purposes during the progress
of the works. We give the account as related by Arthur
Young:—"In carrying on the navigation," he observes,
"a vast quantity of masonry was necessary for building
aqueducts, bridges, warehouses, wharves, &c., and the
want of lime was felt severely. The search that was
made for matters that would burn into lime was for a
long time fruitless. At last Mr. Brindley met with a
substance of a chalky kind, which, like the rest, he tried;
but found (though it was of a limestone nature—lime-

---

[1] A writer in the 'St. James's
Chronicle,' under date the 30th of
September, 1763, gives the following
account of this apparatus, long since
removed:—"At the mouth of the
cavern is erected a water-bellows,
being the body of a tree, forming a
hollow cylinder, standing upright.
Upon this a wooden bason is fixed, in
the form of a funnel, which receives a
current of water from the higher
ground. This water falls into the
cylinder, and issues out at the bottom
of it, but at the same time carries a
quantity of air with it, which is received into the pipes and forced to
the innermost recesses of the coal-
pits, where it issues out as if from a
pair of bellows, and rarefies the body
of thick air, which would otherwise
prevent the workmen from subsisting
on the spot where the coals are dug."

[2] Young's 'Six Months' Tour,' vol.
iii., p. 278.

marl, which was found along the sides of the canal, about a foot below the surface) that, for want of adhesion in the parts, it would not make lime. This most inventive genius happily fell upon an expedient to remedy this misfortune. He thought of tempering this earth in the nature of brick-earth, casting it in moulds like bricks, and then burning it; and the success was answerable to his wishes. In that state it burnt readily into excellent lime; and this acquisition was one of the most important that could have been made. I have heard it asserted more than once that this stroke was better than twenty thousand pounds in the Duke's pocket; but, like most common assertions of the same kind, it is probably an exaggeration. However, whether the discovery was worth five, ten, or twenty thousand, it certainly was of noble use, and forwarded all the works in an extraordinary manner." [1]

It has been stated that Brindley's nervous excitement was so great on the occasion of the letting of the water into the canal, that he took to his bed at the Wheat-sheaf, in Stretford, and lay there until all cause for apprehension was over. The tension on his brain must have been great, with so tremendous a load of work and anxiety upon him; but that he "ran away," [2] as some of

---

[1] 'Six Months' Tour,' vol. iii., p. 270–1. Mr. Hughes, C.E., says of this discovery : " The lime thus made would appear to be the first cement of which we have any knowledge in this country; since the calcareous marl here spoken of would probably produce, when burnt, a lime of strong hydraulic properties."

[2] This story was first set on foot, we believe, by the Earl of Bridgewater, in his singularly incoherent publication entitled, ' A Letter to the Parisians and the French Nation upon Inland Navigation, containing a defence of the public character of His Grace Francis Egerton, late Duke of Bridgewater. By the Hon. Francis Henry Egerton.' The first part of this curious book (published at Paris) was dated " Hôtel Egerton, Paris, 21st Dec., 1818;" the second part was published two years later; and a third part, consisting entirely of a note about Hebrew interpretations, was published subsequently. He had in the mean time become Earl of Bridgewater, in October, 1823, having formerly been prebendary of Durham and rector of Whitchurch in Shropshire. The late Earl of Ellesmere, in his ' Essays on History, Biography,' &c., says of this nobleman that " he died at Paris in the odour of eccentricity." But this is a mild description of his lordship, who had at least a dozen distinct crazes—about canals, the Jews, punctuation, the wonderful

his detractors have alleged, is at variance with the whole character and history of the man.

The Duke's canal, when finished, was for a long time regarded as the wonder of the neighbourhood. Strangers flocked from a distance to see Brindley's " castle in the air ;" and contemporary writers spoke in glowing terms of the surprise with which they saw several barges of great burthen drawn by a single mule or horse along " a river hung in the air," and over another river underneath, by the side of which some ten or twelve men might be seen slowly hauling a single barge against the stream. A lady who writes a description of the work in 1765, speaks of it as " perhaps the greatest artificial curiosity in the world ;" and she states that " crowds of people, including those of the first fashion, resort to it daily." [1] The chief value of the work, however, consisted in its uses. Manchester was now regularly and cheaply supplied with coals. The average price was at once reduced by one-half—from 7$d$. the cwt. to 3$\frac{1}{2}d$. (six score being given to the cwt.)—and the supply was regular instead of intermitting, as it had formerly been. But the full advantages of this improved supply of coals were not experienced until many years after the opening of the canal, when the invention of the steam-engine, and its extensive employment as a motive power in all manufacturing operations, rendered a cheap and abundant supply of fuel of such vital importance to the growth and prosperity of Manchester and its neighbourhood.

merits of the Egertons, the proper translation of Hebrew, the ancient languages generally, but more especially about prophecy and poodle-dogs. When he drove along the Boulevards in Paris, nothing could be seen of his lordship for poodle-dogs looking out of the carriage-windows. The poodles sat at table with him at dinner, each being waited on by a special valet. The most creditable thing the Earl did was to leave the sum of 12,000$l$. to the British Museum, and 8000$l$. to meritorious literary men for writing the well-known 'Bridgewater Treatises.' He died in February, 1829.

[1] Mr. Newbery's 'Lady's Pocketbook.'

## CHAPTER V.

### Proposed Extension of the Duke's Canal to the Mersey.

The Canal had scarcely been opened to Manchester when we find Brindley occupied, at the instance of the Duke, in surveying the country between Stretford and the river Mersey, with the object of carrying out a canal in that direction for the accommodation of the growing trade between Liverpool and Manchester. The first boat-load of coals sailed over the Barton viaduct to Manchester on the 17th of July, 1761, and on the 7th of September following we find Brindley at Liverpool,[1] " rocconitoring ;" and, by the end of the month, he is busily engaged in levelling for a proposed canal to join the Mersey at Hempstones, about eight miles below Warrington Bridge, from whence there was a natural tideway to Liverpool, about fifteen miles distant.

The project in question was a very important one on public grounds. We have seen how the community of Manchester had been hampered by its defective road and water communications, which seriously affected its supplies of food and fuel, and, at the same time, by retarding its trade, hindered to a considerable extent the regular employment of its population. The Duke of

---

[1] It would almost seem as if the extension of the canal to the Mersey had formed part of the Duke's original plan; for Brindley was engaged in making a survey from Longford to Dunham in the autumn of the preceding year, as appears from the following account of Brindley's expenses in making the survey, preserved at the Bridgewater Canal Office at Manchester :—

" Expenses in Surveying from Longford Bridge to Dunham.

Oct<sup>r</sup> 21<sup>st</sup> 1760.

| | | | |
|---|---|---|---|
| Spent at Stretford | .. .. .. | 0 | 6 |
| At Altringham all Night .. .. | | 6 | 0 |
| Gave the Men to Drink that } assisted .. .. .. .. } | | 1 | 0 |
| 22<sup>nd</sup> | | | |
| More at Altringham | .. .. | 2 | 6 |
| | | 10 | 0 |

P<sup>d</sup> Mr. Brinley this."

Bridgewater, by constructing his canal, had opened up
an abundant supply of coal, but the transport of the raw
materials of manufacture was still as much impeded as
before. Liverpool was the natural port of Manchester,
from which it drew its supplies of cotton, wool, silk, and
other produce, and to which it returned them for export
when worked up into manufactured articles.

There were two existing modes by which the communication was kept up between the two places : one
was by the ordinary roads, and the other by the rivers
Mersey and Irwell. From a statement published in
December, 1761, it appears that the quantity of goods
then carried by land from Manchester to Liverpool was
" upwards of forty tons per week," or about two
thousand tons a year. This quantity of goods, insignificant though it appears when compared with the
enormous traffic now passing between the two towns,
was then thought very large, as no doubt it was when
the very limited trade of the country was taken into
account. But the cost of transport was the important
feature ; it was not less than two pounds sterling per ton
—this heavy charge being almost entirely attributable
to the execrable state of the roads. It was scarcely
possible to drive waggons along the ruts and through
the sloughs which lay between the two places at certain
seasons of the year, and even pack-horses had considerable difficulty in making the journey.

The other route between the towns was by the navigation of the rivers Mersey and Irwell. The raw materials used in manufacture were principally transported
from Liverpool to Manchester by this route, at the cost
of about twelve shillings a ton ; the carriage of timber
and such like articles costing not less than twenty per
cent. on their value at Liverpool. But the navigation
was also very tedious and difficult. The boats could only
pass up to the first lock at the Liverpool end with the
assistance of a spring-tide ; and further up the river

there were numerous fords and shallows which the boats could only pass in great freshes, or, in dry seasons, by drawing extraordinary quantities of water from the locks above. Then, in winter, the navigation was apt to be impeded by floods, and occasionally it was stopped altogether. In short, the growing wants of the population demanded an improved means of transit between the two towns, which the Duke of Bridgewater now determined to supply.

The growth of Liverpool as a seaport had been comparatively recent. At a time when Bristol and Hull possessed thriving harbours, resorted to by foreign ships, Liverpool was little better than a fishing-village, its only distinction being that it was a convenient place for setting sail to Ireland. In the war between France and England which broke out in 1347, when Edward the Third summoned the various ports in the kingdom to make contributions towards the naval power according to their means, London was required to provide 25 ships and 662 men; Bristol, 22 ships and 608 men; Hull, 16 ships and 466 men; whilst Liverpool was only asked to find 1 bark and 6 men! In Queen Elizabeth's time, the burgesses presented a petition to Her Majesty, praying her to remit a subsidy which had been imposed upon the seaport and other towns, in which they styled their native place "Her Majesty's poor decayed town of Liverpool." Chester was then of considerably greater importance as a seaport. In 1634-5, when Charles I. made his unconstitutional levy of ship-money throughout England, Liverpool was let off with a contribution of 15*l.*, whilst Chester paid 100*l.*, and Bristol not less than 1000*l.* The channel of the Dee, however, becoming silted up, the trade of Chester decayed, and that of Liverpool rose upon its ruins. In 1699 the excavation of the old dock was commenced; but it was used only as a tidal harbour (being merely an enclosed space with a small pier) until the year 1709, when an Act was

obtained enabling its conversion into a wet dock; since
which time a series of docks have been constructed,
extending for about five miles along the north shore
of the Mersey, which are among the greatest works
of modern times, and afford an unequalled amount of
shipping accommodation.

LIVERPOOL IN 1650.
[From Troughton's History of Liverpool.]

From that time forward the progress of the port of
Liverpool kept steady pace with the trade and wealth
of the country behind it, and especially with the manufacturing activity and energy of the town of Manchester.
Its situation at the mouth of a deep and navigable river,
in the neighbourhood of districts abounding in coal and
iron, and inhabited by an industrious and hardy population, were unquestionably great advantages.  But these
of themselves would have been insufficient to account for
the extraordinary progress made by Liverpool within
the last century, without the opening up of the great
system of canals, which brought not only the towns of
Yorkshire, Cheshire, and Lancashire into immediate
connection with that seaport, but also the manufacturing

districts of Staffordshire, Warwickshire, and the other central counties of England situated at the confluence of these various navigations.[1] Liverpool thus became the great focus of import and export for the northern and western districts. The raw materials of commerce were poured into it from Ireland, America, and the Indies; and from thence they were distributed along the canals amongst the various seats of manufacturing industry, returning mostly by the same route to the same port for shipment to all parts of the world.

At the time of which we speak, however, it will be observed that the communication between Liverpool and Manchester was as yet very imperfect. It was not only difficult to convey goods between the two places, but it was also difficult to convey persons. In fine weather, those who required to travel the thirty miles which separated them, could ride or walk, resting at Warrington for the night. But in winter the roads, like most of the other country roads at the time, were simply impassable. Although an Act had been passed as early as the year 1726 for repairing and enlarging the road from Liverpool to Prescott, coaches could not come nearer to the town than Warrington in 1750, the road being impracticable for such vehicles even in summer.[2]

A stage-coach was not started between Liverpool and Manchester until the year 1767, performing the journey only three times a-week. It required six and sometimes eight horses to draw the lumbering vehicle and its load along the ruts and through the sloughs,—the whole day being occupied in making the journey. The coach was accustomed to start early in the morning from Liverpool;

---

[1] Progress of Liverpool.

| Years. | Vessels entered. | Tonnage. | Duties Paid. |
|---|---|---|---|
| 1701 | 102 | 8,619 | .. |
| 1760 | 1,245 | .. | £2,330 |
| 1800 | 4,746 | 450,060 | 23,379 |
| 1858 | 21,352 | 4,441,943 | 347,889 |

[2] Mr. Baines says: " Carriages were then very rare, and it is mentioned as a singular fact that at the period in question (1750) there was but one *gentleman's* carriage in the town of Liverpool, and that carriage was kept by a *lady* of the name of Clayton."—' History of Lancashire,' vol. iv., p. 90.

it breakfasted at Prescott, dined at Warrington, and
arrived at Manchester usually in time for supper. On
one occasion, at Warrington, the coachman intimated his
wish to proceed, when the company requested him to
take another pint, as they had not finished their wine,
asking him at the same time if he was in a hurry? " Oh,"
replied the driver, " I'm not partic'lar to an hour or so!"
As late as 1775, no mail-coach ran between Liverpool
and any other town, the bags being conveyed to and
from it on horseback ; and one letter-carrier was found
sufficient for the wants of the place. A heavy stage then
ran, or rather crawled, between Liverpool and London,
making only four journeys a-week in the winter time.
It started from the Golden Talbot, in Water-street, and
was three days on the road. It went by Middlewich,
where one of its proprietors kept the White Bear inn ;
and during the Knutsford race-week the coach was sent
all the way round by that place, in order to bring customers to the Bear.

We have said that Brindley was engaged upon the
preliminary survey of a canal to connect Manchester with
the Mersey, immediately after the original Worsley line
had been opened, and before its paying qualities could as
yet be ascertained. But the Duke, having once made up
his mind as to the expediency of carrying out this larger
project, never halted nor looked back, but made arrangements for prosecuting a bill for the purpose of enabling
the canal to be made in the very next session of Parliament. We find that Brindley's first visit to Liverpool and the intervening district on the business of the
survey was made early in September, 1761. During
the remainder of the month he was principally occupied
in Staffordshire, looking after the working of his fire-
engine at Fenton Vivian, carrying out improvements in
the silk-manufactory at Congleton, and inspecting various
mills at Newcastle-under-Lyme and the neighbourhood.

His only idle day during that month seems to have been
the 22nd, which was a holiday, for he makes the entry in
his book of " crounation of Georg and Sharlot," the new
King and Queen of England. By the 25th we find him
again with the Duke at Worsley, and on the 30th he
makes the entry, " set out at Dunham to Level for Liverpool." The work then went on continuously ; the survey
was completed ; and on the 19th of November he set out
for London, with 7*l.* 18*s.* in his pocket.

In the course of his numerous journeys, we find Brindley
carefully noting down the various items of his expenses,
which were curiously small. Although he was four or
five days on the road to London, and stayed eight days in
town, his total expenses, both going and returning,
amounted to only 4*l.* 8*s.* ; though it is most probable that
he lived at the Duke's house whilst in town. On the
1st of December we find him, on his return journey to
Worsley, resting the first night at a place called Brick-
hill ; the next at Coventry, where he makes the entry,
" Moy mar had a bad fall in the frasst;" the third at
Sandon ; the fourth at Congleton ; and the fifth at Worsley. He had still some inquiries to make as to the depth
of water and the conditions of the tide at Hempstones ;
and for three days he seems to have been occupied in
traffic-taking, with a view to the evidence to be given
before Parliament ; for on the 10th of December we find
him at Stretford, " to count the caridgos," and on the
12th he is at Manchester for the same purpose, " counting
the loded caridgos and horses." The following bill refers
to some of the work done by him at this time, and is a
curious specimen of an engineer's travelling charges in
those days—the engineer himself being at the same time
paid at the rate of 3*s.* 6*d.* a day :—

*Expenses for His Grace the Duk of Bridgwator to pay for traveling Chareges by James Brindley.*

18 Novem—1761.

18 No  masuring a Cros from Dunham to Warbuton Mercey and
Thalwall, 3s - 11d Dunham for 2 diners 1s - 3d for the man
1s - 0d at Thalwall 1s - 2d all Night Warington .. .. .. 0 7 4

19 Novem Sat out from Chester for London & at Worsley Septm 5
Retorned back  going to London and at London & hors back
to Worsley  Charged Hors & my salf .. .. .. .. .. 4 8 0

9 december  Coming back from Ham Stone  Charges at Wilders-
pool all Night .. .. .. .. .. .. .. .. .. .. 0 8 0

at Warington to meet M$^r$ Ashley    dining .. .. .. 0 4 2

10 to  ataind the Turn pike Rode 2s  6d & againe on t$^e$ 12 D$^e$
Rode 3s - 6d .. .. .. .. .. .. .. .. .. .. .. 0 6 0

21 Decm  to inspect t$^e$ flux and Reflux at Ham Stone 2 dayes
Charges .. .. .. .. .. .. .. .. .. .. .. 0 6 6

26 Dec$^r$ 1761.  Rec$^d$ the Contents of the above Bill by the Hands of
John Gilbert.    James Brindley .. .. .. .. .. .. £6 00 0

In the early part of the month of January, 1762, we
find Brindley busy measuring soughs, gauging the
tides at Hempstones, and examining and altering the
Duke's paper-mills and iron slitting-mills at Worsley ;
and on the 7th we find this entry : " to masuor the
Duks pools I and Smeaton." On the following day he
makes " an ochilor survey from Saldnoor [Sale Moor] to
Stockport," with a view to a branch canal being carried
in that direction. On the 14th, he sets out from Congleton, by way of Ashbourne, Northampton, and Dunstable, arriving in London on the fifth day. Immediately
on his arrival in town we find him proceeding to rig
himself out in a new suit of clothes. His means were
small, his habits thrifty, and his wardrobe scanty ; but
as he was about to appear in an important character, as
the principal engineering witness before a Parliamentary
Committee in support of the Duke's bill, he felt it necessary to incur an extra expenditure on dress for the occasion. Accordingly, on the morning of the 18th we find
him expending a guinea—an entire week's pay—in the
purchase of a pair of new breeches ; two guineas on a
coat and waistcoat of broadcloth, and six shillings for a

pair of new shoes. The subjoined is a facsimile of the
entry in his pocketbook.

FAC-SIMILE OF BRINDLEY'S HAND-WRITING.

It will be observed that an expenditure is here entered
of nine shillings for going to "the play." It would
appear that his friend Gilbert, who was in London with
him on the canal business, prevailed on Brindley to
go with him to the theatre to see Garrick in the play of
'Richard III.,' and he went. He had never been to an
entertainment of the kind before; but the excitement
which it caused him was so great, and it so completely
disturbed his ideas, that he was unfitted for business for
several days after. He then declared that no consideration
should tempt him to go a second time, and he held to

his resolution. This was his first and only visit to the play. The following week he enters himself in his memorandum-book as ill in bed, and the first Sunday after his recovery we find him attending service at "Sant Mary's Church." The service did not make him ill, as the play had done, and on the following day he attended the House of Commons on the subject of the Duke's bill.

The proposed canal from Manchester to the Mersey at Hempstones stirred up an opposition which none of the Duke's previous bills had encountered. Its chief opponents were the proprietors of the Mersey and Irwell navigation, who saw their monopoly assailed by the measure; and, unable though they had been satisfactorily to conduct the then traffic between Liverpool and Manchester, they were unwilling to allow of any additional water service being provided between the two towns. Having already had sufficient evidence of the Duke's energy and enterprise, from what he had been able to effect in so short a time in forming the canal between Worsley and Manchester, they were not without reason alarmed at his present project. At first they tried to buy him off by concessions. They offered to reduce the rate of 3s. 4d. per ton of coals, timber, &c., conveyed upon the Irwell between Barton and Manchester, to 6d. if he would join their navigation at Barton and abandon the part of his canal between that point and Manchester : but he would not now be diverted from his plan, which he resolved to carry into execution if possible. Again they tried to conciliate his Grace by offering him certain exclusive advantages in the use of their navigation. But it was again too late; and the Duke, having a clear idea of the importance of his project, and being assured by his engineer of its practicability and the great commercial value of the undertaking, determined to proceed with the measure. It offered to the public the advantages of a shorter line of navigation, not liable to be interrupted by floods on the one hand or droughts on the other, and, at the same time, a much lower rate of freight, the maximum charge proposed in the bill being 6s. a ton against 12s., the rate charged by the Mersey and Irwell navigation between Liverpool and Manchester.

The opposition to the bill was led by Lord Strange, son of the Earl of Derby, one of the members for the county of Lancaster, who took the part of the "Old Navigators," as they were called, in resisting the bill. The question seems also to have been treated as a political one; and, the Duke and his friends being Whigs, Lord Strange mustered the Tory party strongly against him. Hence we find this entry occurring in Brindley's note-book, under date the 16th of February: "The Toores [Tories] mad had [made head] agane ye Duk." The principal objections put forward to the proposed canal were, that the landowners would suffer by it from having their lands cut through and covered with water, by which a great number of acres would be for ever lost to the public; that there was no necessity whatever for the canal, the Mersey and Irwell navigation being sufficient to carry more goods than the then trade could supply; that the new navigation would run almost parallel with the old one, and offered no advantage to the public which the existing river navigation did not supply; that the canal would drain away the waters which supplied the rivers, and be very prejudicial to, if not a total obstruction of them in dry seasons; that the proprietors of the old navigation had invested their money on the faith of Parliament, and to permit the new canal to be established would be a gross interference with their vested rights; and so on. To these objections there were very sufficient answers. The bill provided for full compensation being made to the owners of lands through which the canal passed, and, in addition, it was provided that all sorts of manure should be carried for them tonnage free. It was also shown that the Duke's canal could not abstract water from either the Mersey or the Irwell, as the level of both rivers was considerably below that of the intended canal, which would be supplied almost entirely from the drainage of his own coal-mines at Worsley; and with respect to the plea of vested rights set up, it was shown that Parliament, in granting certain powers to the old navigators, had regard mainly to the convenience and advantage of the public; and they were not precluded from empowering a new navigation to be formed if it could be proved to present a more convenient and advantageous mode of conveyance. And on this ground the Duke was strongly supported by the inhabitants of the locality proposed to be served by the intended canal. The "Junto of Old Navigators," as they were termed,[1] had for many years carried things with a very high hand, extorted the highest rates, and, in cases of loss by delay or damage to goods in transit, refused all redress. A feeling very hostile to them and their monopoly had accordingly grown up, which now exhibited itself in a powerful array of petitions to Parliament in favour of the Duke's bill.

On the 17th of February, 1762, the bill came before the Committee of the House of Commons, and Brindley gave his evidence in its support. We regret that no copy of this evidence now exists[2] from which we might have formed an opinion of the engineer's abilities as a witness. Some curious anecdotes have, however, been preserved of his appearance as a witness on canal bills before Parliament. When asked, on one occasion, to produce a drawing of an intended

---

[1] Letter of John Hart to the Gentlemen and Tradesmen at Warrington, Dec. 21st, 1761.

[2] Search has been made at the Bridgewater Estate Offices at Manchester, and in the archives of the Houses of Parliament, but no copy can be found. It is probable that the Parliamentary papers connected with this application to Parliament were destroyed by the fire which consumed so many similar documents some twenty years ago.

bridge, he replied that he had no plan of it on paper,
but he would illustrate it by a model.  He went out and
bought a *large cheese*, which he brought into the room
and cut into two equal parts, saying, " Here is my
model." The two halves of the cheese represented the
semicircular arches of his bridge; and by laying over
them some long rectangular object he could thus readily
communicate to the committee the position of the river
flowing underneath and the canal passing over it.[1]  On
another occasion, when giving his evidence, he spoke so
frequently about " puddling," describing its uses and
advantages, that some of the members expressed a desire
to know what this extraordinary mixture was that could
be applied to so many and important purposes.  Preferring a practical illustration to a verbal description,
Brindley caused a mass of clay to be brought into the
committee-room, and, moulding it in its raw untempered
state into the form of a trough, he poured into it some
water, which speedily ran through and disappeared.
He then worked the clay up with water to imitate the
process of puddling, and again forming it into a trough,
filled it with water, which was now held in without a
particle of leakage.  " Thus it is," said Brindley, " that
I form a water-tight trunk to carry water over rivers
and valleys, wherever they cross the path of the canal." [2]
On another occasion, when Brindley was giving evidence before a committee of the House of Peers as to the
lockage of his proposed canal, one of their Lordships
asked him, " But what is a lock ? " on which the engineer
took a piece of chalk from his pocket and proceeded to
explain it by means of a diagram which he drew upon
the floor, and made the matter clear at once.[3]

---

[1] Stated by Mr. Hughes, in his
' Memoir of Brindley,' as having been
communicated by James Loch, Esq.,
M.P., the agent for the Duke's Trustees.
[2] ' Memoir of Brindley,' by S.

Hughes, C.E. in ' Weale's Papers on
Civil Engineering.'
[3] As the reader may possibly desire information on the same point,
we may here briefly explain the nature of a Canal Lock.  It is employed

On the day following Brindley's examination before
the Committee on the Duke's bill, that is, on the 18th

---

as a means of carrying navigations
through an uneven country, and
raising the boats from one water level
to another, or *vice versâ*. The lock
is a chamber formed of masonry,
occupying the bed of the canal where
the difference of level is to be overcome. It is provided with two pairs
of gates, one at each end; and the
chamber is so contrived that the level
of the water which it contains may be
made to coincide with either the
higher level above, or the lower level
below it. The following diagrams
will explain the form and construction
of the lock. A represents what is
called the upper pond, B the lower, C
is the left wall, and DD side culverts.
When the gates at the lower end of
the chamber (E) are opened, and those
at the upper end (F) are closed, the
water in the chamber will stand at
the lower level of the canal; but
when the lower gates are closed, and
the upper gates are opened, the water
will naturally coincide with that in
the upper part of the canal. In the
first case, a boat may be floated into
the lock from the lower part, and
then, if the lower gates be closed and

water is admitted from the upper
level, the canal-boat is raised, by the
depth of water thus added to the
lock, to the upper level, and on the
complete opening of the gates it is
thus floated onward. By reversing
the process, it will readily be understood how the boat may, in like
manner, be lowered from the higher
to the lower level. The greater the
lift or the lowering, the more water is
consumed in the process of exchange
from one level to another; and where
the traffic of the canal is great, a
large supply of water is required to
carry it on, which is usually provided
by capacious reservoirs situated above
the summit level. Various expedients
are adopted for economising water:
thus, when the width of the canal
will admit of it, the lock is made in
two compartments, communicating
with each other by a valve, which
can be opened and shut at pleasure;
and by this means one-half of the
water which it would otherwise be
necessary to discharge to the lower
level may be transferred to the other
compartment.

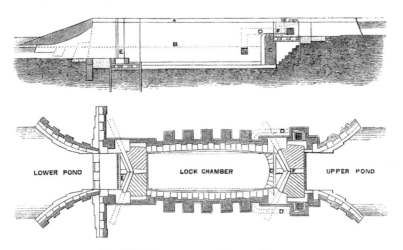

LONGITUDINAL SECTION AND PLAN OF LOCK.

of February, we find him entering in his note-book that the Duke sent out "200 leators" to members—friends of the measure; doubtless containing his statement of reasons in favour of the bill. On the 20th Mr. Tomkinson, the Duke's solicitor, was under examination for four hours and a half. Sunday intervened, on which day Brindley records that he was "at Lord Harrington's." On the following day, the 22nd, the evidence for the bill was finished, and the Duke followed this up by sending out 250 more letters to members, with an abstract of the evidence given in favour of the measure. On the 26th there was a debate of eight hours on the bill, followed by a division, in Committee of the whole House, thus recorded by Brindley :—

"ad a grate Division of 127 fort Duk
98 nos
—
for te Duk 29 Me Jorete "

But the bill had still other discussions and divisions to encounter before it was safe. The Duke and his agents worked with great assiduity. On the 3rd of March he caused 250 more letters to be distributed amongst the members; and on the day after we find the House wholly occupied with the bill. We quote again from Brindley's record : "4 [March] ade bate at the Hous with grate vigor 3 divisons the Duke carred by Numbers evory time a 4 division moved but Noos yelded." On the next day we read "wont thro the closos;" from which we learn that the clauses were settled and passed. Mr. Gilbert and Mr. Tomkinson then set out for Lancashire : the bill was safe. It passed the third reading, Brindley making mention that "Lord Strange" was "sick with geef [grief] on that affair Mr. Wellbron want Rong god,"—which latter expression we do not clearly understand, unless it was that Mr. Wilbraham wanted to wrong God. The bill was carried to the Lords, Brindley on the 10th March making the entry,

"Touk the Lords oath." But the bill passed the Upper House " without opposishin," and received the Royal Assent on the 24th of the same month.

On the day following the passage of the bill through the House of Lords (of which Brindley makes the triumphant entry, " Lord Strange defetted "), he set out for Lancashire, after nine weary weeks' stay in London. To hang about the lobbies of the House and haunt the office of the Parliamentary agent, must have been excessively irksome to a man like Brindley, accustomed to incessant occupation and to see work growing under his hands. During this time we find him frequently at the office of the Duke's solicitor in " Mary Axs;" sometimes with Mr. Tomkinson, who paid him his guinea a-week during the latter part of his stay; and on several occasions he is engaged with gentlemen from the country, advising them about " saltworks at Droitwitch " and mill-arrangements in Cheshire. Many things had fallen behind during his absence and required his attention, so he at once set out home; but the first day, on reaching Dunstable, he was alarmed to find that his mare, so long unaccustomed to the road, had " allmost lost ye use of her Limes " [limbs]. He therefore pushed on slowly, as the mare was a great favourite with him—his affection for the animal having on one occasion given rise to a serious quarrel between him and Mr. Gilbert—and he did not reach Congleton until the sixth day after his setting out from London. He rested at Congleton for two days, during which he " settled the geering of the silk-mill," and then proceeded straight on to Worsley to set about the working survey of the new canal.

## CHAPTER VI.

BRINDLEY CONSTRUCTS THE DUKE'S CANAL TO RUNCORN.

THE course of this important canal, which unites the trade of Manchester with the port of Liverpool, is about twenty-four miles in length.[1] From Longford Bridge, near Manchester, its course lies in a south-westerly direction for some distance, crossing the river Mersey at a point about five miles above its junction with the Irwell. At Altrincham it proceeds in a westerly direction, crossing the river Bollin about three miles further on, near Dunham. After crossing the Bollin, it describes a small semicircle, proceeding onward in the valley of the Mersey, and nearly in the direction of the river as far as the crossing of the high road from Chester to Warrington. It then bends to the south to preserve the high level, passing in a southerly direction as far as Preston, in Cheshire, from whence it again turns round to the north to join the river Mersey.

The canal lies entirely in the lower part of the new red sandstone, the principal earthworks consisting of the clays, marls, bog-earths, and occasionally the sandstones of this formation. The heaviest bog to be crossed was that of Sale Moor, on a bottom of quicksand, west of the Mersey, the construction of the canal at this part being a work of as much difficulty as the laying of the

---

[1] The following statement of the lengths of the different portions of the Duke's canal, including those originally executed, is from the map published by Brindley in 1769 :—

| | Miles. | furl. | chains. | |
|---|---|---|---|---|
| From Worsley to Longford Bridge | 6 | 0 | 0 | Level. |
| „ Longford Bridge to Manchester | 4 | 2 | 0 | „ |
| „ Longford Bridge to Preston Brook | 19 | 0 | 0 | „ |
| „ Preston Brook to upper part of Runcorn | 4 | 4 | 0 | „ |
| „ Upper part of Runcorn to the Mersey | 0 | 5 | 7 | 79 feet fall. |

railroad upon Chat Moss some sixty years later. But Brindley, like Stephenson, looked upon a difficulty as a thing to be overcome; and a difficulty no sooner presented itself, than he at once set his wits to work to study how it was best to be grappled with and surmounted. There were also a large number of brooks to be crossed, and two important rivers, involving a number of aqueducts, bridges, and culverts, to provide for the accommodation of the district. It will, therefore, be obvious that this undertaking was of a much more formidable character—more difficult for the engineer and much more costly to the noble proprietor—than the comparatively limited and inexpensive work between Worsley and Manchester, which we have above described.

The capital idea which Brindley early formed and determined to carry out, was to construct a level of dead water all the way from Manchester to a point as near to the junction of the canal with the Mersey as might be found practicable. Such a canal, he clearly saw, would not be so expensive to work as one furnished with locks at intermediate points. Brindley's practice of securing long levels of water in canals was in many respects similar to that of George Stephenson with reference to flat gradients upon railways; and in all the canals that he constructed, he planned and carried them out upon this leading principle. Hence the whole of the locks on the Duke's canal were concentrated at its lower end near Runcorn, where the navigation descended by a flight of locks into the river Mersey. Lord Ellesmere has observed that this uninterrupted level of the Bridgewater Canal from Leigh and Manchester to Runcorn, and the concentration of its descent to the Mersey at the latter place, have always been considered as among the most striking evidences of the genius and skill of Brindley.

There was, as usual, considerable delay in obtaining possession of the land on which to commence the works. The tenants required a certain notice, which must neces-

sarily expire before the Duke's engineer could take possession ; and many obstacles were thrown in the way both by
tenants and landlords hostile to the undertaking.  In many
cases the Duke had to pay dearly for the land purchased
under the compulsory powers of his Act.  Near Lymm,
the canal passed through a little bit of garden belonging
to a poor man's cottage, the only produce growing
upon the ground being a pear-tree.  For this the Duke
had to pay thirty guineas, and it was thought a very
extravagant price at that time.  Since the introduction
of railways, the price would probably be considered ridi-
culously low.  For the land on which the warehouses
and docks were built at Manchester, the Duke had in all
to pay the much more formidable sum of about forty
thousand pounds.

The Old Quay Navigation (by which the Mersey and
Irwell Company was called), even at this late moment,
thought to delay if not to defeat the Duke's operations,
by lowering their rates nearly one-half.  Only a few days
after the Royal Assent had been given to the bill, they
published an announcement, appropriately dated the 1st
of April, setting forth the large sacrifice they were about
to make, and intimating that " from this Reduction in
the Carriage a real and permanent Advantage will arise
to the Public, and they will experience that Utility so
cried up of late, which has hitherto only existed in promises."  The blow was aimed at the Duke, but he heeded
it not : he was more than ever resolved to go on with his
canal.  He was even offered the Mersey navigation at
the price of thirteen thousand pounds ; but he would not
now have it at any price.

The public spirit and enterprise displayed by many
of the young noblemen of those days was truly admirable.
Brindley had for several years been in close personal
communication with Earl Gower as to the construction
of the canal intended to unite the Mersey with the Trent
and the Severn, and thus connect the ports of Liverpool,

Hull, and Bristol, by a system of inland water-communication.   With this object, as we have seen, he had often
visited the Earl at his seat at Trentham, and discussed
with him the plans by which this truly magnificent enterprise was to be carried out; and he had frequently visited
the Earl of Stamford at his seat at Enville for the same
purpose.   But those schemes were too extensive and
costly to be carried out by the private means of either of
these noblemen, or even by both combined.   They were,
therefore, under the necessity of stirring up the latent
enterprise of the landed proprietors in their respective
districts, and waiting until they had received a sufficient
amount of local support to enable them to act with vigour
in carrying their great design into effect.   The Duke of
Bridgewater's scheme of uniting Manchester and Liverpool by an entirely new line of water-communication,
cut across bogs and out of the solid earth in some places,
and carried over rivers and valleys at others by bridges
and embankments, was scarcely less bold or costly.
Though it was spoken of as another of the Duke's " castles
in the air," and his resources were by no means overflowing at the time he projected it, he nevertheless determined
to enter upon the undertaking, and to go on alone with it
though no one else should join him.   The Duke thus
proved himself a real Dux or leader of the industrial
enterprise of his district; and by cutting his canal, and
providing a new, short, and cheap water-way between
Liverpool and Manchester, which was afterwards extended through the counties of Chester, Stafford, and
Warwick, he thus unquestionably paved the way for the
creation and development of the modern manufacturing
system existing in the north-western counties of England.
We need scarcely say how admirably he was supported
throughout by the skill and indefatigable energy of his
engineer.   Brindley's fertility in resources was the theme
of general admiration.   Arthur Young, who visited the
works during their progress, speaks with enthusiastic

admiration of his "bold and decisive strokes of genius,"
his "penetration which sees into futurity, and prevents
obstructions unthought of by the vulgar mind, merely
by foreseeing them : a man," says he, "with such ideas,
moves in a sphere that is to the rest of the world ima-
ginary, or at best a *terra incognita.*"

It would be uninteresting to describe the works of the
Bridgewater Canal in detail : one part of a canal is
usually so like another, that to do so were merely to
involve a needless amount of repetition of a necessarily
dry description.　We shall accordingly content ourselves
with referring to the original methods by which Brindley
contrived to overcome the more important difficulties of
the undertaking.　From Longford Bridge, where the
new works commenced, the canal, which was originally
about eight yards wide and four feet deep, was carried
upon an embankment of about a mile in extent across
the valley of the Mersey.　One might naturally suppose
that the conveyance of such a mass of earth must have
exclusively employed all the horses and carts in the
neighbourhood for years.　But Brindley, with his usual
fertility in expedients, contrived to make the construction
of the canal itself subservient to the completion of the
remainder.　He had the stuff required to make up the
embankment brought in boats partly from Worsley and
partly from other parts of the canal where the cutting
was in excess ; and the boats, filled with this stuff, were
conducted from the canal along which they had come
into caissons or cisterns placed at the point over which
the earth and clay had to be deposited.

The boats, being double, fixed within two feet of each
other, had a triangular trough supported between them
of sufficient capacity to contain about seventeen tons of
earth.　The bottom of this trough consisted of a line of
trap-doors, which flew open at once on a pin being
drawn, and discharged their whole burthen into the
bed of the canal in an instant.　Thus the level of

the embankment was raised to the point necessary to
enable the canal to be carried forward to the next
length.   Arthur Young was of opinion that the saving
effected by constructing the Stretford embankment in
this way, instead of by carting the stuff, was equivalent
to not less than five thousand per cent.!  The materials of
the caissons employed in executing this part of the work
were afterwards used in forming temporary locks across
the valley of the Bollin, whilst the embankment was
being constructed at that point by a process almost the
very reverse, but of like ingenuity.

BRINDLEY'S BALLAST BOATS.

In the same valley of the Mersey the canal had to be
carried over a large brook subject to heavy floods, by
means of a strong bridge of two arches, adjoining which
was a third, affording provision for a road.   Further on,
the canal was carried over the Mersey itself upon a
bridge with one arch of seventy feet span.   Westward
of this river lay a very difficult part of the work, occasioned by the carrying of the navigation over the Sale
Moor Moss.   Many thought this an altogether impracticable thing ; as not only had the hollow trunk of earth
in which the canal lay to be made water-tight, but to
preserve the level of the water-way it must necessarily be
raised considerably above the level of the Moor across
which it was to be laid.   Brindley overcame the difficulty
in the following manner.   He made a strong casing of
timber-work outside the intended line of embankment on
either side of the canal, by placing deal balks in an erect
position, backing and supporting them on the outside
with other balks laid in rows, and fast screwed together ;
and on the front side of this woodwork he had his earth-

work brought forward, hard rammed, and puddled, to form the navigable canal; after which the casing was moved onward to the part of the work further in advance, and the bottom having previously been set with rubble and gravel, the embankment was thus carried forward by degrees, the canal was raised to the proper level, and the whole was substantially and satisfactorily finished.

A steam-engine of Brindley's contrivance was erected at Dunham Town Bridge to pump the water from the foundations there. The engine was called a Sawney, for what reason is not stated, and, for long after, the bridge was called Sawney's Bridge. The foundations of the under-bridge, near the same place, were popularly supposed to be set on quicksand; and old Lord Warrington, when he had occasion to pass under it, would pretend cautiously to look about him, as if to examine whether the piers were all right, and then run through as fast as he could. A tall poplar-tree stood at Dunham Banks, on which a board was nailed showing the height of the canal level, and the people long after called the place by the name of "The Duke's Folly," believing his scheme to be altogether impracticable. But the skill of the engineer baffled these and other prophets of evil; and the success of his expedients, in nearly every case of difficulty that occurred, must certainly be regarded as remarkable, considering the novel and unprecedented character of the undertaking.

Brindley invariably contrived to economise labour as much as possible, and many of his expedients with this object were very ingenious. So far as he could, he endeavoured to make use of the canal itself for the purpose of forwarding the work. He had a floating blacksmith's forge and shop, provided with all requisite appliances, fitted up in one barge; a complete carpenter's shop in another; and a mason's shop in a third; all of which were floated on as the canal advanced, and were thus always at hand to supply the requisite facilities for prosecuting the operations with economy and despatch. Where there was a break in the line of work, occasioned, for instance, by the erection of some bridge not yet finished, the engineer had similar barges constructed and carried by land to other lengths of the canal which were in progress, where they were floated and advanced in like manner for the use of the workmen. When the bridge across the Mersey, which was pushed on with all despatch with the object of economising labour and cost of materials, was completed, the stone, lime, and timber were brought along the canal from the Duke's own property at Worsley, as well as supplies of clay for the purpose of puddling the bottom of the water-way; and thus the work rapidly advanced at all points.

As one of the great objections made to the construction of the canal had been the danger threatened to the surrounding districts by the bursting of the embankments, Brindley made it his object to provide against the occurrence of such an accident by an ingenious expedient. He had stops or floodgates contrived and laid in various parts of the bed of the canal, across its bottom, so that, in the event of a breach occurring in the bank and a rush of waters taking place, the current which must necessarily set in to that point should have the effect of immediately raising the valvular floodgates, and so shutting off the stream and preventing the escape of more water than was contained in the division between the two nearest gates on either side of the breach. At the same time, these floodgates might be used for cutting off the waters of the canal at different points, for the purpose of making any necessary repairs in particular lengths; the contrivance of waste tubes and plugs being so arranged that the bed of any part of the canal, more especially where it passed over the bridges, might be laid bare in a few hours, and the repairs executed at once. In devising these ingenious expedients, it ought to be remembered that Brindley had no previous experience

to fall back upon, and possessed no knowledge of the
means which foreign engineers might have adopted to
meet similar emergencies. All had been the result of
his own original thinking and contrivance; and, indeed,
many of these devices were altogether new and original,
and had never before been tried by any engineer.

It is curious to trace the progress of the works by
Brindley's own memoranda, which, though brief, clearly
exhibit his marvellous industry and close application to
every detail of the business. He seems to have settled
with the farmers for their tenant-right, sold and accounted for the wood cut down and the gravel dug out
along the line of the canal, paid the workmen employed,
laid out the work, measured off the quantities done from
time to time, planned and erected the bridges, designed
the canal-boats required for conveying the earth to form
the embankments, and united in himself the varied func-
tions of land-surveyor, carpenter, mason, brickmaker,
boat-builder, paymaster, and engineer. We even find him
descending to count bricks and sell grass. Nothing was
too small for him to attend to, or too bold for him to
attempt when the necessity arose. At the same time we
find him contriving a water-plane for the Duke's collieries
at Worsley, and occasionally visiting Newchapel, Leek,
and Congleton, in Staffordshire, for the purpose of
attending to the business on which he still continued to
be employed at those places.

<hr/>

[1] The following bill is preserved amongst the Bridgewater Canal papers. Simcox was a skilled mechanic, and acted as foreman of the carpenters:—

"His Grace the Duke of Bridgewater to Sam[l] Simcox.  D[r]

|  |  | £. | s. | d. |
|---|---|---|---|---|
| 23 Mar[h] 1760 | To 12 days work at 21[d] per .. .. .. | 1 | 1 | 0 |
| 23 Aug[t] | To 6 days more d[o] at d[o] .. .. .. .. | 0 | 10 | 6 |
| 6 Sep[r] | To 8 days more d[o] at d[o] .: .. .. .. | 0 | 14 | 0 |
|  |  | 2 | 5 | 6 |

1 Nov[r] 1760. Rec[d] the Contents above by the Hands of John Gilbert for the Use of Sam[l] Simcox.          P[d]
JAMES BRINDLEY."

The wages of what was called a "right hand man" at that time were from 14d. to 16d. a day, and of a "left-hand man" from 1s. to 14d.

The heavy works at the crossing of the Mersey occupied him almost exclusively towards the end of the year 1763. He was there making dams and pushing on the building of the bridge. Occasionally he enters the words, "short of men at Cornbrook." Indeed, he seems then to have lived upon the works, for we find the almost daily entry of "dined at the Bull, 8*d.*" On the 10th of November he makes this entry : "Aftor noon sattled about the size of the arch over the river Marsee [Mersey] to be 66 foot span and rise 16·4 feet." Next day he is "landing balk out of the ould river in to the canal." Then he goes on, "I prosceded to Worsley Mug was corking ye boats   the masons woss making the senter of the waire [weir]. Whith$^e$ was osing to put the lator side of the water-wheel srouds on   I orderd the pit for ye spindle of ye morter-mill to be sunk level with ye canal   Mr. Gilbert sade ye 20 Tun Boat should be at ye water mitang [meeting] by 7 o'clock the next morn." Next morning he is on the works at Cornhill, setting "a carpentar to make scrwos" [screws], superintending the gravelling of the towing-path, and arranging with a farmer as to Mr. Gilbert's slack. And so he goes on from day to day with the minutest details of the undertaking.

He was not without his petty "werrets" and troubles either. Brindley and Gilbert do not seem to have got on very well together. They were both men of strong tempers, and neither would tolerate the other's interference. Gilbert, being the Duke's factotum, was accustomed to call Brindley's men from their work, which the other would not brook. Hence we have this entry on one occasion,—"A meshender [messenger] from Mr G I retorned the anser No more sosiety." In fact, they seem to have quarrelled.

---

[1] The Earl of Bridgewater, in his rambling 'Letter to the Parisians,' above referred to, alleges that the   quarrel originated in Gilbert's horse breaking into the field where Brindley's mare was grazing—an animal of

We find the following further entries on the subject
in Brindley's note-book : "Thursday 17 Novr past 7
o'clock at night   M Gilbert and sun Tom caled on mee
at Gorshill and I went with them to ye Coik [sign of
the Cock] tha stade all night and the had balk [blank?]
bill of parsill   18 Fryday November 7 morn I went to
the Cock and Bruckfast with Gilberts he in davred to
imploye ye carpinters at Cornhill in making door and
window frames for a Building in Castle field and shades
for the mynors in Dito and other things   I want them
to Saill Moor   Hee took upon him diriction of ye back
drains and likwaise such Lands as be twixt the 2 hous
and ceep uper side the large farme and was displesed
with such raing as I had pointed out."

Those differences between Brindley and Gilbert seem
eventually to have become reconciled, most probably by
the mediation of the Duke, for the services of both were
alike essential to him ; and we afterwards find them
working cordially together and consulting each other as

LONGFORD BRIDGE.

before on any important part of the undertaking.   At
the end of the year 1763, by dint of steady work, Long-

---

which he seems to have been very
fond,—and the consequence was that
the engineer was for a time prevented using the animal in the pursuit of his business.   The Earl says
Brindley was under the impression
that Gilbert had contrived this out of
spite.

ford Bridge was finished and gravelled over, and the embankment was steadily proceeding beyond the Mersey in the manner above described.

Brindley did not want for good workmen to carry out his plans. He found plenty of labourers in the neighbourhood accustomed to hard work, who speedily became expert excavators; and though there was at first a lack of skilled carpenters, blacksmiths, and bricklayers, they soon became trained into such under the vigilant eye of so expert a master as Brindley was. We find him, in his note-book, often referring to the men by their names, or rather byenames, for in Lancashire proper names seem to have been little used at that time. "Black David" was one of the foremen most employed on difficult matters, and "Bill o Toms" and "Busick Jack" seem also to have been confidential workmen in their respective departments. We are informed by a gentleman of the neighbourhood [1] that most of the labourers employed were of a superior class, and some of them were "wise" or "cunning men," blood-stoppers, herb-doctors, and planet-rulers, such as are still to be found in the neighbourhood of Manchester. [2] Their very super-

---

[1] R. Rawlinson, Esq., C.E., Engineer to the Bridgewater Canal.

[2] Whilst constructing the canal, Brindley was very intimate with one Lawrence Earnshaw, of Mottram, a kindred mechanical genius, though in a smaller way. Lawrence was a very poor man's son, and had served a seven years' apprenticeship to the trade of a tailor, after which he bound himself apprentice to a clothier for seven years; but these trades not suiting his tastes, and being of a strongly mechanical turn, he finally bound himself apprentice to a clockmaker, whom he also served for seven years. This eccentric person invented many curious and ingenious machines, which were regarded as of great merit in his time. One of these was an astronomical and geographical machine, beautifully executed, showing the earth's diurnal and annual motion, after the manner of an orrery. The whole of the calculations were made by himself, and the machine is said to have been so exactly contrived and executed that, provided the vibration of the pendulum did not vary, the machine would not alter a minute in a hundred years; but this might probably be an extravagant estimate on the part of Earnshaw's friends. He was also a musical instrument maker and music teacher, a worker in metals and in wood, a painter and glazier, an optician, a bellfounder, a chemist and metallurgist, an engraver—in short, an almost universal mechanical genius. But though he could make all these things, it is mentioned as a remarkable fact that with all his ingenuity, and after many efforts (for he made many), he never could make

stitions, says our informant, made them thinkers and calculators. The foreman bricklayer, for instance, as his son used afterwards to relate, always " ruled the planets to find out the lucky days on which to commence any important work," and he added, " none of our work ever gave way." The skilled men had their trade-secrets, in which the unskilled were duly initiated, and the following were amongst them,—simple matters in themselves, but not without use :—

A wet embankment can be prevented from slipping by dredging or dusting powdered lime in layers over the wet clay or earth.

Sand or gravel can be made water-tight by shaking it together with flat bars of iron run in some depth, say two feet, and washing down loam or soil as the bars are moved about, thus obviating the necessity for clay-puddle.

Dry-rot can be prevented in warehouses by setting the bricks opposite the ends of the main beams of the warehouse in dry sand.

As to the details of the canal works, Mr. Rawlinson observes, " All the bridges and culverts are set in the best hydraulic mortar. The plans are simple, and have special contrivances to suit peculiarities of situation, &c. As a rule, there is a vertical joint betwixt the spandrils and wing walls of bridges, to prevent injury by unequal settlements or bearings."

---

a wicker-basket! Indeed, trying to be a universal genius was his ruin. He did, or attempted to do, so much, that he never stood still and established himself in any one thing ; and, notwithstanding his great ability, he died " not worth a groat." Amongst Earnshaw's various contrivances was a piece of machinery to raise water from a coal-mine at Hague, near Mottram, and (about 1753) a machine to spin and reel cotton at one operation —in fact, a spinning-jenny—which he showed to some of his neighbours as a curiosity, but, after having convinced them of what might be done by its means, he immediately destroyed it, saying that " he would not be the means of taking bread out of the mouths of the poor." He was a total abstainer from strong drink, long before the days of Teetotal Societies. Towards the end of his life he continued on intimate terms with Brindley, holding frequent meetings with him ; and when they met they did not easily separate. Earnshaw died in 1764, at sixty years of age.

Whilst the works were in full progress, which was during several years, from three to four hundred men were regularly employed upon them, divided into gangs of about fifty, over each of which was appointed a captain and setter-out of the works. One who visited the canal while in progress thus writes to the 'St. James's Chronicle,' under date July 1st, 1765 : " I surveyed the Duke's men for two hours, and think the industry of bees or labour of ants is not to be compared to them. Each man's work seems to depend on and be connected with his neighbour's, and the whole posse appeared as I conceive did that of the Tyrians when they wanted houses to put their heads in at Carthage." At Stretford the visitor found " four hundred men at work, putting the finishing stroke to about two hundred yards of the canal, which reached nearly to the Mersey, and which, on drawing up the floodgates, was to receive a proper quantity of water and a number of loaded barges. One of these appeared like the hull of a collier, with its deck all covered, after the manner of a cabin, and having an iron chimney in the centre ; this, on inquiry, proved to be the carpentry, but was shut up, being Sabbath-day, as was another barge, which contained the smith's forge. Some vessels were loaded with soil, which was put into troughs (see Cut at p. 383) fastened together, and rested on boards that lay across two barges ; between each of these there was room enough to discharge the loading by loosening some iron pins at the bottom of the troughs. Other barges lay loaded with the foundation-stones of the canal bridge, which is to carry the navigation across the Mersey. Near two thousand oak piles are already driven to strengthen the foundations of this bridge. The carpenters on the Lancashire side were preparing the centre frame, and on the Cheshire side all hands were at work in bringing down the soil and beating the ground adjacent to the foundations of the bridge, which is designed to be covered with stone in a month, and finished in about ten days more." [1]

By these vigorous measures the works proceeded rapidly towards completion. Before, however, they had made any progress at the Liverpool end, Earl Gower, encouraged and assisted by the Duke, had applied for and obtained an Act to enable a line of navigation to be formed between the Mersey and the Trent; the Duke agreeing with the promoters of the undertaking to vary the course of his canal and meet theirs about midway between Preston-brook and Runcorn, from which point it was to be carried northward towards the Mersey, descending into that river by a flight of ten locks, the total fall being not less than 79 feet from the level of the canal to low-water of spring-tides. When this deviation was proposed, the bold imagination of Brindley projected an aqueduct across the tideway of the Mersey itself, which was there some four hundred and sixty yards wide, with the object of carrying the Duke's navigation directly onward to the port of Liverpool on the Lancashire side of the river. [2] This was an admirable idea, which, if carried out, would probably have redounded more to the fame of Brindley than any other of his works. But the cost of that portion of the canal which had already been executed, had reached so excessive an amount, that the Duke was compelled to stop short at Runcorn, at which place a dock was constructed for the accommodation of the shipping employed in the trade connected with the undertaking.

---

[1] 'A History of Inland Navigations. Particularly those of the Duke of Bridgewater in Lancashire and Cheshire.' 2nd Ed., p. 39.

[2] This bold scheme, which seems to have been earnestly proposed at the time, though never executed, was thus noticed in a Liverpool paper: "On Monday last Mr. Brindley waited upon several of the principal gentlemen of this town and others at Runcorn, in order to ascertain the expense that may attend the building of a bridge over the river Mersey at that place, which is estimated at a sum inferior to the advantages that must arise, both to the counties of Lancaster and Chester, from a communication of this sort."—Williamson's 'Liverpool Advertiser,' July 19, 1768.

From Runcorn, it was arranged that the boats should navigate by the open tideway of the Mersey to the harbour of Liverpool, at which place the Duke made arrangements to provide another dock for their accommodation. Brindley made frequent visits to Liverpool for the purpose of directing its excavation, and it still continues devoted to the purposes of the canal navigation. It lies between the Salthouse and Albert Docks on the north, and the Wapping and King's Docks on the south. The Salthouse was the only public dock near it at the time that Brindley excavated this basin. There were only three others in Liverpool to the north, and not one to the south; but the Duke's Dock is now the centre of about five miles of docks, extending from it on either side along the Lancashire shore of the Mersey.

THE DUKE'S DOCK, LIVERPOOL

[By E. M. Wimperis.]

## CHAPTER VII.

THE DUKE'S DIFFICULTIES—GROWTH OF MANCHESTER.

LONG before the Runcorn locks were constructed, and before the canal from Longford Bridge to the Mersey could be made available for purposes of traffic, the Duke found himself reduced to the greatest straits for want of money. Numerous unexpected difficulties had occurred, so that the cost of the works considerably exceeded his calculations; and though the engineer carried on the whole operations with the strictest regard to economy, the expense was nevertheless almost more than any single purse could bear. The execution of the original canal from Worsley to Manchester did not cost more than about a thousand guineas a mile, to which was to be added the cost of the terminus at Manchester. There was also the outlay which had to be incurred in building the requisite boats for the canal, in opening out the underground workings of the collieries at Worsley, and in erecting various mills, workshops, and warehouses for carrying on the new business.

The Duke was enabled to do all this without severely taxing his resources, and he even entertained the hope of being able to grapple with the still greater under-taking of cutting the twenty-four miles of new canal from Longford Bridge to the Mersey. But before these works were half finished, and whilst the large amount of capital invested in them was lying entirely unproductive, he found that the difficulties of the undertaking were likely to prove almost too much for him. Indeed, it seemed an enterprise beyond the means of any private person—more like that of a monarch with State resources at his command, than of a young English nobleman.

But the Duke was possessed by a brave spirit. He had put his hand to the work, and he would not look back. He had become thoroughly inspired by his great idea, and determined to bend his whole energies to the task of carrying it out. He was only thirty years of age—the owner of several fine mansions in different parts of the country, surrounded by noble domains—he had a fortune sufficiently ample to enable him to command the pleasures and luxuries of life, so far as money can secure them; yet all these he voluntarily denied himself, and chose to devote his time to consultations with an unlettered engineer, and his whole resources to the cutting of a canal to unite Liverpool and Manchester.

Taking up his residence at the Old Hall at Worsley— a fine specimen of the old timbered houses so common

WORSLEY OLD HALL.

[By Percival Skelton, after his original Drawing.]

in South Lancashire and the neighbouring counties,—he cut down every unnecessary personal expense; denied himself every superfluity, except perhaps a pipe of tobacco; paid off his following of servants; put down his carriages and town house; and confined himself and his Ducal establishment to a total expenditure of 400*l*.

a-year. A horse was, however, a necessity, for the purpose of enabling him to visit the canal works during their progress at distant points; and he accordingly continued to maintain one horse for himself and another for his groom.

Notwithstanding this rigid economy, the Duke still found his resources inadequate to the heavy cost of vigorously carrying on the works, and on Saturday nights he was often put to the greatest shifts to raise the requisite money to pay his large staff of craftsmen and labourers. Sometimes their payment had to be postponed for a week or more, until the cash could be raised by sending round for contributions among the Duke's tenantry. Indeed, his credit fell to the lowest ebb, and at one time he could not get a bill for 500*l.* cashed in either Liverpool or Manchester.[1]

When Mr. George Rennie, the engineer, was engaged, in 1825, in making the revised survey of the Liverpool and Manchester Railway, he lunched one day at Worsley Hall with Mr. Bradshaw, manager of the Duke's property, then a very old man. He had been a contemporary of the Duke, and knew of the monetary straits to which his Grace had been reduced during the construction of the works. Whilst at table, Mr. Bradshaw pointed to a small whitewashed cottage on the Moss, about a mile and a half distant, and said that in that cottage, formerly a public-house, the Duke, Brindley, and Gilbert had spent many an evening discussing the prospects of the canal when in progress. One of the principal topics of conversation on those occasions was the means of raising the necessary funds against the next pay-night. " One evening in particular," said Mr.

---

[1] There is now to be seen at Worsley, in the hands of a private person, a promissory note given by the Duke, bearing interest, for as low a sum as five pounds. Amongst the persons known to be lenders of money, to whom the Duke applied at the time, was Mr. C. Smith, a merchant at Rochdale; but he would not lend a farthing, believing the Duke to be engaged in a perfectly ruinous undertaking.

Bradshaw, "the party was unusually dull and silent. The Duke's funds were exhausted; the canal was by no means nearly finished; his Grace's credit was at the lowest ebb; and he was at a loss what step to take next. There they sat, in the small parlour of the little public-house, smoking their pipes, with a pitcher of ale before them, melancholy and silent. At last the Duke broke the silence by asking, in a querulous tone, 'Well, Brindley, what's to be done now? How are we to get at the money for finishing this canal?' Brindley, after a few long puffs, answered through the smoke, 'Well, Duke, I can't tell; I only know that if the money can be got, I can finish the canal, and that it will pay well.' 'Ay,' rejoined the Duke, 'but *where* are we to get the money?' Brindley could only repeat what he had already said; and thus the little party remained in moody silence for some time longer, when Brindley suddenly started up and said, 'Don't mind, Duke; don't be cast down; we are sure to succeed after all!' The party shortly after separated, the Duke going over to Worsley to bed, to revolve in his mind the best mode of raising money to complete his all-absorbing project."

Still undaunted by the difficulties that beset them, the Duke and his agents exerted themselves to the utmost to find the requisite means for completing the works. Gilbert was employed to ride round among the tenantry of the neighbouring districts, and raise five pounds here and ten pounds there, until he had gathered together enough to pay the week's wages. Whilst travelling about among the farmers on one of such occasions, Gilbert was joined by a stranger horseman, who entered into conversation with him; and it very shortly turned upon the merits of their respective horses. The stranger offered to swap with Gilbert, who, thinking the other's horse better than his own, agreed to the exchange. On afterwards alighting at a lonely village inn, which he had not before frequented, Gilbert was surprised to be greeted by the landlord with mysterious marks of recognition, and still more so when he was asked if he had got a good booty. It turned out that he had exchanged horses with a highwayman, who had adopted this expedient for securing a nag less notorious than the one which he had exchanged with the Duke's agent.[1]

At length, when the tenantry could furnish no further advances, and loans were not to be had on any terms in Manchester or Liverpool, and the works must needs come to a complete stand unless money could be raised to pay the workmen, the Duke took the road to London on horseback, attended only by his groom, to try what could be done with his London bankers. The house of Messrs. Child and Co., Temple Bar, was then the principal banking-house in the metropolis, as it is the oldest; and most of the aristocratic families kept their accounts there. The Duke had determined at the outset of his undertaking not to mortgage his landed property, and he had held to this resolution. But the time arrived when he could not avoid borrowing money of his bankers on such other security as he could offer them. He had already created a valuable and lucrative property, which was happily available for the purpose. The canal from Worsley to Manchester had proved remunerative in an extraordinary degree, and was already producing a large annual income. He had not the same scruples at pledging the revenues of his canal that he had to mortgage his lands; and an arrangement was concluded with the Messrs. Child under which they agreed to advance the Duke sums of money from time to time, by means of which he was eventually enabled to finish the entire canal. The books of the firm show that he obtained his first advance from them of 3800*l.* about the middle of the year 1765, at which time he was in the greatest difficulty; shortly after a further sum of 15,000*l.*; then 2000*l.*,

---

[1] The Earl of Ellesmere's ' Essays on History, Biography,' &c., p. 236.

and various other sums, making a total of 25,000*l.*;
which remained owing until the year 1769, when the
whole was paid off—doubtless from the profits of the
canal traffic as well as the economised rental of the
Duke's unbur-
thened estates.

The entire
level length of
the new canal
from Longford
Bridge to the
upper part of
Runcorn, near-
ly twenty-eight
miles in extent, was
finished and opened for
traffic in the year 1767,
after the lapse of about
five years from the pass-
ing of the Act. The for-
midable flight of locks,

THE LOCKS AT RUNCORN.

[By Percival Skelton, after his original Drawing.]

from the level part of the canal down to the waters of
the Mersey at Runcorn, were not finished for several
years later, by which time the receipts derived by the

Duke from the sale of his coals and the local traffic of
the undertaking, enabled him to complete them with comparatively little difficulty. Considerable delay was occasioned by the resistance of an obstinate landowner
near Runcorn, Sir Richard Brooke, who interposed
every obstacle which it was in his power to offer; but
his opposition too was at length overcome, and the
new and complete line of water-communication between
Manchester and Liverpool was finally opened throughout. In a letter written from Runcorn, dated the
1st January, 1773, we find it stated that "yesterday
the locks were opened, and the *Heart of Oak*, a vessel
of 50 tons burden, for Liverpool, passed through them.
This day, upwards of six hundred of his Grace's workmen were entertained upon the lock banks with an ox
roasted whole and plenty of good liquor. The Duke's
health and many other toasts were drunk with the
loudest acclamations by the multitude, who crowded from
all parts of the country to be spectators of these astonishing works. The gentlemen of the country for a long
time entertained a very unfavourable opinion of this
undertaking, esteeming it too difficult to be accomplished,
and fearing their lands would be cut and defaced without producing any real benefit to themselves or the
public; but they now see with pleasure that their fears
and apprehensions were ill-grounded, and they join with
one voice in applauding the work, which cannot fail to
produce the most beneficial consequences to the landed
property, as well as to the trade and commerce of this
part of the kingdom." [1]

Whilst the canal works had been in progress, great
changes had taken place at Worsley. The Duke had
year by year been extending the workings of the coal;
and when the King of Denmark, travelling under the
title of Prince Travindahl, visited the Duke in 1768, the

---

[1] Griswell's ' Account of Runcorn and its Environs,' pp. 63-5.

tunnels had already been extended for nearly two miles under the hill. When the Duke began these works, he possessed only such of the coal-mines as belonged to the Worsley estate, but from time to time he purchased the adjoining lands containing seams of coal which run under the high ground between Worsley, Bolton, and Bury; and in course of time the underground canals connecting the different workings extended for a distance of nearly forty miles.[1] Both the hereditary and the purchased mines are worked upon two main levels, though in all there are four different levels, the highest being a hundred and twenty yards above the lowest. The coals worked out of the higher seams are shot into the boats placed below at certain appointed places; the whole of the subterranean produce being dragged to the light through the tunnels, the entrances to which are shown in our illustration. In opening up these underground workings the Duke is said to have expended about 168,000*l.*; but the immense revenue derived from the sale of the coals by canal rendered this an exceedingly productive outlay. Besides the extension of the canal along these tunnels, the Duke subsequently carried a branch by the edge of Chat-Moss to Leigh, by which means new supplies of coal were introduced to Manchester from that district, and the traffic was still further increased. It was a saying of the Duke's, that " a navigation should always have coals at the heels of it."

The total cost of completing the canal from Worsley to Manchester, and from Longford Bridge to the Mersey at Runcorn, amounted to 220,000*l.* A truly magnificent undertaking, nobly planned and nobly executed. The power conferred by wealth was probably never more munificently exercised than in this case; for, though the traffic proved a source of immense wealth to the Duke, it conferred incalculable blessings upon the

---

[1] See the cut at p. 306, showing the entrances to these underground canals.

population of the district. It greatly added to their comforts, increased their employment, and facilitated the operations of industry in all ways. The canal was no sooner opened than its advantages were at once felt. The charge for water-carriage between Liverpool and Manchester was lowered one-half. All sorts of produce were brought to the latter town, at moderate rates, from the farms and gardens adjacent to the navigation, whilst the value of agricultural property was immediately raised by the facilities afforded for the conveyance of lime and manure, as well as by reason of the more ready access to good markets which it provided for the farming classes. The Earl of Ellesmere has not less truly than elegantly observed, that "the history of Francis Duke of Bridgewater is engraved in intaglio on the face of the country he helped to civilize and enrich."

Probably the most remarkable circumstance connected with the money history of the enterprise is this : that although the canal yielded an income which eventually reached about 80,000*l*. a year, it was planned and executed by Brindley at a rate of pay considerably less than that of an ordinary mechanic of the present day. The highest wage he received whilst in the employment of the Duke was 3*s*. 6*d*. a day. For the greater part of the time he received only half-a-crown. Brindley, no doubt, accommodated himself to the Duke's pinched means, and the satisfactory completion of the canal was quite as much a matter of character with him as of pay. Indeed, he seems to have studied economy in everything, and strove to keep down his expenses to the very lowest point. Whilst superintending the works at Longford Bridge, we find him making an entry of his day's personal expenses at only 6*d*. for " ating and drink." On other days his expenditure was confined to 2*d*. for the turnpike. When living at " The Bull," near the works at Throstle Nest, we find his dinner costing 8*d*. and his breakfast 6*d*. His entire expenses were on an equally low scale, for he studied in all ways to economize the Duke's means, that every available shilling might be expended on the prosecution of the works.

The inadequate character of his remuneration was doubtless well enough known to Brindley himself, and rendered him very independent in his bearing towards the Duke.[1] They had frequent differences as to the

---

[1] The Earl of Bridgewater, in his singular publication, the 'Letter to the Parisians,' above referred to, states that "Brindley offered to stay entirely with the Duke, and do business for no one else, if he would give him a guinea a week;" and this statement is repeated by the late Earl of Ellesmere in his 'Essays on History, Biography,' &c. But, on the face of it, the statement looks untrue; and we have since found, from Brindley's own note-book, that, on the 25th of May, 1762, he was receiving a guinea a day from the Earl of Warrington for performing services for that nobleman; nor is it at all likely that he would prefer the Duke's three-and-sixpence a day to the more adequate rate of payment which he was accustomed to charge and to receive from other employers. It is quite true, however—and the fact is confirmed by Brindley's own record—that he received no more than a guinea a week whilst in the Duke's service; which only affords an illustration of the fact that eminent constructive genius may be displayed and engineering greatness achieved in the absence of any adequate material reward. We regret to have to add that Brindley's widow (afterwards the wife of Mr. Williamson, of Longport) in vain petitioned the Duke and his representatives, as well as the above Earl of Bridgewater, for payment of a balance said to have been due to Brindley for services, at the time of the engineer's death. In her letter to Robert Bradshaw, M.P., dated the 2nd May, 1803, Mrs. Williamson says: "It will doubtless appear to you extraordinary that so very late an application should now be made

. . . . but I must beg leave to state that repeated applications were made by me (after Mr. Brindley's sudden and unexpected death) to the late Mr. Thomas Gilbert and also to his brother, but without any other effect than that of constant promises to lay the matter before His Grace; and I conceive it owing to this channel of application that no settling ever took place. A letter was also written to His Grace on this subject so late as the year 1801, but no answer was received. From the year 1765 to 1772, Mr. Brindley received no money on account of his salary. At that time he was frequently in very great want, and made application to the Duke, whose answer (to use the Duke's expression) was, 'I am much more distressed for money than you; however, as soon as I can recover myself, your services shall not go unrewarded.' In consequence of this, Mr. Brindley was under the necessity of borrowing several sums to make good engagements he was then under to various canal companies. In the year 1774, two years after Mr. Brindley's death, the late Mr. John Gilbert paid my brother, Mr. Henshall, the trifling sum of 100l. on account of Mr. Brindley's time, which is all that has been received. I beg leave to suggest how small and inadequate a return this is for his services during a period of seven years. Mr. B.'s travelling expenses on His Grace's account during that time were considerable, towards which, when he had not sufficient money to carry him the whole journey, he now and then received a small sum. How far his plans and undertakings have been beneficial to His Grace's interest is well known." In

proper mode of carrying on the works; but Brindley was quite as obstinate as the Duke on such occasions, and when he felt convinced that his own plan was the right one he would not yield an inch. It is said that, after long evening discussions at the hearth of the old timbered hall at Worsley, or at the Duke's house at Liverpool, while the works there were in progress, the two would often part at night almost at daggers-drawn. But next morning, on meeting at breakfast, the Duke would very frankly say to his engineer, " Well, Brindley, I have been thinking over what we were talking about last night. I find you may be right after all; so just finish the work in your own way."

The Duke himself, to the end of his life, took the greatest personal interest in the working of his coal-mines, his canals, his mills, and his various branches of industry. These were his hobbies, and he took pleasure in nothing else. He was utterly lost to the fashionable world, and, as some thought, to a sense of its proprieties. Shortly after his canal had been opened for the convey-ance of coals, the Duke established a service of passage-boats between Manchester and Worsley, and between Manchester and a station within two miles of Warring-ton, by which passengers were conveyed at the rate of a penny a mile. The boats were fitted up like the Dutch treckschuyts, and, being found cheap as well as con-venient, were largely patronized by the public.[1] This

---

a statement of the claims of Mr. Brindley's representatives, forwarded to the Earl of Bridgewater on the 3rd of November, 1803, it is further stated that " during the period of his em-ploy under His Grace, many highly advantageous and lucrative offers were made to him, particularly one from the Prince of Hesse, in 1766, who at that time was meditating a canal through his dominions in Germany, and who offered to subscribe to any terms Mr. Brindley might stipulate. To this engagement his family strongly urged him, but the solicitation of the Duke, in this as in every other in-stance, to remain with him, out-weighed all pecuniary considerations; relying upon such a remuneration from His Grace as the profits of his work might afterwards justify."

[1] When the Duke had put on the boats and established the service, he offered to let them for 60l. a year; but not being able to find any person to take them at that price, he was under the necessity of conducting the service himself, by means of an agent.

service was afterwards extended to Runcorn, and from thence to Liverpool. The Duke delighted to travel by his own boats, preferring them to any more stately and aristocratic method. He often went by them to Manchester to watch how the coal-trade was going on. When the passengers alighted at the coal-wharf, there were usually many poor people about, wheeling away their barrow-loads of coals. One of the Duke's regulations was, that whenever any deficiency in the supply was apprehended, those people who came with their wheelbarrows, baskets, and aprons for small quantities, should be served first, and waggons, carts, and horses sent away until the supply was again abundant. The numbers of small customers who thus resorted to the Duke's coal-yard rendered it a somewhat busy scene, and the Duke liked to look on and watch the proceedings. One day a customer, of the poorer sort, having got his sack filled, looked about for some one to help it on to his back. He observed a stoutish man standing near, dressed in a spencer, with dark drab smallclothes. "Heigh! mester!" said the man, "come, gie me a lift wi' this sack o' coal on to my shouder." Without any hesitation, the person in the spencer gave the man the required "lift," and off he trudged with the load. Some one near, who had witnessed the transaction, ran up to the man, and asked, "Dun yo know who's that yo've been speaking tull?"  "Naw! who is he?"  "Why, it's th' Duke his-sen!"  "The Duke!" exclaimed the man, dropping the bag of coals from his shoulder, "Hey! what'll he do at me? Maun a goo an ax his pardon?" But the Duke had disappeared.

He was very fond of watching his men at work, especially when any new enterprise was on foot. When they

---

In the course of a short time the boats were found to yield a clear profit of 1600*l.* a year. (See 'Political Magazine,' published by John Murray, 1786. Vol. xi., 314.)

were boring for coal at Worsley, the Duke came every
morning and looked on for a long time together. The
men did not like to leave off work whilst he remained
there, and they became so dissatisfied at having to work
so long beyond the hour at which the bell rang, that
Brindley had difficulty in getting a sufficient number of
hands to continue the boring. On inquiry, he found out
the cause and communicated it to the Duke, who from
that time made a point of immediately walking off when
the bell rang, returning when the men had resumed
work, and remaining with them usually until six o'clock.
He observed, however, that though the men dropped
work promptly as the bell rang, when he was not by,
they were not nearly so punctual in resuming work,
some straggling in many minutes after time. He asked
to know the reason, and the men's excuse was, that
though they could always hear the clock when it struck
twelve, they could not so readily hear it when it struck
only one. On this, the Duke had the mechanism of
the clock altered so as to make it strike *thirteen* at one
o'clock; which it continues to do until this day.

His time was very fully occupied with his various
business concerns, to which he gave a great deal of personal attention. Habit made him a business man—
punctual in his appointments, precise in his arrangements, and economical both of money and time. When
it was necessary for him to see any persons about matters
of business, he preferred going to them instead of letting
them come to him; " for," said he, " if they come to me,
they may stay as long as they please; if I go to them, I
stay as long as *I* please." His enforced habits of
economy during the construction of the canal had fully
impressed upon his mind the value of money. Yet,
though " near," he was not penurious, but was usually
liberal, and sometimes munificent. When the Loyalty
Loan was raised, he contributed to it no less a sum than

100,000*l.* in cash. He was thoroughly and strongly national, and a generous patron of most of our public benevolent institutions.

The employer of a vast number of workpeople, he exercised his influence over them in such a manner as to extort their gratitude and blessings. He did not " lord it " over them, but practically taught them, above all things, to help themselves. He was the pattern employer of his neighbourhood. With a kind concern for the welfare of his colliery workmen—then a half-savage class —he built comfortable dwellings and established shops and markets for them ; by which he ensured that at least a certain portion of their weekly earnings should go to their wives and families in the shape of food and clothing, instead of being squandered in idle dissipation and drunkenness. In order to put a stop to idle Mondays, he imposed a fine of half-a-crown on any workman who did not go down the pit at the usual hour on that morning ; and hence the origin of what is called Half Crown Row at Worsley, as thus described by one of the colliers :— " T'ould dook fined ony mon as didn't go daown pit o' Moonday mornin auve a craown, and abeaut thot toime he made a new road to t'pit, so t'colliers caw'd it Auve Craown Row." Debts contracted by the men at public-houses were not recognised by the pay-agents. The steadiest workmen were allowed to occupy the best and pleasantest houses as a reward for their good conduct. The Duke also bound the men to contribute so much of their weekly earnings to a general sick club ; and he encouraged a religious tone of character amongst his people by the establishment of Sunday schools, which were directly superintended by his agents, selected from the best available class. The consequence was, that the Duke's colliers soon held a higher character for sobriety, intelligence, and good conduct, than the weavers and other workpeople of the adjacent country.

He did not often visit London, where he had long

ceased to maintain a house; but when he went there he made an arrangement with one of his friends, who undertook for a stipulated sum to provide a daily dinner for His Grace and a certain number of guests whilst he remained in town. He also made occasional visits to his fine estate of Ashridge, in Buckinghamshire, taking the opportunity of spending a few days, going or coming, with Earl Gower and his Countess, the Duke's only sister, Lady Louisa Egerton. The Countess Gower seems to have borne considerable resemblance to her brother in force of character as well as in other respects. It is related of her that when her husband was ambassador at the French Court, during the outbreak of the Revolution in 1792, the Countess was courageously kind to the then members of the French King's family who were confined in the Temple, and consequently exposed her lord to the fury of the mob. A Swiss had been the medium of these kindly acts, and the mob sought the Swiss at the Earl's hotel for the purpose of cutting off his head. The authorities offered the ambassador a guard, which he refused, on the high ground of being protected by his character; but he took care to write in large letters over his door *Hôtel de l'Ambassadeur d'Angleterre.* The Countess, describing these alarming circumstances to a friend in England at the time, concluded thus: "Now we have done all we can; and if the mob attacks us now, it is their concern, not ours." [1]   During his visits at Trentham, the Duke would get ensconced on a sofa in some distant corner of the room in the evenings, and discourse earnestly to those who would listen to him about the extraordinary advantages of canals. There was a good deal of fun made on these occasions about "the Duke's hobby." But he was always like a fish out of water until he got back to

---

[1] See the 'Auckland Journal and Correspondence, 1860.' The noble lady here referred to was grandmother of the late Earl of Ellesmere and the late Duke of Sutherland.

Worsley, to John Gilbert, his coal-pits, his drainage, his mills, and his canals.

No wonder he was fond of Worsley. It had been the scene of his triumphs, and the foundation of his greatness. Illustrious visitors from all parts resorted thither to witness Brindley's "castle in the air," and to explore the underground excavations beneath Worsley-hill. Frisi, the Italian, spoke of the latter with admiration when they were only a mile and a half in length; since then they have been extended to nearly forty miles. Among the numerous visitors entertained by the Duke was Fulton, the American artist, with whose speculations he was much interested. Fulton had given a good deal of attention to the subject of canals, and was then speculating on the employment of steam power for the propulsion of canal boats. The Duke was so much impressed with Fulton's ingenuity, that he urged him to give up the profession of a painter and devote himself to that of a civil engineer. Fulton acted on his advice, and shortly after we find him residing at Birmingham—the central workshop of England—studying practical mechanics, and fitting himself for superintending the construction of canals, on which he was afterwards employed in the midland counties.[1] The Duke did not forget the idea which Fulton had communicated to him as to the employment of steam as a motive power for boats, instead of horses; and when he afterwards heard that Symington's steam-boat, *The Dundas*, had been tried successfully on the Forth and Clyde Canal, he arranged to have six canal boats constructed after Symington's model; for he was a man to shrink from no expense in carrying out an enterprise which, to use his own words, had " utility

---

[1] The treatise which Fulton afterwards published, entitled 'A Treatise on Canal Navigation, exhibiting the numerous advantages to be derived from small Canals, &c., with a description of the machinery for facilitating conveyance by water through the most mountainous countries, independent of Locks and Aqueducts,' (London, 1796,) is well known amongst engineers.

at the heels of it." But the Duke dying shortly after, the
trustees refused to proceed with the experiment, and the
project consequently fell through.[1] Had the Duke lived,
canal steam-tugs would doubtless have been fairly tried;
and he might thus have initiated the practical introduction of steam-navigation in England, as he unquestionably
laid the foundations of the canal system. He lived long
enough, however, to witness the introduction of tram-
roads, and he saw considerable grounds for apprehension
in them. "We may do very well," he once observed
to Lord Kenyon, "if we can keep clear of these ——
tram-roads."

He was an admirable judge of character, and was rarely
deceived as to the men he placed confidence in. John
Gilbert was throughout his confidential adviser—a practical out-doors man, full of energy and perseverance.
When any proposal was made to the Duke, he would
say, "Well, thou must go to Gilbert and tell him all
about it; I'll do nothing without I consult him." From
living so much amongst his people, he had contracted
their style of speaking, and "thee'd" and "thou'd" those
whom he addressed, after the custom of the district.
He was rough in his speech, and gruff and emphatic
in his manner, like those amidst whom he lived; but
with the rough word he meant and did the kindly
act. His early want of education debarred him in a
measure from the refining influences of letters; for
he read little, except perhaps an occasional newspaper,
and he avoided writing whenever he could. He also
denied himself the graces of female society; and the

---

[1] The Earl of Ellesmere, in his
'Essay on Aqueducts and Canals,'
states that the Duke made actual
experiment of a steam-tug, and quotes
the following from the communication
of one of the Duke's servants, alive in
1844: "I well remember the steam-
tug experiment on the canal. It was
between 1796 and 1799. Captain
Shanks, R.N., from Deptford, was
at Worsley many weeks preparing it,
by the Duke's own orders and under
his own eye. It was set going, and
tried with coal-boats; but it went
slowly, and the paddles made sad
work with the bottom of the canal,
and also threw the water on the bank.
The Worsley people called it Bona-
parte."—Lord Ellesmere's 'Essays,'
p. 241.

seclusion which his early disappointment in love had
first driven him to, at length grew into a habit.  He
lived wifeless and died childless.  He would not even
allow a woman servant to wait upon him.

In person he was large and corpulent ; and the slim
youth on whom the bet had been laid that he would

FRANCIS DUKE OF BRIDGEWATER (AET. 70)

[By T. D. Scott, after Picart's engraving ]

be blown off his horse when riding the race in Trentham
Park so many years before, had grown into a bulky and
unwieldy man.  His features strikingly resembled those
of George III. and other members of the royal family.
He dressed carelessly, and usually wore a suit of brown

—something of the cut of Dr. Johnson's—with dark drab breeches, fastened at the knee with silver buckles. At dinner he rejected, with a kind of antipathy, all poultry, veal, and such like, calling them " white meats," and wondered that everybody, like himself, did not prefer the brown. He was a great smoker, and smoked far more than he talked. Smoking was his principal evening's occupation when Brindley and Gilbert were pondering with him over the difficulty of raising funds to complete the navigation, and the Duke continued his solitary enjoyment through life. One of the droll habits to which he was addicted was that of rushing out of the room every five minutes, with the pipe in his mouth, to look at the barometer. Out of doors he snuffed, and he would pull huge pinches out of his right waistcoat pocket and thrust the powder up his nose, accompanying the operation with sundry strong short snorts.

He would have neither conservatory, pinery, flower-garden, nor shrubbery at Worsley; and once, on his return from London, finding some flowers which had been planted in his absence, he whipped their heads off with his cane, and ordered them to be rooted up. The only new things introduced about the place were some Turkey oaks, with which his character seemed to have more sympathy. But he took a sudden fancy for pictures, and with his almost boundless means the purchase of pictures was easy.[1] Fortunately, he was well advised as to the paintings selected by him, and the numerous fine

---

[1] Lord Ellesmere says : " An accident laid the foundation of the Bridgewater collection. Dining one day with his nephew, Lord Gower, afterwards Duke of Sutherland, the Duke saw and admired a picture which the latter had picked up a bargain, for some 10*l.*, at a broker's in the morning. ' You must take me,' he said, ' to that d——d fellow to-morrow.' Whether this impetuosity produced any immediate result we are not informed, but plenty of such ' fellows' were doubtless not wanting to cater for the taste thus suddenly developed ; and such advisers as Lord Farnborough and his nephew lent him the aid of their judgment. His purchases from Italy and Holland were judicious and important, and, finally, the distractions of France forcing the treasures of the Orleans Gallery into this country, he became a principal in the fortunate speculation of its purchase."—' Essays on History, Biography,' &c.

works of art which were pouring into the country at
the time, occasioned by the disturbances prevailing on
the Continent, enabled him to lay the foundation of
the famous Bridgewater Gallery, now reckoned to be
one of the finest private collections in Europe. At his
death, in 1803, its value was estimated at not less than
150,000*l.*

The Duke very seldom took part in politics, but
usually followed the lead of his relative Earl Gower,
afterwards Marquis of Stafford, who was a Whig. In
1762, we find his name in a division on a motion to
withdraw the British troops from Germany, and on the
loss of the motion he joined in a protest on the subject.
When the repeal of the American Stamp Act was under
discussion, his Grace was found in the ranks of the
opposition to the measure. He strongly supported Mr.
Fox's India Bill, and generally approved the policy of
that statesman.

The title of Duke of Bridgewater died with him.[1] The
Earldom went to his cousin General Egerton, seventh
Earl of Bridgewater, and from him to his brother the

---

[1] In 1720, when Scroop Egerton, Earl of Bridgewater, had obtained the promise of a dukedom, he acquainted his brother with the circumstance, and told him, moreover, he had so much interest he could get the dukedom settled collaterally upon him and his heirs male, in case there should happen a failure of males in his own line direct, provided his brother would pay the additional office-fees, which amounted to less than 320*l.*, for extending the patent. His brother, then Bishop of Hereford (and who, if he had lived, was to have been Archbishop of York), replied that if the Duke would consent to entail the old family estates upon the dukedom, he would consent to discharge the additional fees. To this the Duke answered that he himself had no immediate concern, and no particular interest, in the above proposal. He made it solely because he conceived it might be acceptable to his brother; he would bind himself, however, by no promise or condition in a matter which regarded the Bishop alone. If the Bishop thought it worth while to give about three hundred guineas for the chance, well; otherwise, the patent would stand as it was already directed to be made out. Hence the patent was not extended, and now there is a failure of males in the Duke's own line direct. The dukedom of Bridgewater is consequently become extinct, in the branch of the family of Egerton, by the death of Francis, the late Duke; and the Earldom of Bridgewater is devolved to the direct heir of the above-mentioned Henry Bishop of Hereford.—'Gentleman's Mag.' for June, 1807. Vol. 77, p. 499.

crazed Francis Henry, eighth Earl; and on his death at
Paris, some thirty years since, that title too became
extinct. The Duke bequeathed about 600,000*l.* in
legacies to his relatives, General Egerton, the Countess
of Carlisle, Lady Anne Vernon, and Lady Louisa Mac-
donald. He devised most of his houses, his pictures,
and his canals, to his nephew George Granville (son of
Earl Gower), second Marquis of Stafford and first Duke
of Sutherland, with reversion to his second son, Lord
Francis Egerton, first Earl of Ellesmere, who thus succeeded to the principal part of the vast property created
by the Duke of Bridgewater. The Duke was buried in
the family vault at Little Gaddesden, Hertfordshire, in
the plainest manner, without any state, at his own express
request.

The Duke was a great public benefactor. The boldness of his enterprise, and the salutary results which
flowed from its execution, entitle him to be regarded as
one of the most useful men of his age. A Liverpool
letter of 1765 says, " The services the Duke has rendered
to the town and neighbourhood of Manchester have en-
deared him to the country, more especially to the poor,
who, with grateful benedictions, repay their noble benefactor." [1]   If he became rich through his enterprise, the
public grew rich with him and by him; for his undertaking was no less productive to his neighbours than it was
to himself. His memory was long venerated by the people
amongst whom he lived,—a self-reliant, self-asserting
race, proud of their independence, full of persevering
energy, and strong in their attachments. The Duke
was a man very much after their own hearts, and a good
deal after their own manners. In respecting him, they
were perhaps but paying homage to those qualities which
they most cherished in themselves. Long after the Duke
had gone from amongst them, they spoke to each other

---

[1] 'History of Inland Navigation,' p. 76.

of his rough words and his kindly acts, his business zeal and his indomitable courage. He was the first great "Manchester man." His example deeply penetrated the Lancashire character, and his presence seems even yet to hover about the district. The Duke's canal still carries a large proportion of the merchandise of Manchester and the neighbouring towns ; the Duke's horses [1] still draw the Duke's boats ; the Duke's coals still issue from the Duke's levels ; and when any question affecting the traffic of the district is under consideration, the questions are still asked of " What will the Duke say ? " " What will the Duke do ? "

Manchester men of this day may possibly be surprised to learn that they owe so much to a Duke, or that the old blood has helped the new so materially in the development of England's modern industry. But it is nevertheless true that the Duke of Bridgewater, more than any other single man, contributed to lay the foundations of the prosperity of Manchester, Liverpool, and the surrounding districts. The cutting of the canal from Worsley to Manchester conferred upon that town the immediate benefit of a cheap and abundant supply of coal ; and when Watt's steam-engine became the great motive power in manufactures, such supply became absolutely essential to its existence as a manufacturing town. Being the first to secure this great advantage, Manchester thus got the start forward which she has never since lost.[2]

---

[1] The Duke at first employed mules in hauling the canal-boats, because of the greater endurance and freedom from disease of those animals, and also because they could eat almost any description of provender. The Duke's breed of mules was for a long time the finest that had been known in England. The popular impression in Manchester is, that the Duke's Acts of Parliament authorising the construction of his canals, forbade the use of horses, in order that men might be employed ; and that the Duke consequently dodged the provisions of the Acts by employing mules. But this is not the case, there being no clause in any of them prohibiting the use of horses.

[2] The cotton trade was not of much importance at first, though it rapidly increased when the steam-engine and spinning-jenny had become generally adopted. It may be interesting to know that sixty years since it was considered satisfactory if

But, besides being a waterway for coal, the Duke's canal, when opened out to Liverpool, immediately conferred upon Manchester the immense advantage of direct connection with an excellent seaport. New canals, supported by the Duke and constructed by the Duke's engineer, grew out of the original scheme between Manchester and Runcorn, which had the further effect of placing the former town in direct water-communication with the rich districts of the north-west of England. Then the Duke's canal terminus became so important, that most of the new navigations were laid out so as to join it; those of Leigh, Bolton, Stockport, Rochdale, and the West Riding of Yorkshire, being all connected with the Duke's system, whose centre was at Manchester. And thus the whole industry of these districts was brought, as it were, to the very doors of that town.

But Liverpool was not less directly benefited by the Duke's enterprise. Before his canal was constructed, the small quantity of Manchester woollens and cottons manufactured for exportation, was carried on horses' backs to Bewdley and Bridgenorth on the Severn, from whence they were floated down that river to Bristol, then the chief seaport on the west coast. No sooner, however, was the new water-road opened out than the Bridgenorth pack-horses were taken off, and the whole export trade of the district concentrated on Liverpool. The additional accommodation required for the increased business of the port was promptly provided as occasion required. New harbours and docks were built, and before many years had passed Liverpool had shot far ahead of Bristol, and became the chief port on the west coast, if not in all England. Had Bristol been blessed with a Duke of Bridgewater, the

---

one cotton-flat a day reached Manchester from Liverpool. In the Duke's time the flats always "cast anchor" on their way, or at least laid up for the night, at six o'clock precisely, starting again at six o'clock on the following morning.

result might have been altogether different; and the valleys of Wilts, the coal and iron fields of Wales, and the estuary of the Severn, might have been what South Lancashire and the Mersey are now. Were statues any proof of merit, the Duke would long since have had the highest statue in Manchester as well as Liverpool erected to his memory, and that of Brindley would have been found standing by his side; for they were both heroes of industry and of peace, though even in commercial towns men of war are sometimes more honoured.

We can only briefly glance at the extraordinary growth of Manchester since the formation of the Duke's canal, as indicated by the annexed plan.

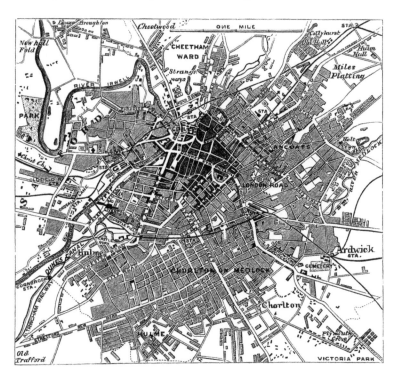

PLAN OF MANCHESTER, SHOWING ITS EXTENT AT THREE PERIODS.

(The parts printed black ⌐ represent Manchester in 1770; those dark-shaded ▙ show its extent in 1804; and the light-shaded parts ▥ Manchester at the present day.)

Though Manchester was a place of some importance about the middle of last century, it was altogether insignificant in extent, trade, and population compared with what it is now.   It consisted of a few principal streets—narrow, dark, and tortuous—one of them, leading from the Market Place to St. Ann's Square, being very appropriately named " Dark Entry."   Deansgate was the principal original street of the town, and so called because of its leading to the dean or valley along which it partly extended.   From thence a few streets diverged in different directions into the open country.   St. Ann's Square, the fashionable centre of modern Manchester, was in 1770 a corn-field surrounded with lofty trees, and known by the name of " Acre's Field."   The cattle-fairs of the town were held there, the entrance from Deansgate being by Toll Lane, a narrow, dirty, unpaved way, so called because toll was there levied on the cattle proceeding towards the fair.   The ancient seat of the Radcliffe family still stood at Pool Fold, close to the site of the modern Cross Street, and the water in the moat was used as a ducking-pond for scolds.   When the pool became filled up, the ducking-pond was removed to Daub Holes, then on the outskirts of the town, where the Infirmary now stands.   The site of King Street, now the very heart of Manchester, was as yet comparatively retired, a colony of rooks having established themselves in the tall trees at its upper end, from which they were only finally expelled about thirty-five years ago.   Cannon Street was the principal place of business, the merchants and their families living in the comparatively humble tenements fronting the street, the equally humble warehouses in which their business was done standing in the rear.   The ground on which the crowded thoroughfares of Oldham Street, London Road, Mosley Street, and their continuations, now exist, was as yet but garden or pasture-land. Salford itself was only a hamlet contained in the bend of the Irwell.   It consisted of a double line of mean houses, extending from the Old Bridge (now Victoria Bridge) to about the end of Gravel Lane—a country road containing only a few detached cottages. The comparatively rural character of Manchester may be inferred from the circumstance that the Medlock and the Irk, the Tib and Shooter's Brook, were favourite fishing streams. Salmon were caught in the Medlock and at the mouth of the Irk; and the others were well stocked with trout. The Medlock and the Irk are now as black as old ink, and as thick; but the Tib and Shooter's Brook are entirely lost,—having been absorbed, like the London Fleet, in the sewage system of the town. Tib Street and Tib Lane indicate the former course of the Tib; but of Shooter's Brook not a trace is left.

The townships of Ardwick Green, Hulme, and Chorlton-upon-Medlock (formerly called Chorlton Row), were purely rural. The old rate-books of Chorlton Row exhibit some curious facts as to the transformations effected in that township. In 1720, a "lay" of 14d. in the pound produced a sum of 26l. 18s., the whole disbursements for the year amounting to 28l. 8s. 5d. From the highway-rate laid in 1722, it appears that the contributors were only twenty persons in all, whose payments ranged from 8d. to 1l. 13s. 4d., producing a total levy of 6l. 18s. 10d. for the year. From the disbursements, it appears that the regular wage paid to the workmen employed was a shilling a-day. In 1750, a lay of 3d. in the pound produced only 6l. 2s. 1½d.; so that the population and value of property in Chorlton Row had not much increased during the thirty years that had passed. In 1770, two brought in 57l. 8s. 6d.; and in 1794, four levies made in that year produced 208l. 2s. 4d.[1] Among the list of contributors in the latter year we find "Mrs. Quincey 16s. 6d."—the mother of De Quincey,

---

[1] Recent "poor-lays" exhibit a very different result from what they did in former years. In 1860-1 the poor-rate levied in Chorlton-upon-Medlock yielded (at 2s. 10d. in the pound), 18,798l.; the property in the township being at the rateable value of 145,844l.

the English opium-eater, who was brought up in Chorlton Row. De Quincey describes the home of his childhood as a solitary house, " beyond which was nothing but a cluster of cottages, composing the little hamlet of Greenhill." It was connected by a winding lane with the Rusholme road. The house, called Greenheys—the nucleus of an immense suburban district—built by De Quincey's father, " was then," he says, " a clear mile from the outskirts of Manchester," Princess Street being then the termination of the town on that side.[1]   Now it is enveloped by buildings in all directions, and nothing of the former rural character of the neighbourhood remains but the names of Greenhill, Rusholme, and Greenheys.

Coming down to the second expansion of Manchester, as exhibited on our plan, it will be observed that a considerable increase of buildings had taken place in the interval between 1770 and 1804. The greater part of the town was then contained in the area bounded by Deansgate, the crooked lanes leading to Princess Street, Bond Street, and Dand Street, to the Rochdale Canal, and round by Ancoats Lane (now Great Ancoats Street) and Swan Street, to Long Millgate, then a steep narrow lane forming the great highway into North Lancashire. Very few buildings existed outside the irregular quadrangle indicated by the streets we have named. The straggling houses of Deansgate, which were principally of timber, ended at Knott Mill. A few dye-works stood at intervals along the Medlock, now densely occupied with buildings for miles along both banks. Salford had not yet extended to St. Stephen's Street in one direction, nor above half way to Broughton Bridge in another.[2] The comparatively limited spaces thus indicated sufficed,

---

[1] De Quincey's 'Autobiographic Sketches,' pp. 34, 48.

[2] The corner of Irwell Street, Salford, as recently as 1828, was occupied by an old canal " flat," tenanted by an eccentric character, after whom it was designated " Bannister's Ship." Opposite it was a row of cottages with gardens in front. Oldfield and Ordsall Lanes were country roads, and the streets adjacent to them were not as yet in existence.

however, for places of business and habitations for the population. Now the central districts are almost exclusively occupied for business purposes, and houses for dwellings have rapidly extended in all directions. The now populous districts of Broughton, Higher and Lower, did not exist thirty-five years ago. They contained no buildings excepting Strangeways Hall and a few cottages which lay scattered beyond the bottom of the workhouse brow.

But pastures, corn-fields, and gardens rapidly disappeared before the advance of streets and factory buildings.[1] The suburban districts of Ardwick, Hulme, and Cheetham, with their hamlets, became altogether absorbed in the great city. Stretford New Road, a broad straight street nearly a mile and a half long, forms the main highway for a district which has sprung up during the life of the present generation, and alone contains a population greater than that of many cities. Not fifty years ago, a few farm-houses and detached dwellings were all the buildings it contained, and Chester road, the principal one in the district, was a narrow winding lane, with hedges on either side. Less than thirty years since, Jackson's Lane was a mere farm-road through corn-fields. Within that time it has been converted into a spacious street, now dignified with the name of Great Jackson Street.[2] The portion of Hulme nearest Manchester was occupied by tea and ginger-beer gardens, the

---

[1] The growth of Manchester and the sister borough of Salford will be more readily appreciated, perhaps, by a glance at the population at different periods than by any other illustration :—

| In 1774. | 1801. | 1821. | 1861. |
| --- | --- | --- | --- |
| 41,032 | 84,020 | 187,031 | 460,028 |

[2] A relic of rural Hulme is to be seen in a remnant of the buildings formerly known as "Jackson's Farm," which gave a name first to the "lane," and subsequently to the "street." It is a single-storey building, covered with gray flags, and standing in an oblique recess situated on the left-hand side about half-way between Chester Road and the recently formed City Road. Higher up, near the end of Upper Jackson Street, at the junction of Chapman Street with Preston Street, are still standing, but without any field near, the buildings, also covered with gray flags, which were known within a comparatively recent period as "Geary's Farm."

principal of which was at the White House. Late roy-
sterers at that place of resort, unless they could form a
party or secure the services of "the patrol," had frequently to sojourn there all night. The officers consti-
tuting· the patrol[1] carried swords and horn lanterns;
and, clad as they were in heavy greatcoats with many
capes, they were by no means light of foot, or at all
formidable adversaries to the footpads who "worked"
the district.

Among the most remarkable improvements in Manchester of late years, have been the numerous spacious
thoroughfares which have been opened up in all directions. In this respect, the public spirit of Manchester
has not been surpassed by any town in the kingdom,—
the new streets being laid out on a settled plan with
a view to future extension, and executed with admirable
judgment. Narrow, dark, and crooked ways have thus
been converted into wide and straight streets, admitting
light, air, and health to the inhabitants, and affording
spacious highways for the great and growing traffic of
the town and district. The important street-improvements executed in Manchester during the last thirty
years have cost an aggregate of about 800,000*l.* The
central and oldest part of the town has thus undergone
a complete transformation. So numerous are the dark
and narrow entries that have been opened up—the ob-
structive buildings that have been swept away, the projecting angles that have been cut off, and the crooked ways
that have been made straight—that the denizen of a
former age would be very unlikely to recognise the Manchester of to-day, were it possible for him to revisit it.

Some of the street-improvements have their peculiar
social aspects, and call up curious reminiscences. The

---

[1] On the subject of watchmen, it
may be mentioned that the first
watchman was appointed for Chorlton-
on-Medlock in 1814. In 1832 an
Act was obtained for improving and
regulating that township, and so recently as 1833 it was first lighted
with gas. The Police Act for the
Township of Hulme was obtained in
1834.

stocks, pillory, and Old Market Cross, were removed from the Market Place in 1816. The public whipping of culprits on the pillory stage is within the recollection of the elder portion of the present inhabitants. Another "social institution," of a somewhat different character, was extinguished much more recently, by the construction of the splendid piece of terrace-road in front of the cathedral, known as the Hunt's Bank improvement. This road swept away a number of buildings, shown on the old plans of Manchester as standing on the water's edge, close to the confluence of the Irk with the Irwell. They were reached by a flight of some thirty steps, and consisted of a dye-work, employing three or four hands, two public-houses, and about a dozen cottages and other buildings. The public-houses, the 'Ring o' Bells' and the 'Blackamoor,' particularly the former, were famous places in their day. On Mondays, wedding-parties from the country, consisting sometimes of from twenty to thirty couples, accompanied by fiddlers, visited "t' Owd Church" to get married. The 'Ring o' Bells' was the rendezvous until the parties were duly married and ready to form and depart homewards, in a more or less orderly manner, headed by their fiddlers as they had come. The 'Ring o' Bells' was also a favourite resort of the recruiting-serjeant, and more recruits, it is said, were enlisted there than at any other public-house in the kingdom. But these, and many other curious characteristics of old Manchester, have long since passed away; and not only the town but its population have become entirely new.

BRIDGEWATER HALFPENNY TOKEN.

## CHAPTER VIII.

### THE GRAND TRUNK CANAL.

LONG before the Duke's canal was finished, Brindley was actively employed in carrying out a still larger enterprise—no less than the canal to connect the Mersey with the Trent, and both with the Severn; thus uniting by a grand line of water-communication the ports of Liverpool, Hull, and Bristol. He had, indeed, already made a survey of such a canal, at the instance of Earl Gower, before his engagement as engineer for the Bridgewater undertaking. Thus, in the beginning of February, 1758, before the Duke's bill had been even applied for, we find him occupied for days together " a bout the novogation," and he then appears to have surveyed the country between Longbridge, in Stafford-shire, and King's Mills, in Derbyshire. The enterprise, however, seems to have made little progress. It was of too formidable a character, and canals were as yet too untried in England, to be hastily entered upon. But again, in 1759, we find Brindley proceeding with his survey of the Staffordshire Canal; and in the middle of the following year he was occupied about twenty days in levelling from Harecastle, at the summit of the pro-posed. canal, to Wilden, near Derby; and we find him frequently with Earl Gower, at Trentham, and with the Earl of Stamford, at Enville, discussing the project.

The next step taken was the holding of a public meeting at Sandon, in Staffordshire, as to the proper course in which the canal should be constructed. Con-siderable difference of opinion was expressed at that meeting, in consequence of which it was arranged that

Mr. Smeaton should be called upon to co-operate with Brindley in making a joint survey and a joint report. A second meeting was afterwards held at Wolseley Bridge, at which the plans of the two engineers were ordered to be engraved and circulated amongst the landowners and others interested in the project. Here the matter rested for several years more, without any action being taken. Brindley was hard at work upon the Duke's canal, and the Staffordshire projectors were disposed to wait the issue of that experiment; but no sooner had it been opened, and its extraordinary success become matter of fact, than the project of the canal through Staffordshire was again revived. The gentlemen of Cheshire and Staffordshire, especially the salt-manufacturers of the former county and the earthen-ware-manufacturers of the latter, now determined to enter into co-operation with the leading landowners in concerting the necessary measures with the object of opening up a line of water-communication with the Mersey and the Trent.

The earthenware manufacture, though in its infancy, had already made considerable progress, but, like every other branch of industry in England at that time, its further development was greatly hampered by the wretched state of the roads. Throughout Staffordshire they were as yet, for the most part, narrow, deep, cir-cuitous, miry, and inconvenient; barely passable with rude waggons in summer, and almost impassable, even with pack-horses, in winter. Yet the principal materials used in the manufacture of pottery, especially of the best kinds, were necessarily brought from a great distance —flint-stones from the south-eastern ports of England, and clay from Devonshire and Cornwall. The flints were brought by sea to Hull, and the clay to Liverpool. From Hull the materials were brought up the Trent in boats to Willington; and the clay was in like manner brought from Liverpool up the Weaver to Winsford, in

Cheshire. Considerable quantities of clay were also conveyed in boats from Bristol, up the Severn, to Bridgenorth and Bewdley. From these various points the materials were conveyed by land-carriage, mostly on the backs of horses, to the towns in the Potteries, where they were worked up into earthenware and china. The manufactured articles were returned for export in the same rude way. Large crates of pot-ware were slung across horses' backs, and thus conveyed to their respective ports, not only at great risk of breakage and pilferage, but also at a heavy cost. The expense of carriage was not less than a shilling a ton per mile, and the lowest charge was eight shillings the ton for ten miles. Besides, the navigation of the rivers above mentioned was most uncertain, arising from floods in winter and droughts in summer. The effect was, to prevent the expansion of the earthenware manufacture, and very greatly to restrict the distribution of the lower-priced articles in common use.

The same difficulty and cost of transport checked the growth of nearly all other branches of industry, and made living both dear and uncomfortable. The indispensable article of salt, manufactured at the Cheshire Wiches, was in like manner carried on horses' backs all over the country, and reached almost a fabulous price by the time it was sold two or three counties off. About a hundred and fifty pack-horses, in gangs, were also occupied in going weekly from Manchester, through Stafford, to Bewdley and Bridgenorth, loaded with woollen and cotton cloth for exportation;[1] but the cost

---

[1] In a curious book published in 1766, by Richard Whitworth, of Balcham Grange, Staffordshire, afterwards Sir Richard Whitworth, member for Stafford, entitled 'The Advantages of Inland Navigation,' he points to the various kinds of traffic that might be expected to come upon the canal then proposed by him, and amongst other items he enumerates the following :—" There are three pot-waggons go from Newcastle and Burslem weekly, through Eccleshall and Newport to Bridgenorth, and carry about eight tons of pot-ware every week, at 3l. per ton. The same waggons load back with ten tons of close goods, consisting of white clay, grocery,

of the carriage by this mode so enhanced the price, that it is clear that in the case of many articles it must have acted as a prohibition, and greatly checked production and consumption. Even corn, coal, lime, and iron-stone were conveyed in the same way, and the operations of agriculture, as of manufacture, were alike injuriously impeded. There were no shops then in the Potteries, the people being supplied with wares and drapery by packmen and hucksters, or from Newcastle-under-Lyme, which was the only town in the neighbourhood worthy of the name.

The people of the district in question were quite as rough as their roads. Their manners were coarse, and their amusements brutal. Bull-baiting, cock-throwing, and goose-riding were the favourite sports. When Wesley first visited Burslem, in 1760, the potters assembled to jeer and laugh at him. They then proceeded to pelt him. " One of them," he says, " threw a clod of earth which struck me on the side of the head; but it neither disturbed me nor the congregation." About that time the whole population of the Potteries did not amount to more than about 7000. The villages in which they lived were poor and mean, scattered up and down, and the houses were mostly covered with thatch. Hence the Rev. Mr. Middleton, incumbent of Stone—a man of great shrewdness and quaintness, distinguished for his love of harmless mirth and sarcastic humour—when enforcing the duty of humility upon his leading parishioners, took the opportunity, on one occasion,

---

and iron, at the same price, delivered on their road to Newcastle. Large quantities of pot-ware are conveyed on horses' backs from Burslem and Newcastle to Bridgenorth and Bewdley for exportation—about one hundred tons yearly, at 2*l*. 10*s*. per ton. Two broad-wheel waggons (exclusive of 150 pack-horses) go from Manchester through Stafford weekly, and may be computed to carry 312 tons of cloth and Manchester wares in the year, at 3*l*. 10*s*. per ton. The great salt-trade that is carried on at Northwich may be computed to send 600 tons yearly along this canal, together with Nantwich 400, chiefly carried now on horses' backs, at 10*s*. per ton on a medium."

after the period of which we speak, of reminding them of the indigence and obscurity from which they had risen to opulence and respectability. He said they might be compared to so many sparrows, for that all of them had been hatched *under the thatch*. When the congregation of this gentleman, growing rich, bought an organ and placed it in the church, he persisted in calling it the hurdy-gurdy, and often took occasion to lament the loss of his old psalm-singers.

The people further north were no better, nor were those further south. When Wesley preached at Congleton, four years later, he said, "even the poor potters [though they had pelted him] are a more civilized people than the better sort (so called) at Congleton." Arthur Young visited the neighbourhood of Newcastle-under-Lyme in 1770, and found poor-rates high, wages low, and employment scarce. "Idleness," said he, "is the chief employment of the women and children. All drink tea, and fly to the parishes for relief at the very time that even a woman for washing is not to be had. By many accounts I received of the poor in this neighbourhood, I apprehend the rates are burthened for the spreading of laziness, drunkenness, tea-drinking, and debauchery,—the general effect of them, indeed, all over the kingdom." [1] Hutton's account of the population inhabiting the southern portion of the same county is even more dismal. Between Hales Owen and Stourbridge was a district usually called the Lie Waste, and sometimes the Mud City. Houses stood about in every direction, composed of clay scooped out into a tenement, hardened by the sun, and often destroyed by the frost. The males were half-naked, the children dirty and hung over with rags. "One might as well look for the moon in a coal-pit," says Hutton, "as for stays or white linen in the City of Mud. The principal tool in

---

[1] Young's 'Six Months' Tour.   Ed. 1770.   Vol. iii., p. 317.

business is the hammer, and the beast of burden the ass." [1]

The district, however, was not without its sprinkling of public-spirited men, who were actively engaged in devising new sources of employment for the population ; and, as one of the most effective means of accomplishing this object, opening up the communications, by road and canal, with near as well as distant parts of the country. One of the most zealous of such workers was the illustrious Josiah Wedgwood. He was one of those indefatigable workers who from time to time spring from the ranks of the common people, and by their energy, skill, and enterprise, not only practically educate the working population in habits of industry, but, by the example of diligence and perseverance which they set before them, largely influence the public activity in all directions, and contribute in a great measure to form the national character. Josiah Wedgwood was born in a very humble position in life ; and though he rose to eminence as a man of science as well as a manufacturer, he possessed no greater advantages at starting than Brindley himself did. His grandfather and granduncle were both potters, as was also his father Thomas, who died when Josiah was a mere boy, the youngest of a family of thirteen children. He began his industrial life as a thrower in a small pot-work, conducted by his elder brother ; and he might have continued working at the wheel but for an attack of virulent small-pox, which, being neglected, led to a disease in his right leg, which ended in his suffering amputation, when he became unfitted for following even that humble employment. During his illness he took to reading and thinking, and turned over in his mind the various ways of making a living by his trade, now that he could no longer work at the potter's wheel. When sufficiently recovered, he began making fancy articles out of potter's clay,

---

[1] 'History of Birmingham.' Ed. 1836, p. 24.

such as knife-hafts, boxes, and sundry curious little
articles for domestic use.    He joined in several successive
partnerships with other workmen, but made comparatively small progress until he began business on his
own account, in 1759, in a humble cottage near the
Market House in Burslem, known by the name of the
Ivy House.    He there pursued his manufacture of knife-

IVY HOUSE, BURSLEM, WEDGWOOD'S FIRST POTTERY. [1]

[From Ward's ' History of Stoke-upon-Trent ']

handles and other small wares, striving at the same time
to acquire such a knowledge of practical chemistry as
might enable him to improve the quality of his work in
respect of colour, glaze, and durability.    Success attended
Wedgwood's diligent and persistent efforts, and he proceeded from one stage of improvement to another, until
at length, after a course of about thirty years' labour, he
firmly established a new branch of industry, which not
only added greatly to the conveniences of domestic life,
but proved a source of remunerative employment to

---

[1] The Ivy House, in which Wedgwood began business on his own account, is the cottage shown on the
right-hand of the engraving.   The
other house is the old " Turk's Head."

many thousand families throughout England. His trade having begun to extend, a demand for his articles sprang up not only in London but in foreign countries. But there was this great difficulty in his way,—that the roads in his neighbourhood were so bad that he was at the same time prevented from obtaining a sufficient supply of the best kinds of clay and also from disposing of his wares in distant markets. This great evil weighed heavily upon the whole industry of the district, and Wedgwood accordingly appears to have bestirred himself at an early period in his career to improve the local communications. In conjunction with several of the leading potters he promoted an application to Parliament for powers to repair and widen the road from the Red Bull at Lawton, in Cheshire, to Cliff Bank, in Staffordshire. This line, if formed, would run right through the centre of the Potteries, open them to traffic, and fall at either end into a turnpike road. The measure was, however, violently opposed by the people of Newcastle-under-Lyme, on the ground that the proposed new road would enable waggons and packhorses to travel north and south from the Potteries without passing through their town. The Newcastle innkeepers acted as if they had a vested interest in the bad roads ; but the bill passed, and the new line was made, stopping short at Burslem. This was, no doubt, a great advantage, but it was not enough. The heavy carriage of clay, coal, and earthenware needed some more convenient means of transport than waggons and roads ; and, when the subject of water communication came to be discussed, Josiah Wedgwood at once saw that a canal was the very thing for the Potteries. Hence he immediately entered with great spirit into the movement again set on foot for the construction of Brindley's Grand Trunk Canal.

The field was not, however, so clear now as it had been before. The success of the Duke's canal led to the projection of a host of competing schemes in the county of

Cheshire, and it appeared that Brindley's Grand Trunk project would now have to run the gauntlet of a powerful local opposition. There were two other projects besides his, which formed the subject of much pamphleteering and controversy at the time, one entering the district by the river Weaver, and another by the Dee. Neither of these proposed to join the Duke of Bridgewater's canal, whereas the Grand Trunk line was laid out so as to run into his at Preston-on-the-Hill near Runcorn. As the Duke was desirous of placing his navigation —and through it Manchester, Liverpool, and the intervening districts—in connection with the Cheshire Wiches and the Staffordshire Potteries, he at once threw the whole weight of his support upon the side of Brindley's Grand Trunk. Indeed, he had himself been partly at the expense of its preliminary survey, as we find from an entry in Brindley's memorandum-book, under date the 12th of April, 1762, as follows: " Worsley—Rec$^d$ from Mr Tho Gilbert for ye Staffordshire survey, on account, 33$l$. 16$s$. 11$d$." The Cheshire gentlemen protested against the Grand Trunk scheme, as calculated to place a monopoly of the Staffordshire and Cheshire traffic in the hands of the Duke; but they concealed the fact, that the adoption of their respective measures would have established a similar monopoly in the hands of the Weaver Navigation Company, whose line of navigation, so far as it went, was tedious, irregular, and expensive. Both parties mustered their forces for a Parliamentary struggle, and Brindley exerted himself at Manchester and Liverpool in obtaining support and evidence on behalf of his plan. The following letter from him to Gilbert, then at Worsley, relates to the rival schemes.

" 21 Decr. 1765

" On Tusdey Sr Georg [Warren] sent Nuton in to Manchester to make what intrest he could for Sir Georg and to gather ye old Navogtors togather to meet Sir Georg at Stoperd to make Head a ganst His Grace

"I sawe Docter Seswige who sese Hee wants to see you about pamant of His Land in Cheshire

"On Wednesday ther was not much transpired but was so dark I could carse do aneything

"On Thursdey Wadgwood of Burslam came to Dunham & sant for mee and wee dined with Lord Gree [Grey] & Sir Hare Mainwering and others   Sir Hare cud not ceep His Tamer [temper] Mr. Wedgwood came to seliset Lord Gree in faver of the Staffordshire Canal & stade at Mrs Latoune all night & I whith him & on frydey sat out to wate on Mr Edgerton to seliset Him   Hee sase Sparrow and others are indavering to gat ye Land owners consants from Hare Castle to Agden

"I have ordered Simcock to ye Langth falls of Sanke Navegacion.

"Ryle wants to have coals sant faster to Alteringham that Hee may have an opertunety dray of ye sale Moor Canal in a bout a weeks time.

"I in tend being back on Tusdy at fardest."

The first public movement was made by the supporters of Brindley's scheme. They held an open meeting at Wolseley Bridge, Staffordshire, on the 30th of December, 1765, at which the subject was fully discussed. Earl Gower, the lord-lieutenant of the county, occupied the chair; and Lord Grey and Mr. Bagot, members for the county,—Mr. Anson, member for Lichfield,—Mr. Thomas Gilbert, the agent for Earl Gower, then member for Newcastle-under-Lyme,—Mr. Wedgwood, and many other influential gentlemen, were present to take part in the proceedings. Mr. Brindley was called upon to explain his plans, which he did to the satisfaction of the meeting; and these having been adopted, with a few immaterial alterations, it was determined that steps should be taken to apply for a bill conferring the necessary powers in the ensuing session of Parliament. Mr. Wedgwood put his name down for a thousand pounds towards the preliminary expenses, and promised to subscribe largely for shares besides.[1] The promoters of the

---

[1] Wedgwood even entered the lists as a pamphleteer in aid of the Grand | Trunk project, and, in 1765, he and his partner, Mr. Bentley (son of Dr.

JOSIAH WEDGWOOD.

[By T. D. Scott, after Reynolds.]

measure proposed to designate the undertaking "The
Canal from the Trent to the Mersey;" but Brindley,
with sagacious foresight, urged that it should be called
The Grand Trunk, because, in his judgment, numerous
other canals would branch out from it at various points
of its course, in like manner as the arteries of the human
system branch out from the aorta; and we need scarcely

Bentley, archdeacon of Ely), drew up
a very able statement, showing the
advantages likely to be derived from
the construction of the proposed canal,
under the title of 'A View of the
Advantages of Inland Navigation,
with a plan of a Navigable Canal intended for a communication between
the ports of Liverpool and Hull.' It

pointed out in glowing language the
advantages to be derived from opening
up the internal communications of a
country by means of roads, canals,
&c.; and showed how the comfort and
even the necessity of all classes must
be so much better provided for by a
reduction in the cost of carriage of
useful and necessary commodities.

add that, before many years had passed, our engineer's anticipations of the progress of canal enterprise were fully justified. The Staffordshire potters were so much rejoiced at the decision of the meeting that on the following evening they assembled round a large bonfire at Burslem, and drank the healths of Lord Gower, Mr. Gilbert, and the other promoters of the scheme, with fervent demonstrations of joy.

The opponents of the measure also held meetings at which they strongly declaimed against the Duke's proposed monopoly, and set forth the superior merits of their respective schemes. One of these was a canal from the river Weaver, by Nantwich, Eccleshall, and Stafford, to the Trent at Wilden Ferry, without touching the Potteries at all. Another was for a canal from the Weaver at Northwich, passing by Macclesfield and Stockport, round to Manchester, thus completely surrounding the Duke's navigation, and preventing its extension southward into Staffordshire or any other part of the Midland districts. But there was also a strong party opposed to all canals whatever—the party of croakers, who are always found in opposition to improved communications, whether in the shape of turnpike roads, canals, or railways. These proclaimed that if the proposed canals were made, the country would be ruined, the breed of English horses would be destroyed, the innkeepers would be made bankrupts, and the pack-horses and their drivers would be deprived of their subsistence. It was even said that the canals, by putting a stop to the coasting-trade, would destroy the race of seamen. It is a fortunate thing for England that it has contrived to survive these repeated prophecies of ruin. But the manner in which our countrymen contrive to grumble their way along the high road of enterprise, thriving and grumbling, is one of the peculiar features in our character which perhaps Englishmen only can understand and appreciate.

It is a curious illustration of the timidity with which
the projectors of those days entered upon canal enterprise, that one of their most able advocates, in order to
mitigate the opposition of the pack-horse and waggon
interest, proposed that " no main trunk of a canal should
be carried nearer than within four miles of any great
manufacturing and trading town; which distance from
the canal would be sufficient to maintain the same number of carriers and to employ almost the same number of
horses as before."[1]  But as none of the towns in the
Potteries were as yet large manufacturing or trading
places, this objection did not apply to them, nor prevent the canals being carried quite through the centre
of what has since become a continuous district of populous manufacturing towns and villages.  The vested
interests of some of the larger towns were, however,
for this reason preserved, greatly to their own ultimate
injury; and when the canal, to conciliate the local opposition, was so laid out as to pass them at a distance,
not many years passed before they became clamorous
for branches to join the main trunk—but not until the
mischief had been done, and a blow dealt to their own
trade, in consequence of being left so far outside the
main line of water communication, from which many of
them never after recovered.

It is not necessary to describe the Parliamentary contest upon the Grand Trunk Canal Bill.  There was the
usual muster of hostile interests,—the river navigation
companies uniting to oppose the new and rival company; the array of witnesses on both sides,—Brindley,
Wedgwood, Gilbert, and many more, giving their evidence in support of their own scheme, and a powerful
array of the Cheshire gentry and Weaver Navigation
Trustees appearing on behalf of the others; and the
whipping-up of votes, in which the Duke of Bridgewater

---

[1] ' The Advantages of Inland Navigation,' by R. Whitworth.  1766.

and Earl Gower worked their influence with the Whig party to good purpose. Brindley's plan was, on the whole, considered the best. It was the longest and the most circuitous, but it appeared calculated to afford the largest amount of accommodation to the public. It would pass through important districts, urgently in need of an improved communication with the port of Liverpool on the one hand, and with Hull on the other. But it was not so much the connection of those ports with each other that was needed, as a more convenient means of communication between them and the Staffordshire manufacturing districts; and the Grand Trunk system—somewhat in the form of a horse-shoe, with the Potteries lying along its extreme convex part—promised effectually to answer this purpose, and to open up a ready means of access to the coast on both sides of the island.

A glance at the course of the proposed line will show its great importance. Starting from the Duke's canal at Preston-on-the-Hill, near Runcorn, it passed southwards by Northwich and Middlewich, through the great salt-manufacturing districts of Cheshire, to the summit at Harecastle. It was alleged that the difficulties presented by the long tunnel at that point were so great that it could never be the intention of the projectors of the canal to carry their "chimerical idea," as it was called, into effect. Brindley however insisted, not only that the tunnel was practicable, but that, if the necessary powers were granted, he would certainly execute it.[1]

---

[1] In one of the many angry pamphlets published at the time, the 'Supplement to a pamphlet entitled Seasonable Considerations on a Navigable Canal intended to be cut from the Trent to the Mersey,' &c., the following passage occurs : "When our all is at stake, these gentlemen [the promoters of the Grand Trunk Canal] must not be surprised at bold truths. We conceive more favourably of their *understanding* than of their *motive ;* we cannot suspect them of entertaining the chimerical idea of cutting through *Hare Castle !* We rather believe that they are desirous of cutting their canal at both ends, and of leaving the middle for the project of a future day. Are these projectors *jealous* of their *honour ?* Let them adopt a clause (which reason and justice strongly enforce) to restrain them from meddling with *either end* till they have finished the *great trunk.* This, and this alone, will shield them from suspicion."

near Derby. From thence there was a clear line of navigation, by Nottingham, Newark, and Gainsborough, to the Humber. Provided this admirable project could be carried out, it appeared likely to meet all the necessities of the case. Ample evidence was given in support of the allegations of its promoters; and the result was, that Parliament threw out the bills promoted by the Cheshire gentlemen on behalf of the old river navigation interest, and the Grand Trunk Canal Act passed into law. At the same time another important Act was passed, empowering the construction of the Wolverhampton Canal, from the river Severn, near Bewdley, to the river Trent, near Haywood Mill; thus uniting the navigation of the three rivers which had their termini at the ports of Liverpool, Hull, and Bristol, on the opposite sides of the island.

There was great rejoicing at Burslem on the news arriving at that place of the passing of the bill; and very shortly after, on the 26th of July, 1766, the first sod of the canal was cut by Josiah Wedgwood on the declivity of Bramhills, in a piece of land within a few yards of the bridge which crosses the canal at that place. Brindley was present on the occasion, and due honours were paid to him by the assembled potters. After Mr. Wedgwood had cut the first sod, many of the leading persons of the neighbourhood followed his example, putting their hand to the work by turns, and each cutting a sod or wheeling a barrow of earth in honour of the occasion. It was, indeed, a great day for the Potteries, as the event proved. In the afternoon a sheep was roasted whole in Burslem marketplace, for the good of the poorer class of potters; a *feu de joie* was performed in front of Mr. Wedgwood's house, and sundry other demonstrations of local rejoicing followed the auspicious event.

Wedgwood was of all others the most strongly impressed with the advantages of the proposed canal.

He knew and felt how much his trade had been hindered by the defective communications of the neighbourhood, and to what extent it might be increased provided a ready means of transit to Liverpool, Hull, and Bristol could be secured ; and, confident in the accuracy of his anticipations, he proceeded to make the purchase of a considerable estate in Shelton, intersected by the canal, on the banks of which he built the celebrated Etruria—the finest manufactory of the kind up to that time erected in England, alongside of which he built a mansion for himself and cottages for his workpeople. He removed his works thither from Burslem, partially in 1769, and wholly in 1771, shortly before the works of the canal had been completed.

The Grand Trunk was the most formidable undertaking of the kind that had yet been attempted in England. Its whole length, including the junctions with the Birmingham Canal and the river Severn, was $139\frac{1}{2}$ miles. In conformity with Brindley's practice, he laid out as much of the navigation as possible upon a level, concentrating the locks in this case at the summit, near Harecastle, from which point the waters fell in both directions, north and south. Brindley's liking for long flat reaches of dead water made him keep clear of rivers as much as possible. He likened water in a river flowing down a declivity, to a furious giant running along and overturning everything ; whereas (said he) " if you lay the giant flat upon his back, he loses all his force, and becomes completely passive, whatever his size may be." Hence he contrived that from Middlewich, a distance of seventeen miles, to the Duke's canal at Preston Brook, there should not be a lock ; but goods might be conveyed from the centre of Cheshire to Manchester, for a distance of about seventy miles, along the same uniform water level. He carried out the same practice, in like manner, on the Trent side of Harecastle, where he laid out the canal in as many long lengths of dead water as possible.

The whole rise of the canal from the level of the Mersey, including the Duke's locks at Runcorn, to the summit at Harecastle, is 395 feet; and the fall from thence to the Trent at Wilden is 288 feet 8 inches. The locks on the Grand Trunk proper, on the northern side of Harecastle, are thirty-five, and on the southern side forty. The dimensions of the canal, as originally constructed, were twenty-eight feet in breadth at the top, sixteen at the bottom, and four and a-half feet in depth; but from Wilden to Burton, and from Middlewich to Preston-on-the-Hill, it was thirty-one feet broad at the top, eighteen at the bottom, and five and a-half feet deep, so as to be navigable by large barges; and the locks at those parts of the canal were of correspondingly large dimensions. The width was afterwards made uniform throughout. The canal was carried over the river Dove on an aqueduct of twenty-three arches, approached by an embankment on either side—in all a mile and two furlongs in length. There were also aqueducts over the Trent, which it crosses at four different points—one of these being of six arches of twenty-one feet span each—and over the Dane and other smaller streams. The number of minor aqueducts was about 160, and of road-bridges 109.

But the most formidable works on the canal were the tunnels, of which there were five—the Harecastle, 2880 yards long; the Hermitage, 130 yards; the Barnton, 560 yards; the Saltenford, 350 yards; and the Preston-on-the-Hill, 1241 yards. The Harecastle tunnel (subsequently duplicated by Telford) was constructed only nine feet wide and twelve feet high;[1] but the others

---

[1] Brindley's tunnel had only space for a narrow canal-boat to pass through, and it was propelled by the tedious and laborious process of what is called "legging." It still continues to be worked in the same way, while horses haul the boats through the whole length of Telford's wider tunnel. The men who "leg" the boat, literally kick it along from one end to

were seventeen feet four inches high, and thirteen feet six inches wide. The most extensive ridge of country to be penetrated was at Harecastle, involving by far the most difficult work in the whole undertaking. This ridge is but a continuation of the high ground, forming what is called the " back-bone of England," which extends in a south-westerly direction from the Yorkshire moun-

NORTHERN ENTRANCE OF HARECASTLE TUNNELS.[1]

[By E. M. Wimperis, after a Sketch by the Author.]

tains to the Wrekin in Shropshire. The flat county of Cheshire, which looks almost as level as a bowling-green when viewed from the high ground near New Chapel, seems to form a deep bay in the land, its innermost point being almost immediately under the village of Hare-

the other. They lie on their backs on the boat-cloths, with their shoulders resting against some package, and propel it along by means of their feet pressing against the top or sides of the tunnel.

[1] The smaller opening into the hill on the right-hand of the view is Brindley's tunnel; that on the left is Telford's, executed some forty years since. Harecastle church and village occupy the ground over the tunnel entrances.

castle; and from thence to the valley of the Trent the ridge is at the narrowest. That Brindley was correct in determining to form his tunnel at this point has since been confirmed by the survey of Telford, who there constructed his parallel tunnel for the same canal, and still more recently by the engineers of the North Staffordshire Railway, who have also formed their railway tunnel almost parallel with the line of both canals.

When Brindley proposed to cut a navigable way under this ridge, it was declared to be chimerical in the extreme. The defeated promoters of the rival projects continued to make war upon it in pamphlets, and in the exasperating language of mock sympathy proclaimed Brindley's proposed tunnel to be " a sad misfortune,"[1] inasmuch as it would utterly waste the capital raised by the subscribers, and end in the inevitable ruin of the concern. Some of the small local wits spoke of it as another of Brindley's " Air Castles ;" but the allusion was not a happy one, as his first " castle in the air," despite all prophecies to the contrary, had been built, and continued to stand firm at Barton ; and judging by the issue of that undertaking, it was reasonable to infer that he might equally succeed in this, difficult though it was on all hands admitted to be.

The Act was no sooner passed than Brindley set to work to execute the impossible tunnel. Shafts were sunk from the hill-top at different points down to the level of the intended canal. The stuff was drawn out of

LONGITUDINAL SECTION OF TUNNEL, SHOWING THE STRATA.

the shafts in the usual way by horse-gins ; and so long as the water was met with in but small quantities, the power

---

[1] ' Seasonable Considerations,' &c. ; Canal pamphlet dated 1766.

of windmills and watermills working pumps over each shaft was sufficient to keep the excavators at work. But as the miners descended and cut through the various strata of the hill on their downward progress, water was met with in vast quantities; and here Brindley's skill in pumping-machinery proved of great value. The miners were often drowned out, and as often set to work again by his mechanical skill in raising water. He had a fire-engine, or atmospheric steam-engine, of the best construction possible at that time, erected on the top of the hill, by the action of which great volumes of water were pumped out night and day. This abundance of water, though it was a serious hinderance to the execution of the work, was a circumstance on which Brindley had calculated, and .indeed depended, for the supply of water for the summit level of his canal. When the shafts had been sunk to the proper line of the intended waterway, the excavation then proceeded in opposite directions, to meet the other driftways which were in progress. The work was also carried forward at both ends of the tunnel, and the whole line of excavation was at length united by a continuous driftway —it is true, after long and expensive labour—when the water ran freely out at both ends, and the pumping-apparatus on the hilltop was no longer needed. At a general meeting of the Company, held on the 1st October, 1768, after the works had been in progress about two years, it appeared from the report of the Committee that four hundred and nine yards of the tunnel were cut and vaulted, besides the vast excavations at either end for the purpose of reservoirs; and the Committee expressed their opinion that the work would be finished without difficulty.

Active operations had also been in progress at other parts of the canal. About six hundred men in all were employed, and Brindley went from point to point super-intending and directing their labours. A Burslem correspondent, in September, 1767, wrote to a distant friend

thus :—" Gentlemen come to view our eighth wonder of
the world, the subterraneous navigation, which is cutting
by the great Mr. Brindley, who handles rocks as easily
as you would plum-pies, and makes the four elements
subservient to his will. He is as plain a looking man
as one of the boors of the Peak, or as one of his own
carters; but when he speaks, all ears listen, and every
mind is filled with wonder at the things he pronounces
to be practicable. He has cut a mile through bogs,
which he binds up, embanking them with stones which
he gets out of other parts of the navigation, besides about
a quarter of a mile into the hill Yelden, on the side of
which he has a pump worked by water, and a stove, the
fire of which sucks through a pipe the damps that would
annoy the men who are cutting towards the centre of
the hill. The clay he cuts out serves for bricks to arch
the subterraneous part, which we heartily wish to see
finished to Wilden Ferry, when we shall be able to
send Coals and Pots to London, and to different parts
of the globe."

In the course of the first two years' operations, twenty-
two miles of the navigation had been cut and finished,
and it was expected that before eighteen months more
had elapsed the canal would be ready for traffic by water
between the Potteries and Hull on the one hand, and
Bristol on the other. It was also expected that by the
same time the canal would be ready for traffic from the
north end of Harecastle Tunnel to the river Mersey.
The execution of the tunnel, however, proved so tedious
and difficult, and the excavation and building went on
so slowly, that the Committee could not promise that it
would be finished in less than five years from that time.
As it was, the completion of the Harecastle Tunnel
occupied nine years more; and it was not finally completed until the year 1777, by which time the great
engineer had finally rested from his labours.

It is scarcely necessary to describe the benefits which

the canal conferred upon the inhabitants of the districts through which it passed. As we have already seen, Staffordshire and the adjoining counties had been inaccessible during the chief part of each year. The great natural wealth which they contained was of little value, because it could with difficulty be got at; and even when reached, there was still greater difficulty in distributing it. Coal could not be worked at a profit, the price of land-carriage so much restricting its use, that it was placed altogether beyond the reach of the great body of consumers. It is difficult now to realise the condition of poor people situated in remote districts of England less than a century ago. In winter time they shivered over scanty wood-fires, for timber was almost as scarce and as dear as coal. Fuel was burnt only at cooking-times, or to cast a glow about the hearth in the winter evenings. The fireplaces were little apartments of themselves, sufficiently capacious to enable the whole family to ensconce themselves under the chimney, to listen to stories or relate to each other the events of the day. Fortunate were the villagers who lived hard by a bog or a moor, from which they could cut peat or turf at will. They ran all risks of ague and fever in summer, for the sake of the ready fuel in winter. But in places remote from bogs, and scantily timbered, existence was scarcely possible; and hence the settlement and cultivation of the country were in no slight degree retarded until comparatively recent times, when better communications were opened up.

So soon as the canals were made, and coals could be readily conveyed along them at comparatively moderate rates, the results were immediately felt in the increased comfort of the people. Employment became more abundant, and industry sprang up in their neighbourhood in all directions. The Duke's canal, as we have seen, gave the first great impetus to the industry of Manchester and that district. The Grand Trunk had precisely the same effect throughout the Pottery and other districts of Staffordshire ; and their joint action was not only to employ, but actually to civilize the people. The salt of Cheshire could now be manufactured in immense quantities, readily conveyed away, and sold at a comparatively moderate price in all the midland districts of England. The potters of Burslem and Stoke, by the same mode of conveyance, received their gypsum from Northwich, their clay and flints from the seaports now directly connected with the canal, returning their manufactures by the same route. The carriage of all articles being reduced to about one-fourth of their previous rates,[1] articles of necessity and comfort, such as had formerly been unknown except amongst the wealthier classes, came into common use amongst the people. Existence ceased to be difficult, and came to be easy. Led by the enterprise of Wedgwood and others like him, new branches of industry sprang up, and the manufacture of earthenware, instead of being insignificant and comparatively unprofitable, which it was before his time, became a staple branch of English trade. Only about ten years after the Grand Trunk Canal had been opened, Wedgwood stated in evidence before the House of Commons, that from 15,000 to 20,000

---

[1] The following comparison of the rates per ton at which goods were conveyed by land-carriage before the opening of the Grand Trunk Canal, and those at which they were subsequently carried by it, will show how great was the advantage conferred on the country by the introduction of navigable canals :—" The cost of carrying a ton of goods from Liverpool to Etruria, the centre of the Staffordshire potteries, by land-carriage, was 50s.; the Trent and Mersey reduced it to 13s. 4d. The land-carriage from Liverpool to Wolverhampton was 5l. a ton; the canal reduced it to 1l. 5s. The land-carriage from Liverpool to Birmingham, and also to Stourport, was 5l. a ton; the canal reduced both to 1l. 10s. . . . . . Thus the cost of inland transport was reduced, on the average, to about one-fourth of the rate paid previous to the introduction of canal-navigation. The advantages were enormous : wheat, for example, which formerly could not be conveyed a hundred miles, from corn-growing districts to the large towns and manufacturing districts, for less than 20s. a quarter, could be conveyed for about 5s. a quarter. These facts show how great was the service conferred on the country by Brindley and the Duke of Bridgewater."— Baines's 'History of the Commerce and Town of Liverpool.'

persons were then employed in the earthenware-manu-
facture alone, besides the large number of others employed in digging coals for their use, and the still larger
number occupied in providing materials at distant parts,
and in the carrying and distributing trade by land and
sea. The annual import of clay and flints into Staffordshire at that time was from fifty to sixty thousand tons;
and yet, as Wedgwood truly predicted, the trade was
but in its infancy. The outwards and inwards tonnage
to the Potteries is now upwards of three hundred thousand
tons a-year.

The moral and social influences exercised by the canals
upon the Pottery districts were not less remarkable.
From a half-savage, thinly-peopled district of some 7000
persons in 1760, partially employed and ill-remunerated,
we find them increased, in the course of some twenty-five
years, to about treble the population, abundantly employed, prosperous, and comfortable.[1] Civilization is
doubtless a plant of very slow growth, and does not
necessarily accompany the rapid increase of wealth.
On the contrary, higher earnings, without improved
morale, may only lead to wild waste and gross indulgence.
But the testimony of Wesley to the improved character
of the population of the Pottery district in 1781, within
a few years after the opening of Brindley's Grand Trunk
Canal, is so remarkable, that we cannot do better than
quote it here; and the more so, as we have already given
the account of his first visit in 1760, on the occasion of
his being pelted. "I returned to Burslem," says Wesley;
" how is the whole face of the country changed in about
twenty years! Since which, inhabitants have continually
flowed in from every side. Hence the wilderness is
literally become a fruitful field. Houses, villages, towns,
have sprung up, and the country is not more improved
than the people."

---

[1] The population of the same district in 1861 was found to be upwards
of 120,000.

## CHAPTER IX.

### Brindley's Last Canals.

It is related of Brindley that, on one occasion, when giving evidence before a committee of the House of Commons, in which he urged the superiority of canals to rivers for purposes of inland navigation, the question was asked by a member, " Pray, Mr. Brindley, what then do you think is the use of navigable rivers ? " " To make canal navigations, to be sure," was his instant reply. It is easy to understand the gist of the engineer's meaning. For purposes of trade he regarded regularity and certainty of communication as essential conditions of any inland navigation; and he held that neither of these could be relied upon in the case of rivers, which are in winter liable to interruption by floods, and in summer by droughts. In his opinion, a canal, with enough of water always kept banked up, or locked up where the country would not admit of the level being maintained throughout, was absolutely necessary to satisfy the requirements of commerce. Hence he held that one of the great uses of rivers was to furnish a supply of water for canals. It was only another illustration of the " nothing like leather " principle; Brindley's head being so full of canals, and his labours so much confined to the making of canals, that he could think of little else.

In connection with the Grand Trunk—which proved, as Brindley had anticipated, to be the great aorta of the canal system of the midland districts of England— numerous lines were projected and afterwards carried out under our engineer's superintendence. One of the most important of these was the Wolverhampton Canal,

connecting the Trent with the Severn, and authorised
in the same year as the Grand Trunk itself. It is now
known as the Staffordshire and Worcestershire Canal,
passing close to the towns of Wolverhampton and Kid-
derminster, and falling into the Severn at Stourport. This
branch opened up several valuable coal-fields, and placed
Wolverhampton and the intermediate districts, now
teeming with population and full of iron manufactories,
in direct connection with the ports of Liverpool, Hull,
and Bristol. Two years later, in 1768, three more
canals, laid out by Brindley, were authorised to be constructed : the Coventry Canal to Oxford, connecting the
Grand Trunk system by Lichfield with London and the
navigation of the Thames; the Birmingham Canal, which
brought the advantages of inland navigation to the very
doors of the central manufacturing town in England;
and the Droitwich Canal, to connect that town by a
short branch with the river Severn. In the following
year a further Act was obtained for a canal, laid out by
Brindley, from Oxford to the Coventry Canal at Longford, eighty-two miles in length.

These were highly important works; and though they
were not all carried out strictly after Brindley's plans,
they nevertheless formed the groundwork of future Acts,
and laid the foundations of the midland canal system.
Thus, the Coventry Canal was never fully carried out
after Brindley's designs; a difference having arisen
between the engineer and the Company during the progress of the undertaking, in consequence, as is supposed,
of the capital provided being altogether inadequate to
execute the works considered by Brindley as indispensable. He probably foresaw that there would be nothing
but difficulty, and very likely there might be discredit
attached to himself by continuing connected with an
undertaking the proprietors of which would not provide
him with sufficient means for carrying it forward to
completion; and though he finished the first fourteen

miles between Coventry and Atherstone, he shortly after gave up his connection with the undertaking, and it remained in an unfinished state for many years, in consequence of the financial difficulties in which the Company had become involved through the insufficiency of their capital. The connection of the Coventry Canal with the Grand Trunk was afterwards completed, in 1785, by the Birmingham and Fazeley and Grand Trunk Companies conjointly, and the property eventually proved of great value to all parties concerned.

The Droitwich Canal, though only a short branch five and a half miles in length, was a very important work, opening up as it did an immense trade in coal and salt between Droitwich and the Severn. The works of this navigation were wholly executed by Brindley, and are considered superior to those of any others on which he was engaged. Whilst residing at Droitwich, we find our engineer actively engaged in pushing on the subscription to the Birmingham Canal, of which the capital was taken slowly. Matthew Boulton, of Birmingham, was one of the most active promoters of the scheme, and Josiah Wedgwood was also bestirring himself in its behalf. In a letter written by him about this time, we find him requesting one of his agents to send out plans to gentlemen whom he names, in the hope of completing the subscription-list.[1] Brindley did not live to finish the Birmingham

---

[1] The letter is so characteristic of Josiah Wedgwood that we here insert it at length, as copied from the original in the possession of Mr. Mayer of Liverpool :—

"Burslem, 12th July, 1769.

"Dear Sir,—I should have wrote to you about young Wilson, but the multiplicity of branches you wrote me he was expected to learn, made me despair of teaching him any. Pray give my compliments to his father, and if he chooses to have his son to learn to be a warehouseman and book-keeper, which is quite sufficient and better than more for any one person, I will learn him those in the best manner; but, even then, Mr. Wilson must not expect him to

be set on the top of a ladder without setting his feet upon the lowermost steps; and unless he will let the Boy pursue that method, I would not be concerned with him on any account. I will not attempt to teach him any more trades; it would injure the Boy, and do me no good. If he has a mind at his leisure time to amuse himself with drawing I have no objection, and would encourage him in it, as an innocent amusement, and what may be of use to him, but would not make this a branch of his business. If the business I propose is too humble for Mr. Wilson's son, I would not by any means have him accept of it.

"Mr. Brindley desires you'll send 30 plans to each of the undermentioned Gentlemen by the first Waggons, and let us know

Canal ; it was carried out by his successors,—partly by his pupil, Mr. Whitworth, and partly by Smeaton and Telford. Brindley's plan was, as usual, to cut the level as flat as possible, in order to avoid lockage ; but his successors, in order to relieve the capital expenditure, as they supposed, constructed it with a number of locks to carry it over the summit at Smithwick. Shortly after its opening, however, the Company found reason to regret their rejection of Mr. Brindley's advice, and they lowered the summit by cutting a tunnel, as he had originally proposed, and thereby incurred an extra expense of about 30,000*l.*

Another of Brindley's canals, authorised in 1769, was that between Chesterfield and the river Trent, at Stockwith, about forty-six miles in length, intended for the transport of coal, lime, and lead from the rich mineral districts of Derbyshire, and the return trade of deals, corn, and groceries to the same districts. It would appear that Mr. Grundy, another engineer, of considerable reputation in his day, was consulted about the project, and that he advised a much more direct route than that pointed out by Brindley, who looked to the accommodation of the existing towns, rather than shortness of route, as the main thing to be provided for. Brindley, in this respect, took very much the same view in laying out his canals as was afterwards taken by George Stephenson—a man whom he resembled in many respects—in laying out railways. He would rather go *round* an obstacle in the shape of an elevated range of country, than go *through* it, especially if in going round and avoiding expense he could accommodate a number

---

when they are sent, as we shall advertise them in several of the Country papers: Mr. Walker, of Oxford, Steward to D. of Marlbro'—you may perhaps get them sent from Marlbro' house; Mr. Dudley, Attorney-at-Law, Coventry ; Mr. Richardson, Silversmith, in Chester: Mr. Perry, Wolverhampton.

"We shall send you this week end, double salts, creams, pott'y potts, table plates of all sorts, sallad dishes, covered do., pierced desert plates, &c. We cannot get Sadler to send us any pierced desert ware, &c. Pay Addison 5 or 6 guineas.

"Yours, &c., J. W."

of towns and villages.  Besides, by avoiding the hills
and following the course of the valleys, along which
population usually lies, he avoided expense of construction and secured flatness of canal; just as Stephenson
secured flatness of railway gradient.  Although the
length of canal to be worked was longer, yet the cost
of tunnelling and lockage was avoided.  Besides, the
population of the district was fully accommodated, which
could not have been accomplished by the more direct
route through unpopulated districts or under barren
hills.  The proprietors of the Chesterfield Canal concurred in Brindley's view, adopting his plan in preference to Grundy's, and it was accordingly carried into
effect.  This navigation was, nevertheless, a work of
considerable difficulty, proceeding, as it did, across a
very hilly country, the summit tunnel at Hartshill being
2850 yards in extent.  Like many of Brindley's other
works projected about this time, it was finished by his
brother-in-law, Mr. Henshall, and opened for traffic
several years after the great engineer's death.[1]

The whole of these canals were laid out by Brindley,
though they were not all executed by him, nor precisely
after his plans.  No record of any kind has been preserved of the manner in which these works were carried
out.  Brindley himself made few reports, and these
merely stated results, not methods ; yet he had doubtless
many formidable difficulties to encounter, and must have
overcome them by the adoption of those ingenious expedients, varying according to the circumstances of each

---

[1] The following were the canals laid out and principally executed by
Brindley :—

|  |  | Miles. | furl. | chains. |
|---|---|---|---|---|
| The Duke's Canals { Worsley to Manchester .. .. .. .. | | 10 | 2 | 0 |
| Longford Bridge to the Mersey below Runcorn | | 24 | 1 | 7 |
| Grand Trunk Canal Proper, from Wilden Ferry to Preston Brook | | 88 | 7 | 9 |
| Wolverhampton Canal .. .. .. .. .. .. .. .. | | 46 | 4 | 0 |
| Coventry ,, .. .. .. .. .. .. .. .. | | 36 | 7 | 8 |
| Birmingham ,, .. .. .. .. .. .. .. .. | | 24 | 2 | 0 |
| Droitwich ,, .. .. .. .. .. .. .. .. | | 5 | 4 | 9 |
| Oxford ,, .. .. .. .. .. .. .. .. | | 82 | 7 | 3 |
| Chesterfield ,, .. .. .. .. .. .. .. .. | | 46 | 0 | 0 |

case, in which he was always found so fertile. He had
no treasury of past experience, as recorded in books, to
consult, for he could scarcely read English; and certainly he could neither read French nor Italian, in
which languages the only engineering works of eminence were then written; nor had he any store of English experience to draw from, he himself being the first
of the English engineers of eminence, and having all
his experience to make for himself. It would doubtless
have been most interesting could we have had some
authentic record of this strong original man's struggles
with opposition and difficulty, and the methods by which
he contrived not only to win persons of high station to
support him with their influence but also with their
purses, at a time when money was comparatively a much
rarer commodity than it is now. " That want of records,
journals, and memoranda," says Mr. Hughes, "which is
ever to be deplored when we seek to review the progress of engineering works, is particularly felt when we
have to look back upon those undertakings which first
called for the exercise of engineering skill in many new
and untried departments. In Brindley's day, the entire
absence of experience derived from former works, the
obscure position which the engineer occupied in the
scale of society, the imperfect communication between
the profession in this country and the engineers and
works of other countries, and, lastly, the backward condition of all the mechanical arts and of the physical
sciences connected with engineering, may all be ranked
in striking contrast with the vast appliances which are
placed at the command of modern engineers." [1]

Moreover, the great canal works upon which Brindley
was engaged during the later part of his career, were
as yet scarcely appreciated as respects the important
influences which they were calculated to exercise upon

---

[1] Memoir, by Samuel Hughes, C.E. Weale's 'Quarterly Papers,' 1844.

society at large. The only persons who seem to have
regarded them with interest were far-sighted men like
Josiah Wedgwood, who saw in them the means not
only of promoting the trade of his own county, but
of opening up the rich natural resources of the kingdom, and diffusing amongst the people the elements of
comfort, intelligence, and civilization. The literary and
scientific classes as yet took little or no interest in them.
The most industrious and observant literary man of the
age, Dr. Johnson, though he had a word to say upon
nearly every subject, never so much as alluded to them,
though all Brindley's canals were finished in Johnson's
lifetime, and he must have observed the works in progress when passing on his various journeys through the
midland districts. The only reference which he makes
to the projects set on foot for opening up the country by
means of better roads, was to the effect, that whereas
there were before cheap places and dear places, now all
refuges were destroyed for elegant or genteel poverty.

Before leaving this part of the subject, it is proper to
state that during the latter part of Brindley's life, whilst
canals were being projected in various directions, he
was, on many occasions, called upon to give his opinion
as to the plans which other engineers had prepared.
Among the most important of the new projects on
which he was thus consulted were, the Leeds and Liverpool Canal; the improvement of the navigation of the
Thames to Reading; the Calder Navigation; the Forth
and Clyde Canal; the Salisbury and Southampton Canal;
the Lancaster Canal; and the Andover Canal. Many
of these schemes were of great importance in a national
point of view. The Leeds and Liverpool Canal, for
instance, opened up the whole manufacturing district
of Yorkshire along the valley of the Aire to Liverpool
and the intermediate districts of Lancashire. The advantages of this navigation to Leeds, Bradford, Keighley,
and the neighbouring towns, are felt to this day, and

their extraordinary prosperity is doubtless in no small degree attributable to the facilities which the canal has provided for the ready conveyance of raw materials and manufactured produce between those places and the towns and sea-ports of the west. Brindley surveyed and laid out the whole line of this navigation, 130 miles in length, and he framed the estimate on which the Company proceeded to Parliament for their bill. On the passing of the Act in 1768-9, the directors appointed Brindley their engineer; but, being almost overwhelmed with other business at the time, and feeling that he could not give the proper degree of personal attention to carrying out so extensive an undertaking, he was under the necessity of declining the appointment.[1]

Brindley being now the recognised head of his

---

[1] The works were immediately commenced at both ends of the canal, and portions were speedily opened out; but the difficulty and expensiveness of the remaining works greatly delayed their execution, and the canal was not finished until the year 1816. Twenty miles, extending from near Bingley to the neighbourhood of Bradford, were opened on 21st March, 1774; and the opening was thus described by a correspondent of 'Williamson's Liverpool Advertiser': "From Bingley and about three miles down, the noblest works of the kind that perhaps are to be found in the universe are exhibited, namely, a five-fold, a three-fold, a two-fold, and a single lock, making together a fall of 120 feet; a large aqueduct-bridge of seven arches over the river Aire, and an aqueduct on a large embankment over Shipley valley. Five boats of burden passed the grand lock, the first of which descended through a fall of sixty-six feet in less than twenty-nine minutes. This much-wished-for event was welcomed with ringing of bells, a band of music, the firing of guns by the neighbouring militia, the shouts of spectators, and all the marks of satisfaction that so important an acquisition merits. On the 21st Octo-

ber of the same year the following further paragraph appears: "The Liverpool end of the Leeds and Liverpool Canal was opened from Liverpool to Wigan on Wednesday, the 19th October, 1774, with great festivity and rejoicings. The water had been led into the basin the evening before. At nine A.M. the proprietors sailed up the canal in their barge, preceded by another filled with music, with colours flying, &c., and returned to Liverpool about one. They were saluted with two royal salutes of twenty-one guns each, besides the swivels on board the boats, and welcomed with the repeated shouts of the numerous crowds assembled on the banks, who made a most cheerful and agreeable sight. The gentlemen then adjourned to a tent on the quay, where a cold collation was set out for themselves and their friends. From thence they went in procession to George's coffee-house, where an elegant dinner was provided. The workmen, 215 in number, walked first, with their tools on their shoulders, and cockades in their hats, and were afterwards plentifully regaled at a dinner provided for them. The bells rang all day, and the greatest joy and order prevailed on the occasion."

profession, and the great authority on all questions of navigation, he was, in 1770, employed by the Corporation of London to make a survey of the Thames above Battersea, with the object of having it improved for purposes of navigation.  As usual, Brindley strongly recommended the construction of a canal in preference to carrying on the navigation by the river, where it was liable to be interrupted by the tides and floods, or by the varying deposits of silt in the shallow places.  In his first report to the Common Council, dated the 16th of June, 1770, he pointed out that the cost of hauling the barges was greatly in favour of the canal.  For example, he stated that the expense of taking a vessel of 100 or 120 tons from Isleworth to Sunning, and back again to Isleworth, was 80*l.*, and sometimes more ; whilst the cost by the canal would only be 16*l.*  The saving in time would be still greater, for the double voyage might easily be performed in fifteen hours ; whereas by the river the boats were sometimes three weeks in going up, and almost as much in coming down.  He estimated that there would be a saving to the public of at least 64*l.* on every voyage, besides the saving of time in performing it.  After making a further detailed examination of the district, and maturing his views on the whole subject, he sent in a report, accompanied by a profile of the river about seven feet long.  We quote the document, which is little known, as a specimen of Brindley's reports, though doubtless written by a more practised hand :—

" To the Committee of the Common Council of the City of London.

" Gentlemen,—Pursuant to your instructions, dated 27th July, 1770, I have made a survey of the River Thames from Boulter's Lock to Mortlake in Surrey, and have made a plan and profile of the same, with a level and fall line, remarking the different falls from one town to another ; and likewise have examined the most material obstructions and inconveniences that attend the present navigation,

which are considerable and many; for it hath been found by long
experience to be impassable for barges in time of flood, which in
most years continues several months during the winter, and is out
of the power of art to remedy. It likewise is impassable in time
of long droughts, for want of a sufficient depth of water; but this
difficulty may be removed, and the most effectual way to do it
would be by making dams and (cistern) locks, the dams to
pound up one to another; the number of which may be ascertained
by the profile, which, I suppose, will be about twelve; for if they
be made to pound more than five or six feet, some of the adjacent
lands will be laid under water, or be subject to be soon flooded, as
may be seen by the surface of it in the profile. But the expense of
improving so large a river in this way will be so great that I
suppose it will not be put in practice.

"It is impossible for me to tell what the expense would be, but
dare make bold to say it will be five or six times the expense of
making a canal, and, when done, will be far from being so safe and
speedy a conveyance; yet the river may be made better than it
is, and that at no very extraordinary expense. The method that I
would propose is, to contract the channel in the shallow and broad
places, most of which are marked in the plan. By this means a
sufficient depth of water, I suppose, may be obtained in all places,
or at least may be made much better than it is; but the fall will
remain the same, and the current increase by the increased depth
of water, and consequently it will require more strength of men
and horses to draw the barges against the stream; yet by this
means it may be rendered much more certain than it is, and an
easy navigation downward (except in time of flood); but the great
labour and expense of taking a vessel upwards cannot be taken away
by any method that I know of but by making a canal, by means
of which most places upon the banks of the river may be supplied
by collateral cuts; for it is practicable to make branches from the
main canal to fall into the river wheresoever it may be most useful
to the country, viz. one may be made to communicate with the
river above Windsor Bridge; the length will be about a mile, and
the fall or lockage ten feet. Another cut may very easily be made
from near West Bedford to the lower end of Staines: the length
will be about two miles and seven furlongs, and the fall twenty-six
feet. And whenever it is thought proper, another may be made to
communicate with the river at or near Shepperton: this would be
convenient to the Guildford navigation and the several places down
the stream to Kingston. By these collateral cuts all the places

upon the river (at least above Kingston) may be much better, cheaper, and more certainly supplied with their commodities than ever they were, or ever can be, by the river navigation alone, supposing the shallow places to be contracted (as mentioned before) so that the vessels have a sufficient depth of water to carry them down the stream from the ends of the cuts to the several places where they are designed.

" When I made my former report about the canal, I was informed that the vessels came very well up to Isleworth, but now understand it was generally at two tides; yet notwithstanding I take Isleworth to be the most proper place for the canal to communicate with the river, for the reasons there given.

" With respect to the latter part of my instructions, directing me to make an estimate of the expense of improving the river, it is impossible for me to make an estimate with any degree of accuracy. I can only say, that whatever sum is thought proper to be laid out —suppose 10,000*l.* in contracting, &c.—in the best manner, the river may be made much better than it is, or perhaps ever was; but if twenty times that sum be laid out, it would not make it a good and permanent navigation.

" Upon examining that part of the river between Mortlake and Richmond Gardens, I find much time is lost, particularly in neap tides, owing to the shoals and sand-banks arising on or towards the towing-side between those two places, which cannot be passed over for a great part of the tide of ebb, so that vessels not being able to reach this place before high water must remain till the next tide; which they need not do, as the towing part begins at Mortlake, could they have a sufficient depth of water to float them up. To remedy this inconvenience, I would propose a dam to be made across the river, somewhere between Mortlake and Kew Bridge, with a lock at each end. This would receive all vessels from London that could save their tide to this place, and by being come to the towing-path would proceed up to the canal or towards Kingston without hindrance or delay.

" Also all vessels coming down would, for the whole tide, have deep water and an easy passage into the tideway. This would also be very useful in another respect, as it would make Kew Bridge a convenient landing-place for all passengers, and would preserve a pool of water large enough to receive commodiously all vessels which must wait during the flood-tide before they can proceed towards London. The river being thus sufficiently raised, will render the navigation to the mouth of the canal and towards Kingston

extremely easy, and be of great utility to the several towns of
Brentford, Isleworth, Richmond, and Twickenham, and render that
part of the country in general perhaps the most delightful spot in
Europe.

" The expense of this dam and two locks (from the best accounts
I have been able to get of the price of materials) will be about
17,500*l.*

"*December 12th,* 1770."                    " JAMES BRINDLEY.

This plan, however, was never carried out; the proposal to construct a canal parallel with the Thames
having been abandoned so soon as the Grand Junction
Canal was undertaken.

These and numerous other schemes in various parts of
the country—at Stockton, at Leeds, at Cambridge, at
Chester, at Salisbury and Southampton, at Lancaster, and
in Scotland—fully occupied the attention of Brindley; in
addition to which, there was the personal superintendence which he must necessarily give to the canals in
active progress, and for the execution of which he was
responsible. In fact, there was scarcely a design of
canal navigation set on foot throughout the kingdom,
during the later years of his life, on which he was not
consulted, and the plans of which he did not entirely
form, or revise and improve.

In addition to his canal works, Brindley was also consulted as to the best mode of draining the low lands
in different parts of Lincolnshire, and the Great Level
in the neighbourhood of the Isle of Ely. He supplied
the corporation of Liverpool with a plan for cleansing
the docks and keeping them clear of mud, which is said
to have proved very. effective; and he pointed out to
them an economical method of building walls against the
sea without mortar, which long continued to be employed
with complete success. In such cases he laid his plans
freely open to the public, seeking to secure them by
no patent, nor shrouding his proceedings in any mystery.
He was free and open with professional men, harbouring

no petty jealousy of rivals. His pupils had free access
to all his methods, and he took a pride in so training
them that they should be an honour to the profession
then rising into importance, and enabled, after he had
left the scene, to carry on those great industrial undertakings, which he probably foresaw clearly enough in the
future of England.

Before dismissing the subject of Brindley's canals, we
may briefly allude to the influence which they exercised
upon the enterprise as well as the speculation of the time.
" When these fellows," says Sheridan in the ' Critic,'
" have once got hold of a good thing, they do not know
when to stop." This might be said of the speculative
projectors of canals, as afterwards of railways. The
commercial success which followed the opening of the
Duke's Canal, and shortly after it the Grand Trunk,
soon infected the whole country, and canal schemes
were projected in great numbers for the accommodation
even of the most remote and unlikely places. In those
districts where the demand for improved water-communication grew out of an actual necessity—as, for instance,
where a large coal-field required to be opened up for the
supply of a population urgently in need of fuel—or
where two large towns, such as Manchester and Liverpool, required to be provided with a more cheap and
convenient means of trading intercourse than had formerly existed—or where districts carrying on extensive
and various manufactures, such as Birmingham, Wolverhampton, and the Potteries, needed a more ready means
of communication with other parts of the kingdom—there
was no want of trade for the canals ; and those constructed
for such purposes very soon had as much traffic as they
could carry. The owners of land discovered that their
breed of horses was not destroyed, and that their estates
were not so cut up as to be rendered useless, as many of
them had prognosticated. On the contrary, the demand

for horses to carry coals, lime, manure, and goods to and from the canal depôts, rapidly increased. The canals meandering through their green fields were no such unsightly objects after all, and they very soon found that inasmuch as the new water-ways readily enabled agricultural produce to reach good markets in the large towns, they were likely even to derive considerable pecuniary advantages from their formation.

Another objection alleged against canals, on public grounds, was alike speedily disproved. It was said that inland navigation, by reason of its greater cheapness, ease, and certainty, must necessarily diminish the coasting trade, and consequently discourage the training of seamen, the constitutional bulwark of the kingdom. But the extraordinarily rapid growth of the shipping-trade of Liverpool, and the vastly increased number of seagoing vessels required to accommodate the traffic now converging on that seaport, very soon showed that canals, instead of diminishing, were calculated immensely to promote the naval power and resources of England. Thus it was found that in the thirty years which elapsed subsequent to the opening of the Duke's Canal between Worsley and Manchester,—during which time the navigation had also been opened to the Mersey, and the Grand Trunk and other main canals had been constructed, connecting the principal inland towns with the sea-ports,—the tonnage of English ships had increased threefold, and the number of sailors been more than doubled.

So great an impulse had thus been given to the industry of the country, and it had become so clear that facility of communication must be an almost unmixed good, that a desire for the extension of canals sprang up in all districts; and instead of being resisted and denounced, they everywhere became the rage. They were advocated in pamphlets, in newspapers, and at public meetings. One enthusiastic pamphleteer, advocating the formation of a canal between Kendal and Manchester, denounced the wretched state of the turnpike-roads, which were maintained by "an enormous tax," and exclaimed, "May we all scorn to plod through the dirt as we long have done at so large an expense; and for the support of our drooping manufactories, let canals be made through the whole nation as common as the public highways."[1]

There seemed, indeed, to be every probability that this desire would be shortly fulfilled; for so soon as the canals which had been made began to pay dividends, the strong motive of personal gain became superadded to that of public utility.    The rapid increase of wealth which they promoted served to stimulate the projection of new schemes; and in a very few years after Brindley's death we find an immense number of Navigation Acts receiving the sanction of the legislature, and canal works in progress in all parts of the country.    The shares were quoted upon 'Change, when they became the subject of commerce, and very shortly of wild speculation.    By the year 1792, the country was in a perfect ferment about canal shares. Notices of eighteen new canals were published in the ' Gazette ' of the 18th August in that year.    The current premiums of single shares in those companies for which Acts had been obtained were as follows : Grand Trunk, 350l.; Birmingham and Fazeley, 1170l.; Coventry, 350l.; Leicester, 155l.; and so on.    There was a rush to secure shares in the new schemes, and the requisite capitals were at once eagerly subscribed.    At the first meeting, held in 1790, of the promoters of the Ellesmere Canal, 112 miles in extent, to connect the Mersey, the Dee, and the Severn, applications were made for four times the disposable number of shares.    A great number of worthless and merely speculative schemes were

---

[1] ' A Cursory View of a proposed Canal from Kendal to the Duke of Bridgewater's Canal, leading to the great manufacturing town of Manchester.' 1785.

thus set on foot, which brought ruin upon many, and led to waste both of labour and capital. But numerous sound projects were at the same time launched, and an extraordinary stimulus was thus given to the prosecution of measures, too long delayed, for effectually opening up the communications of the country. The movement extended to Scotland, where the Forth and Clyde Canal, and the Crinan Canal, were projected; and to Ireland, where the Grand Canal and Royal Canal were undertaken. But, as Arthur Young pithily remarked, in reference to these latter projects, " a history of public works in Ireland would be a history of jobs."

In the course of the four years ending in 1794, not fewer than eighty-one Canal and Navigation Acts were obtained : of these, forty-five were passed in the two latter years, authorising the expenditure of not less than 5,300,000*l.* As in the case of the railways at a subsequent period, works which might, without pressure upon the national resources, easily have been executed if spread over a longer period, were undertaken all at once; and the usual consequences followed, of panic, depreciation, and loss. But though individuals lost, the public were eventually the gainers. Many projects fell through, but the greater number were commenced, and after passing through the usual financial difficulties, were finished and opened for traffic. The country became thoroughly opened up in all directions by about 2600 miles of navigable canals in England, 276 miles in Ireland, and 225 miles in Scotland. The cost of executing these great water ways is estimated to have amounted to about fifty millions sterling. There was not a place in England, south of Durham, more than fifteen miles from water communication; and most of the large towns, especially in the manufacturing dictricts, were directly accommodated with the means of easy transport of their goods to the principal markets. " At the beginning of the present century," says Dr. Aiken, writing in 1795, "it was thought a most arduous task to make a high road practicable for carriages over the hills and moors which separate York-shire from Lancashire, and now they are pierced through by three navigable canals!"

Many of these great works were executed by the engineers whose biographies will form the subject of succeeding chapters; but in this place we may take the opportunity of stating, that notwithstanding the great additional facilities for conveyance of merchandise which have been provided of late years by the construction of railways, a very large proportion of the heavy carrying trade of the country still continues to be conducted upon canals.   It was indeed at one time proposed, during the railway mania, and that by a somewhat shrewd engineer, to fill up the canals and make railways of them.   It was even predicted, during the construction of the Liverpool and Manchester Railway, that " within twelve months of its opening, the Bridgewater Canal would be closed, and the place of its waters be covered over with rushes." But canals have stood their ground, even against rail-ways; and the Duke's Canal, instead of being closed, con-tinues to carry as much traffic as ever.   It has lost the conveyance of passengers by the fly-boats,[1] it is true; but it has retained and in many instances increased its traffic in minerals and merchandise.   The canals have stood the competition of railways far more successfully than the old turnpike-roads, though these too are still, in their way, as indispensable as canals and railways themselves. Not less than twenty millions of tons of traffic are esti-

---

[1] The following curious paragraph is from the 'Times' of the 19th December, 1806. It relates to the despatching of troops from London for Ireland, during a time of great excitement: —" The first division of the troops that are to proceed by Paddington Canal for Liverpool, and thence by transports for Dublin, will leave Paddington to-day, and will be fol-lowed by others to-morrow, and on Sunday. By this mode of conveyance the men will be *only seven days* in reaching Liverpool, and with com-paratively little fatigue, and it would take them above fourteen days to march that distance. Relays of fresh horses for the canal-boats have been ordered to be in readiness at all the stations."

mated to be carried annually upon the canals of England alone, and this quantity is steadily increasing.[1] In 1835, before the opening of the London and Birmingham Railway, the through tonnage carried on the Grand Junction Canal was 310,475 tons; and in 1845, after the railway had been open for ten years, the tonnage carried on the canal had increased to 480,626 tons. At a meeting of proprietors of the Birmingham Canal Navigations, held in October, 1860, the chairman said, " the receipts for the last six months were, with one exception, the largest they had ever had."

Railways are a great invention, but in their day canals were as highly valued, and indeed quite as important; and it is fitting that the men by whom they were constructed should not be forgotten. We may be apt to think lightly of the merits and achievements of the early engineers, now that works of so much greater magnitude are accomplished without difficulty. The appliances of modern mechanics enable men of this day to dwarf by comparison the achievements of their predecessors, who had obstructions to encounter which modern engineers know nothing of. The genius of the older men now seems slow, although they were the wonder of their own age. The canal, and its barges tugged along by horses, now appears a cumbersome mode of communication, beside the railway and the locomotive with its power and speed. Yet canals still are, and will long continue to form, an essential part of our great system of commercial communication,—as much as roads, railways, or the ocean itself.

---

[1] Braithwaite Poole's 'Statistics of British Commerce.' London, 1852.

# CHAPTER X.

## Brindley's Death—Characteristics.

It will be observed that Brindley's employment as an engineer extended over a wide district. Even before his employment by the Duke of Bridgewater, he was under the necessity of travelling great distances to fit up water-mills, pumping-engines, and manufacturing machinery of various kinds, in the counties of Stafford, Cheshire, and Lancashire. But when he had been appointed to superintend the construction of the Duke's canals, his engagements necessarily became of a still more engrossing character, and he had very little leisure left to devote to the affairs of private life. He lived principally at inns, in the immediate neighbourhood of his work; and though his home was at Leek, he sometimes did not visit it for weeks together. He had very little time for friendship, and still less for courtship. Nevertheless, he did contrive to find time for marrying, though at a comparatively advanced period of his life. In laying out the Grand Trunk Canal, he was necessarily brought into close connection with Mr. John Henshall, of the Bent, near New Chapel, land-surveyor, who assisted him in making the survey. We find him visiting his house in September, 1762, and settling with him as to the preliminary operations. At these visits Brindley seems to have taken a special liking for Mr. Henshall's daughter Anne, then a girl at school, and when he went to see her father, he was accustomed to take a store of gingerbread for Anne in his pocket. She must have been a comely girl, judging by the portrait of her as a woman, which we have seen. In course of time, the liking ripened into an attachment; and shortly after the girl had left school, at the age of only nineteen, Brindley proposed to her, and was accepted. By this time he was close upon his fiftieth year, so that the union may possibly have been quite as much a matter of convenience as of love on his part. He had now left the Duke's service for the purpose of entering upon the construction of the Grand Trunk Canal, and with that object resolved to transfer his home to the immediate neighbourhood of Harecastle, as well as of his colliery at Golden Hill. Shortly after the marriage, the old mansion of Turnhurst fell vacant, and Brindley became its tenant, with his young wife. The marriage took place on the 8th December, 1765, in the parish church of Wolstanton, Brindley being described in the register as " of the parish of Leek, engineer;" but from that time until the date of his death his home was at Turnhurst.

The house at Turnhurst was a comfortable, roomy, old-fashioned dwelling, with a garden and pleasure-ground behind, and a little lake in front. It was formerly the residence of the Bellot family, and is said to have been the last house in England in which a family fool was maintained. Sir Thomas Bellot, the last of the name, was a keen sportsman, and the panels of several of the upper rooms contain pictorial records of some of his exploits in the field. In this way Sir Thomas seems to have befooled his estate, and it shortly after became the property of the Alsager family, from whom Brindley rented it. A little summer-house, standing at the corner of the outer courtyard, is still pointed out as Brindley's office, where he sketched his plans and prepared his calculations. As for his correspondence, it was nearly all conducted, subsequent to his marriage, by his wife, who, notwithstanding her youth, proved a most clever, useful, and affectionate partner.

Turnhurst was conveniently near to the works then

BRINDLEY'S HOUSE AT TURNHURST.

in progress at Harecastle Tunnel, which was within easy walking distance, whilst the colliery at Golden Hill was only a few fields off. From the elevated ground at Golden Hill, the whole range of high ground may be seen under which the tunnel runs—the populous pottery towns of Tunstall and Burslem filling the valley of the Trent towards the south. At Golden Hill, Brindley carried out an idea which he had doubtless brought with him from Worsley. He and his partners had an underground canal made from the main line of the Harecastle Tunnel into their coal-mine, about a mile and a half in length ; and by that tunnel the whole of the coal above that level was afterwards worked out, and conveyed away for sale in the Pottery and other districts, to the great profit of the owners as well as to the equally great convenience of the public.

These various avocations involved a great amount of labour as well as anxiety, and probably considerable tear and wear of the vital powers. But we doubt whether mere hard work ever killed any man, or whether Brindley's labours, extraordinary though they were, would have shortened his life, but for the far more trying condition of the engineer's vocation—irregular living, exposure in all weathers, long fasting, and then, perhaps, heavy feeding when the nervous system was exhausted, together with habitual disregard of the ordinary conditions of physical health. These are the main causes of the shortness of life of most of our eminent engineers, rather than the amount and duration of their labours. Thus the constitution becomes strained, and is ever ready to break down at the weakest place. Some violation of the natural laws more flagrant than usual, or a sudden exposure to cold or wet, merely presents the opportunity for an attack of disease which the ill-used physical system is found unable to resist. Such an accidental exposure unhappily proved fatal to Brindley. While engaged one day in surveying a branch canal between Leek and Froghall, he got drenched near Ipstones, and went about for some time in his wet clothes. This he had often before done with impunity, and he might have done so again ; but, unfortunately, he was put into a damp bed in the inn at Ipstones, and this proved too much for his constitution, robust though he naturally was. He became seriously ill, and was disabled from all further work. Diabetes shortly developed itself, and, after an illness of some duration, he expired at his house at Turnhurst, on the 27th of September, 1772, in the fifty-sixth year of his age, and was interred in the burying-ground at New Chapel, a few fields distant from his dwelling.

James Brindley was probably one of the most remarkable instances of self-taught genius to be found in the whole range of biography. The impulse which he gave to social activity, and the ameliorative influence which he exercised upon the condition of his countrymen, seem out of all proportion to the meagre intellectual culture which he had received in the course

of his laborious and active career. We must not, however, judge him merely by the literary test. It is true,
he could scarcely read, and he was thus cut off, to his
own great loss, from familiar intercourse with a large
class of cultivated minds, living and dead ; for he could
not share in the conversation of educated men, nor enrich his mind by reading the stores of experience found
treasured up in books. Neither could he write, except
with difficulty and inaccurately, as we have shown from
the extracts above quoted from his note-books still
extant.

Brindley was, nevertheless, a highly-instructed man
in many respects. He was full of the results of careful observation, ready at devising the best methods
of overcoming material difficulties, and possessed of a
powerful and correct judgment in matters of business.
Where any emergency arose, his quick invention and
ingenuity, cultivated by experience, enabled him almost
at once unerringly to suggest the best means of providing for it. His ability in this way was so remarkable, that those about him attributed the process by
which he arrived at his conclusions rather to instinct
than reflection—the true instinct of genius. " Mr.
Brindley," says one of his contemporaries, " is one of
those great geniuses whom Nature sometimes rears by
her own force, and brings to maturity without the necessity of cultivation. His whole plan is admirable, and
so well concerted that he is never at a loss ; for, if any
difficulty arises, he removes it with a facility which
appears so much like inspiration, that you would think
Minerva was at his fingers' ends."

His mechanical genius was indeed most highly cultivated. From the time when he bound himself apprentice to the trade of a millwright—impelled to do so
by the strong bias of his nature—he had been under-
going a course of daily and hourly instruction. There
was nothing to distract his attention, or turn him from

pursuing his favourite study of practical mechanics.
The training of his inventive faculty and constructive
skill was, indeed, a slow but a continuous process; and
when the time and the opportunity arrived for turning
these to account—when the silk-throwing machinery of
the Congleton mill, for instance, had to be perfected and
brought to the point of effectively doing its intended
duty—Brindley was found able to take it in hand and
finish the work, when even its own designer had given
it up in despair. But it must also be remembered that
this great facility of Brindley had been in a great measure the result of the closest observation, the most pains-
taking study of details, and the most indefatigable industry.

The same qualities were displayed in his improvements of the steam-engine, and his arrangements to
economise power in the pumping of water from drowned
mines. It was often said of his works, as was said of
Columbus's discovery, "how easy! how simple!" but
this was after the fact. Before he had brought his fund
of experience and clearness of vision to bear upon a
difficulty, every one was equally ready to exclaim, " how
difficult! how absolutely impracticable!" This was the
case with his " castle in the air," the Barton Viaduct—
such a work as had never before been attempted in
England, though now any common mason would undertake it. It was Brindley's merit always to be ready
with his simple, practical expedient; and he rarely failed
to effect his purpose, difficult although at first sight its
accomplishment might seem to be.

Like men of a similar stamp, Brindley had great confidence in himself and in his powers and resources.
Without this, it were impossible for him to have accomplished so much as he did. It is said that the King of
France, hearing of his great natural genius, and the
works he had performed for the Duke of Bridgewater
at Worsley, expressed a desire to see him, and sent a

message inviting him to view the great canal of Languedoc. But Brindley's reply was characteristic : " I will have no journeys to foreign countries," said he, " unless to be employed in surpassing all that has been already done in them."

His observation was remarkably quick. In surveying a district, he rapidly noted the character of the country, the direction of the hills and the valleys, and, after a few journeys on horseback, he clearly settled in his mind the best line to be selected for a canal, which almost invariably proved to be the right one. In like manner he would estimate with great rapidity the fall of a brook or river while walking along the banks, and thus determined the height of his cuttings and embankments, which he afterwards settled by a more systematic survey. In these estimates he was rarely, if ever, found mistaken.

His brother-in-law, Mr. Henshall, has said of him, " when any extraordinary difficulty occurred to Mr. Brindley in the execution of his works, having little or no assistance from books or the labours of other men, his resources lay within himself. In order, therefore, to be quiet and uninterrupted whilst he was in search of the necessary expedients, he generally retired to his bed ;[1] and he has been known to be there one, two, or three days, till he had attained the object in view. He would then get up and execute his design, without any drawing or model. Indeed, it was never his custom to make either, unless he was obliged to do it to satisfy his employers. His memory was so remarkable that he has often declared that he could remember, and execute, all the parts of the most complex machine, provided he had time, in his survey of it, to settle in his mind the

---

[1] The younger Pliny seems to have adopted almost a similar method : " Clausæ fenestræ manent. Mirè enim silentio et tenebris animus alitur. Ab iis quæ avocant abductus, et liber, et mihi relictus, non oculos animo sed animum oculis sequor, qui eadem quæ mens vident quoties non vident alia."—Epist. lib. ix., ep. 36.

several parts and their relations to each other. His method of calculating the powers of any machine invented by him was peculiar to himself. He worked the question for some time in his head, and then put down the results in figures. After this, taking it up again at that stage, he worked it further in his mind for a certain time, and set down the results as before. In the same way he still proceeded, making use of figures only at stated parts of the question. Yet the ultimate result was generally true, though the road he travelled in search of it was unknown to all but himself, and perhaps it would not have been in his power to have shown it to another." [1]

The statement about his taking to bed to study his more difficult problems is curiously confirmed by Brindley's own note-book, in which he occasionally enters the words "lay in bed," as if to mark the period, though he does not particularise the object of his thoughts on such occasions. It was a great misfortune for Brindley, as it must be to every man, to have his mental operations confined exclusively within the limits of his profession. He thought and lived mechanics, and never rose above them. He found no pleasure in anything else; amusement of every kind was distasteful to him; and his first visit to the theatre, when in London, was also his last. Shut out from the humanising influence of books, and without any taste for the politer arts, his mind went on painfully grinding in the mill of mechanics. "He never seemed in his element," said his friend Bentley, "if he was not either planning or executing some great work, or conversing with his friends upon subjects of importance." To the last he was full of projects, and full of work; and then the wheels

---

[1] 'Biographia Britannica,' 2nd Ed. Edited by Dr. Kippis. The materials of the article are acknowledged to have been obtained principally from Mr. Henshall by Messrs. Wedgwood and Bentley, who wrote and published the memoir in testimony of their admiration and respect for their deceased friend, the engineer of the Grand Trunk Canal.

of life came to a sudden stop, when he could work no
longer. It is related of him that, when dying, some
eager canal undertakers insisted on having an interview with him. They had encountered a serious difficulty in the course of constructing their canal, and
they *must* have the advice of Mr. Brindley on the subject. They were introduced to the apartment where he
lay scarce able to gasp, yet his mind was clear. They
explained their difficulty—they could not make their
canal hold water. "Then puddle it," said the engineer.
They explained that they had already done so. "Then
puddle it again—and again." This was all he could
say, and it was enough.

It remains to be added that, in his private character,
Brindley commanded general respect and admiration.
His integrity was inflexible ; his manner, though rough
and homely, was kind ; and his conduct unimpeachable.[1]
He was altogether unassuming and unostentatious, and
dressed and lived with great plainness. He was the
furthest possible from a narrow or jealous temper, and
nothing gave him greater pleasure than to assist others
with their inventions, and to train up a generation of
engineers, in the persons of his pupils, able to carry out
the works he had designed, when no longer able to conduct them. The principal undertakings in which he
was engaged up to the time of his death were carried
on by his brother-in-law, Mr. Henshall, formerly his

---

[1] It has, indeed, been stated in
the crazy publication of the last Earl
of Bridgewater, to which we have
already alluded, that when in the
service of the Duke, Brindley was
"drunken." But this is completely
contradicted by the testimony of
Brindley's own friends; by the evidence of Brindley's note-book, from
repeated entries in which it appears
that his "ating and drink" at dinner
cost no more than 8d.; by the confidence generally reposed in him, and
the friendship entertained for him, by
men such as Josiah Wedgwood; and
by the fact of the vast amount of
work that he subsequently contrived
to get through. No man of "drunken"
habits could possibly have done this.
We should not have referred to this
topic but for the circumstance that the
late Mr. Baines, of Leeds, has quoted
the Earl's statement, without contra-
diction, in his excellent 'History of
Lancashire.'

clerk of the works on the Grand Trunk Canal, and by
his able pupil, Mr. Robert Whitworth, for both of whom
he had a peculiar regard, and of whose integrity and
abilities he had the highest opinion.

Brindley left behind him two daughters, one of whom,
Susannah, married Mr. Bettington, of Bristol, merchant,
afterwards the Honourable Mr. Bettington, of Brindley's
Plains, Van Diemen's Land, where their descendants
still live. His other daughter, Anne, died unmarried,
on her passage home from Sydney, in 1838. His
widow, still young, married again, and died at Longport in 1826. Brindley had the sagacity to invest a
considerable portion of his savings in Grand Trunk
shares, the great increase in the value of which, as well
as of his colliery property at Golden Hill, enabled him
to leave his family in affluent circumstances.

BRINDLEY'S BURIAL-PLACE AT NEW CHAPEL.

# INDEX TO VOL. I.

END OF VOL. I.

LONDON: PRINTED BY WILLIAM CLOWES AND SONS, STAMFORD STREET,
AND CHARING CROSS.